Interconnect Analysis and Synthesis

Interconnect Analysis and Synthesis

Chung-Kuan Cheng
University of California at San Diego

John Lillis
University of Illinois at Chicago

Shen Lin
Hewlett-Packard

Norman H. Chang
Hewlett-Packard

A Wiley-Interscience Publication

JOHN WILEY & SONS, Inc.

New York • Chichester • Weinheim • Brisbane • Singapore • Toronto

This book is printed on acid-free paper. ∞

For ordering and customer service, call 1-800-CALL WILEY.

Library of Congress Cataloging-in-Publication Data:

Interconnect analysis and synthesis / Chung-Kuan Cheng ... [et al.].
p. cm.
ISBN 0-471-29366-0 (cloth : alk. paper)
1. Interface circuits—Computer simulation. 2. Electric network analysis—Data processing. 3. Computer-aided design. I. Cheng, Chung-Kuan.
TK7868.I58I583 2000 99-21918
621.3815—dc21 CIP

Printed in the United States of America

10 9 8 7 6 5 4 3 2 1

Contents

Preface

Our main goal in writing this book was to present a physical-design oriented perspective on contemporary issues in VLSI interconnect. Of particular interest was, in one book, to cover both interconnect analysis and synthesis. Because of the increasing importance of interconnect in modern VLSI systems and the fact that these fields of research have gained a certain degree of maturity in recent years, we felt that the time was right for such a book. We hope that we have succeeded in providing a unique and timely perspective on the topic.

The work presented in this book grew out of research projects initiated at both the University of California at San Diego and at Hewlett Packard Labs. Chapters 2 through 4 were developed largely as the product of graduate courses and research conducted at UCSD by Chung–Kuan Cheng. Chapter 5 embodies research conducted at HP Labs. Chapters 6 through 9 grew largely out of John Lillis' Ph.D. work at UCSD, his later work at UC Berkeley, and continuing work at the University of Illinois at Chicago.

We want to make clear that the book is by no means intended to be comprehensive. The absence of coverage of work by other researchers should by no means diminish their contributions. Several research groups have made significant contributions in the field and the reader is encouraged to investigate their works. Key contributors to progress in interconnect–oriented design automation include Jason Cong and Andrew Kahng (UCLA), Gabriel Robins (U. Virginia), Majid Sarrafzadeh (NWU), Martin Wong (UT Austin), Margaret Marek-Sadowska (UCSB), Wayne Dai (UCSC), Larry Pileggi (CMU), Sachin Sapatnekar (U. Minnesota), Massoud Pedram (USC), Eby Friedman (U. Rochester), Kurt Antreich (Technical Univ. Munich), Pak Chan (UCSC),

Jun-Dong Cho (Sungkyunkwan Univ.), Cheng-Kok Koh (Purdue), Xianlong Hong (Tsinghua Univ), Steve Kang (UIUC), C.L. Liu (Tsing Hua Univ.), Tatsuo Ohtsuki (Waseda), Takayasu Sakurai (Tokyo Univ.), Jacob White (MIT), CK Wong (Chinese Univ. Hong Kong), and Dian Zhou (UNCC).

Acknowledgments

We are first and foremost indebted to Professors Ernest S. Kuh of UC Berkeley and T.C. Hu of UCSD for their inspiration and encouragement of the production of this book. We also thank Professor Kuh for his thoughts and suggestions for improving the book. We are grateful to Professor Jiayuan Fang for his review of the book. We thank Professor Peter Absec for helpful discussions on scaling trends. Alina Deutch, Den Edelstein, Robert Rosenberg, and their colleagues at IBM gave us key insights on electro-migration and transmission line issues.

Special acknowledgement is due to Luca Daniel for his critique and detailed proofreading of the manuscript. Thanks to Frank Liu and Peining Guo for help on several figures. Lung-Tien Liu was also invaluable as a reviewer.

In the review of Chapter 5, we thank Tony Fourcroy, Lei He, Wayne Greene, Dan Krueger, Jon Lachman, Ken Lee, Antonio Martinez, Alan Meyer, Tim Michalka, Ed Miller, Alan Meyer, Eric Fetzer, Dan Robuck, Hans Stork, and David Quint. We also acknowledge the financial support from various sources which made this work possible. We thank Bob Grafton of the NSF in this regard. Support under the California MICRO program matched by AT&T, Cadence, Fujitsu, LSI, Raytheon, Rockwell, and Xilinx must also be recognized. Chung-Kuan Cheng thanks Mentor Graphics for support while on leave from UCSD. Chung-Kuan Cheng thanks his wife Jenny and John Lillis thanks his wife Rebecca for their support and patience.

1

Introduction

1.1 OVERVIEW

It has long been understood that the relentless advance of integrated circuit (IC) fabrication technology would eventually change the nature of IC design. As devices and interconnect are continually scaled down, an increasingly large number of critical design criteria—performance, noise immunity, and so on—are dominated or strongly influenced by interconnect. While there continues to be debate regarding the degree to which interconnect does and will dominate, there is widespread agreement on the overall trend as well as a number of design projects in which unexpected interconnect effects have lengthened the design cycle, which are essentially living proof of the trend.

The modern IC designer is, more than ever, reliant on effective computer-aided design (CAD) tools. From functional simulation to logic synthesis to floorplanning and placement and routing to layout analysis and verification, CAD plays a critical role in each phase of design. Given the trend toward interconnect dominance, a natural and well-founded fear has been that the tools developed for a device-dominated world would become inadequate (or perhaps even hamful) as technology advanced and the objectives and assumptions of those tools drift farther from physical reality. In response to this fear, a great deal of research has been done in the past decade to deal with the effects of interconnect via CAD tools.

Much of this research can be broadly categorized as either *analysis oriented*—given a design, how we can efficiently characterize its properties (performance, noise immunity, etc.)—or *synthesis oriented*—how we should design (place, route, etc.) a circuit so as to maximize the chances of achieving the desired results (as verified by the subsequent analysis phase). These are closely related and complementary issues and our goal in writing this book is to provide some perspective on these relationships. Accordingly, the book progresses from modeling and analysis techniques to synthesis and optimization techniques.

It is the authors' hope that the book will provide both CAD developers and researchers with a powerful set of tools for both analysis and optimization and also will give some perspective on the relationships and boundaries between the two disciplines.

1.2 BOOK ORGANIZATION

1.2.1 Models and Analysis

Chapter 2 through 5 deal with analysis. Scaling trends and their impact on interconnect and device models are discussed as well as modern analysis techniques. The focus of these chapters is summarized as follows.

- Chapter 2 deals with interconnect models. Technology, scaling trends and their impact are first covered: the effects of scaling on capacitive and inductive coupling are discussed. Projections from the semiconductor industry provide insight into the trend. In addition, the situations in which scaling necessitates transmission line modeling are covered. Finally, the effects of scaling on power dissipation and noise are discussed.
- In Chapter 3, device models are covered. Deep Sub-Micron (DSM) technology has impacted device modeling in several ways. The short-channel effect, for example, is studied, which requires more sophisticated modeling and encourages more complex gates (longer pull-down/up chains) than in previous technologies. Another result of DSM is due to highly resistive interconnect—lumped capacitance no longer adequately reflects the load driven by a gate. These effects are captured through a general set of device formats using input transition and output load. Within this context, explicit expressions are derived, a table-lookup strategy is presented, and a method for finding the "effective capacitance" of an interconnect load is given so that the general format may take resistive shielding into account.
- Building on the models developed in Chapters 2 and 3, Chapter 4 develops modern analysis techniques. First, classical time-domain and S-domain analysis techniques are covered. Subsequently, the emphasis is on containing of the complexity of the analysis via matrix reduction and moment matching. Properties of the Elmore delay—the first moment of the impulse response—are discussed particularly in relation to interconnect-driven layout. Additionally, a more general incremental moment computation technique convenient in interconnect-driven layout is covered. Finally, transmission-line analysis is touched on.
- Chapter 5 gives a state-of-the-art overview of the effects of inductance. Both analysis and design techniques are discussed.

1.2.2 Synthesis

Chapter 6 through 9 deal with synthesis techniques — CAD algorithms for synthesizing and optimizing interconnect. Throughout, the emphasis is on establishing flexible algorithmic frameworks that can be generalized and tuned for particular purposes. Further, an emphasis is placed on realistic problem formulations, capturing complex design tradeoffs.

The contents of the synthesis chapters are summarized as follows.

- Chapter 6 presents algorithms for automatic buffer insertion and wire sizing. The focus is on optimizing given (static) interconnect topologies and the development of general algorithmic techniques. Dynamic programming techniques pioneered by van Ginneken and Dhar, which have proved so flexible and powerful in interconnect optimization, are presented. An algorithm for finding an optimal assignment of buffers to the available insertion points in a given multi-pin tree topology is given. Generalizations incorporating wire sizing and improved delay models are also presented, demonstrating the flexibility of the approach. In addition, key data structures are described to ensure efficient implementation.
- Chapter 7 studies a degree of freedom neglected in Chapter 6 — the synthesis of the global routing topology itself. A natural timing-driven Steiner routing problem is formulated: construct a Steiner topology that minimizes cost (e.g., wire length) subject to given delay requirements of the sinks (using an RC delay model). Though this is a computationally difficult problem (it is NP-hard even without delay constraints), an efficient scheme — the P-Tree algorithm — works well in practice. This scheme is presented in detail. Several natural refinements and generalizations of these techniques are discussed.
- Chapter 8 examines the timing optimization of multisource nets (e.g., buses). Given the prevalence of such structures in modern designs, it is of practical interest to characterize the issues involved. In a multisource net, any pin can (in the worst case) act as the driver of the net on a given clock cycle. As a result, the influence of a particular design solution on, for example, clock period, is complex. This relationship is naturally characterized by the *Augmented RC Diameter*. Given this characterization of the problem, synthesis techniques — which may be viewed as generalizations of those presented in Chapter 6 — are illustrated by an algorithm for optimal bidirectional repeater insertion.
- Finally, Chapter 9 presents algorithms for *timing-driven maze routing*. This is a topic that has received relatively little attention in the literature, but nevertheless is of great practical importance. The problem consists of finding connecting paths between pins that are not only short in length, but also have reasonable delay. The routing resources in the problem are now quite complex: not only must we deal with blockages caused by previously routed nets, but it is also desirable to consider sizing options (some routing channels

may allow wider wires), varying parasitics from one layer to the next, the parasitics of vias, the coupling capacitance with previously routed nets, and so forth. Several practical maze-routing formulations resulting from these complexities are presented, targeted at different phases in the design flow (e.g., detailed routing vs. global routing and buffer allocation in early stages of design) and different global optimization strategies (e.g., constraint-driven vs. composite objective strategies). The algorithms presented are based on graph-labeling schemes.

Throughout the book, exercises are provided to shed light on the discussed topics and give insight into application and generalization of the described techniques for specific purposes.

2

Interconnect Models

In this chapter, we talk about the technology trend and scaling effect on timing, power, noise, and reliability. Scaling into deep submicron technologies has shifted the paradigm from device-dominated to interconnect-dominated design methodology. We thus start with discussing the model of resistance, capacitance, and inductance as basic elements for interconnect analysis. We explore the crosstalk due to coupling capacitance and mutual inductance. Finally, we touch on the transmission line, power dissipation, and reliability.

2.1 TECHNOLOGY TRENDS

In the past two decades, we have seen an explosive growth in the worldwide semiconductor market. It is estimated that in the past two decades, the worldwide semiconductor market has been increasing in size at an annual rate of 15%. This phenomenal growth is tied to the rapid rate of cost reduction, which has been at 25% to 30% per function per year. Rapid cost reduction makes high-performance semiconductor products affordable and thus widely used by the mass public [1].

There are three factors to cost reduction:

1. Feature size reduces at 10.5% to 15.5% per year.
2. Chip size increases 12% per year.
3. Design innovation and the above two factors have contributed to memory cell capacity increases of four times every three years, which corresponds to an increase in design complexity of 58% per year.

Increasing circuit sizes on the one hand and shrinking design rules on the other have shifted ICs and MCMs from being device dominated to being interconnect dominated. In deep-submicron technology, wire delay can no longer be ignored, and must be factored prominently into any overall performance metric. Furthermore, the frequency range of synchronous IC designs continues to increase. The design issues focus on timing, power, noise, and reliability.

2.2 DEVICE AND INTERCONNECT SCALING

Table 2.1 shows the effects of device scaling, while Table 2.2 illustrates the effects of interconnect scaling. Note that these two tables are different from the scaling tables in [2, 3]. We use **scaling ratio** to represent the trend of the scale instead of **scaling factor** used in [2, 3]. For example, as dimension W of a device shrinks by 20%, the scaling factor would be $1/0.8 = 1.25$, but according to our notation, the scaling ratio is $S = 0.8$. This ratio notation eliminates the inverse operation, which is needed when we translate the meaning of scaling factor.

In Table 2.1, we introduce scaling ratio f for clock frequency. Frequency ratio f is used to demonstrate the effect of clock rates. Note that in deep-submicron technologies, clock frequency depends not only on the fabrication processes but also on the design styles. In Table 2.2, we use scaling ratio R for metal effective resistivity to reflect the transition of metal material from aluminum to copper. We use E to indicate the scaling ratio of the oxide permittivity. We denote the global interconnect length scaling ratio as D: the ratio of chip semiperimeter scaling.

In Table 2.1, we use the same scaling ratio S for the device dimensions, gate oxide thickness, and supply voltage. Following the formula listed in the left column, we derive that the transistor resistance remains the same. The transistor resistance is unchanged because the supply voltage level scales down with the

TABLE 2.1 Device Scaling

Quantity	Scaling Ratio
Dimensions (W, L)	S
Area $(A = WL)$	S^2
Gate Oxide Thickness (t_{gox})	S
Supply Voltage (v_{dd})	S
Drain-Source Current $\left(i_{DS} \propto W/L(v_{gox}/t_{gox})(v_{dd} - v_T)^2\right)$	S
Gate Capacitance $\left(C_g = \varepsilon_{gox}(WL/t_{gox})\right)$	S
Transistor On-Resistance $(R_{tr} \propto v_{dd}/i_{DS})$	1
Intrinsic Gate Delay $(\tau = R_{tr}C_g)$	S
Clock Frequency	f
Power Dissipation per Gate $\left(P = (f/z)C_g v_{dd}^2\right)$	fS^3
Power Dissipation Density (P/A)	fS

TABLE 2.2 Interconnect Scaling

Quantity	Scaling Ratio
Cross-Sectional Dimensions ($W_{int}, H_{int}, W_{sp}, t_{ox}$)	S
Metal Effective Resistivity	R
Interconnect Oxide Permittivity (ε_{ox})	E
Resistance per Unit Length ($\mathcal{R}_{int} = \rho_{int}(1/W_{int}H_{int})$)	R/S^2
Capacitance per Unit Length ($\mathcal{C}_{int} = \varepsilon_{ox}(W_{int}/t_{ox})$)	E
RC Delay per Unit Length ($\mathcal{R}_{int}\mathcal{C}_{int}$)	ER/S^2
Local Interconnect Length (l_{loc})	S
Local Interconnect RC Delay ($\mathcal{R}_{int}\mathcal{C}_{int}l_{loc}^2$)	ER
Local Interconnect Power Dissipation per wire ($\propto fl_{loc}\mathcal{C}_{int}v_{dd}^2$)	fES^3
Global Interconnect Length (l_{int})	D
Global Interconnect RC Delay ($\mathcal{R}_{int}\mathcal{C}_{int}l_{int}^2$)	ERD^2/S^2
Global Interconnect Power Dissipation per wire ($\propto fl_{int}\mathcal{C}_{int}v_{dd}^2$)	$fDES^2$
Transmission Line Time of Flight (l_{int}/v_c)	$D\sqrt{E}$

device dimensions. If supply voltage v_{dd} is fixed, the resistance would reduce with ratio S. In that case, the rest of the table should be revised as well. Such fluctuation of the scaling effect would occur, if the scaling of device dimensions and the scaling of the voltage level are not synchronized.

The intrinsic gate delay due to driving device resistance and loading gate capacitance scales down with ratio S. The power dissipation per unit area is independent of S. It increases by the scaling ratio f of clock rate.

In Table 2.2, the interconnect resistance per unit length changes with a ratio R/S^2. The interconnect capacitance per unit length scales down with ratio E. Thus, the ratio of RC delay per unit length becomes ER/S^2 Ignoring the effect of driving device and loading capacitance, we derive the local interconnect RC delay ratio to be ER. The local interconnect power dissipation is fES^3, which reduces as ratio S becomes smaller. On the other hand, for the global net that travels across the whole chip, its delay changes by a ratio ERD^2/S^2, which increases as ratio S decreases. The global net power dissipation is $fDES^2$. Note that this is the ratio of a single net. The total number of nets is proportional to the size of the transistors, that is D^2/S^2.

Let us take an example of the technology trend in a four-year span. The scaling ratios are $S = 0.6$, $f = 2$, $D = 1.3$, $R = 2/3$, and $E = 2/3$. As a result, the intrinsic gate delay drops by a ratio of $S = 0.6$, the local interconnect RC delay drops by a ratio of $ER = 0.44$, and the global interconnect delay increases by a ratio of $ERD^2/S^2 = 2.09$. In terms of power dissipation, the gate power dissipation density increases by a ratio of $fS = 1.2$. Below we give an overview of some of the problems brought up by scaling [4].

2.2.1 Timing

1. *Increase of frequency*: As gate delay drops, a circuit's feasible clock frequency increases. The increase in frequency puts stringent timing constraints on high-performance circuit designs.
2. *Gate delay versus interconnect delay*: As can be seen from Tables 2.1 and 2.2, scaling S causes intrinsic gate delay to *decrease* linearly with S, while causing global interconnect delay to *increase* by a ratio of ERD^2/S^2. Although, we reduce the oxide permittivity E and the metal resistance R to alleviate the signal delay, the overall trend is still dominated by the scaling ratio S. Thus, the trend in deep-submicron technologies is toward dominance of interconnect delay over gate delay.
3. *Dominance of coupling capacitances in overall wire capacitance:* As interconnect scales down, the spacing between wires becomes smaller. Thus, coupling capacitances replace parallel-plate and fringing capacitances as the dominant source for overall wire capacitance.
4. *Impact of inductance on wire delay*: Due to increasing clock speeds and decreasing rise times, inductance becomes an important issue in signal integrity and delay. Inductance causes ringing in the signal waveform, which can degrade signal integrity if it exceeds allowed thresholds. Also, with increasing wire lengths, the effective current return loop is increased, making mutual inductances a nonnegligible effect.
5. *Impact of clock skew variations*: As clock rate increases, the clock skew sensitivity becomes an important design factor in consideration of setup time and hold time constraints. Due to parameter changes, the clock skew shifts with fabrication process and operating environment variations. For setup time constraints, we allocate extra intervals in clock cycle time to accommodate the skew changes. However, for hold time constraints, we cannot solve the constraint violation by just making clock cycle time longer. Proper circuit design and clock distribution is essential to satisfy the timing constraints.

2.2.2 Noise

1. *Crosstalk*: With coupling capacitance and inductance taking on more important roles in deep-submicron technologies, crosstalk (the propagation of a signal to a neighboring wire) becomes a serious problem. If not properly handled, crosstalk can degrade signal integrity and can even cause circuit malfunctions when the coupled noise exceeds allowed thresholds.
2. *Voltage fluctuations due to simultaneous-switching gates*: Another potential source of noise is the so-called simultaneous-switching noise. When many gates switch at the same time, a voltage fluctuation on the power ground network will occur due to the *current* (ΔI) drawn by the simultaneously switching gates. If this fluctuation causes the voltage levels on the network to exceed allowed thresholds (e.g. $> 1.1v_{dd}$ or $< 0.9v_{ss}$), malfunctions can occur.

3. *Charge sharing and leakage current:* Sensitivity of circuits to the noise caused by charge sharing and leakage current increases. In deep submicron, the interconnect capacitance becomes comparable or even larger than the gate capacitance. Thus, for dynamic logic, the charge of interconnect may overwrite the content of the gate capacitance. Due to the drop of threshold voltage, leakage current increases. The leakage current causes design problems of dynamic memory.

2.2.3 Power

Reducing power dissipation becomes important for portable system designs. Due to battery limits, power dissipation is a crucial consideration in portable designs. Even in more conventional systems, power dissipation remains an important part of the engineering picture (e.g., heat management).

1. *Power dissipation versus frequency*: As clock frequency increases, power dissipation increases. Reducing the supply voltage can alleviate this problem, However, voltage level is bounded below by threshold voltage. As voltage level drops, the leakage current increases. Consequently, the leakage current will cause static power dissipation problem.
2. *Power dissipation due to interconnect capacitance*: Traditionally, gates are the main sources of on-chip power dissipation. However, with the vast increase in the amount of interconnect, the situation has changed. Today, interconnect capacitance is the cause of most of the on-chip power dissipation. Clock nets in particular are often a severe source of power dissipation, due to the number and length of these nets, and the fact that they switch on every cycle. Thus, minimization of power dissipation in clock nets becomes an important issue.
3. *Power dissipation due to leakage current*: It is common practice to drop the supply voltage to reduce power dissipation. However, the threshold voltage drops as we reduce the voltage source. Due to the threshold voltage drop, leakage current increases.
4. *Power dissipation due to short circuit*: The sharpness of a signal's rising and falling edges strongly affect the short circuit current. In the middle of a transition, both pull-up and pull-down paths may be on. Thus, we want to reduce the transition period to reduce the short circuit current flow. However, coupling noise may change the shape of the signal transition waveform, which causes extra short circuit current flow, further complicating the problem.

2.2.4 Reliability

1. *Thermal*: Power dissipation causes thermal problems in high-frequency designs. The high density of power dissipation creates hot spots that cause

temperature increase. If the temperature rises beyond the operating range, the chip may fail to function.

2. *Electromigration*: As interconnect shrinks, the current density increases. High current density is the cause of electron migration for aluminum. The replacement of aluminum by copper for interconnect may solve this problem.

2.3 INTERCONNECT MODELS

A VLSI interconnect, such as that between nodes *a* and *b* in Figure 2.1, can be modeled in a number of ways with varying accuracy and computational overhead. The RC model is easy to compute, but relatively inaccurate. The RLC model increases accuracy by incorporating inductance. The transmission line model provides the accuracy but with added computational overhead. Furthermore, for on-chip interconnect, the wires are more than simplified transmission lines because of the coupling to adjacent objects. In this section we will give a brief introduction to each of these three models [5].

For CMOS circuits, if inductance is ignored, the interconnect is usually modeld by an RC tree, as shown in Figure 2.2. Inductance becomes important at high frequencies and for long wires. Inductance can be modeled by inductors that are placed in series with interconnect resistances. The transmission line model can be approximated by the RLC model, with a sufficiently large number of RLC segments. As the number of RLC segments increases, the behavior of the RLC model asymptotically approaches that of the transmission line model.

2.3.1 Resistors

The wire resistance remains constant for a wide range of frequency, although skin effect increases the resistance at very high frequencies. In such a case, the wave penetrates in a short-depth of the conductor as it propagates along the surface.

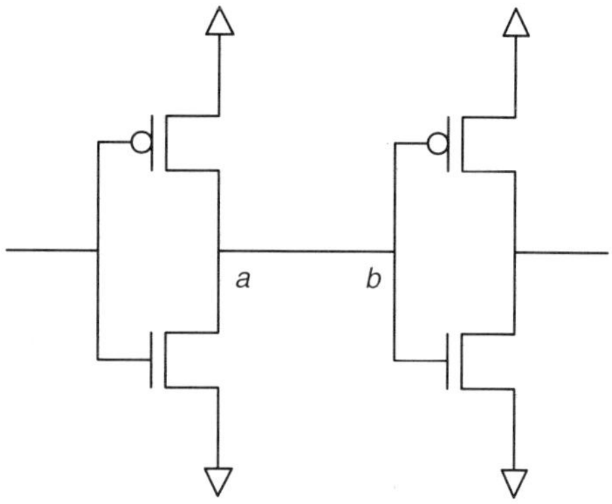

FIGURE 2.1 *A VLSI interconnect.*

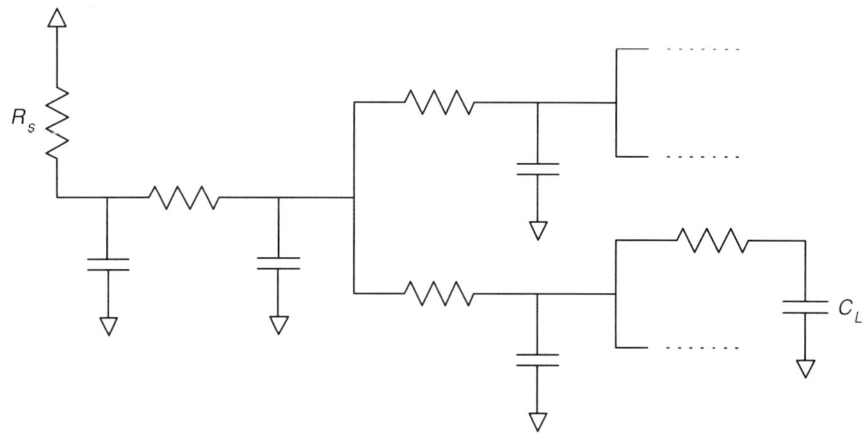

FIGURE 2.2 *A RC tree.*

Thus, current crowds on the surface of the conductor and increases its effective resistance.

The skin depth δ is defined as the penetration distance at which the current density is attenuated by $1/e = -8.7$ decibels:

$$\delta = \sqrt{\frac{\rho}{\pi f \mu}} \tag{2.1}$$

where f, ρ and μ are the frequency, conductor resistivity, and magnetic permeability, respectively. For example, a signal with a $t_r = 100$ pico-seconds rise time has its significant energy at frequency around 3 $GHz \approx 1/\pi t_r$. For 3 GHz signals, the skin depth δ of aluminum is 1.6 μm, which is larger than the line thickness. Thus, the current density can be considered to be uniform.

2.3.2 Capacitors

As long as there is voltage difference across a capacitor, there is charge stored in it. Figure 2.3 shows a simple capacitor. The amount of charge stored in a capacitor has the following relation:

$$Q = Cv \tag{2.2}$$

where Q is the stored charge, C is the capacitance value, and v is the voltage difference across the capacitor.

Differentiating equation (2.2) with respect to time, we get

$$i = C\frac{dv}{dt} \tag{2.3}$$

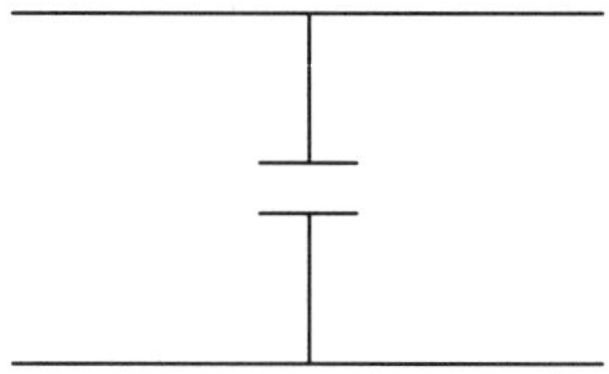

FIGURE 2.3 *A simple capacitor.*

Integrating the equation on both sides, we derive:

$$v(t_2) - v(t_1) = \frac{1}{C} \int_{t_1}^{t_2} i \, dt \tag{2.4}$$

Suppose t_1 and t_2 represent two steady at the start and end of the transition. If the voltage at these two steady states is the same, that is, $v(t_1) = v(t_2)$, then we have

$$\int_{t_1}^{t_2} i \, dt = 0 \tag{2.5}$$

which means the total amount of inflow charge from t_1 to t_2 is equal to outflow charge.

The permittivity of insulator SiO_2 does not change with frequency variations. Thus, we can assume that the interconnect capacitance remains constant.

2.3.3 Inductors

As long as there is current through an inductor (Fig. 2.4), there is magnetic flux stored in it. The amount of magnetic flux stored in an inductor has the following relation:

$$\phi = Li \tag{2.6}$$

where ϕ is the stored magnetic flux, L is the inductance value, and i is the current through the inductor.

Differentiating equation (2.6) with respect to time, we get

$$v = L\frac{di}{dt} \tag{2.7}$$

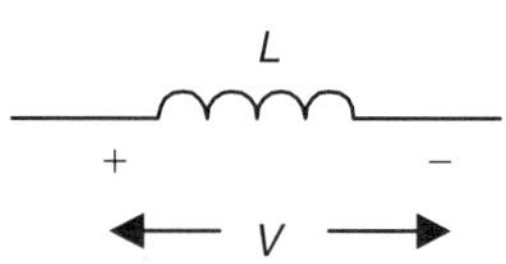

FIGURE 2.4 *A simple inductor.*

Integrating the equation on both sides, we derive:

$$i(t_2) - i(t_1) = \frac{1}{L}\int_{t_1}^{t_2} v\,dt \tag{2.9}$$

Suppose t_1 and t_2 represent two steady states at the start and end of the transition. If the current at these two steady states is zero, that is, $i(t_1) = i(t_2) = 0$, then we have

$$\int_{t_1}^{t_2} v\,dt = 0 \tag{2.10}$$

which means in the voltage versus time curve the area under positive voltage is equal to the area under negative voltage.

Suppose the signal is a sinusoidal wave; Equation (2.8) can be rewritten as $v = j\omega Li$. The impedance of an inductor is $Z_{ind} = j\omega L$. As frequency increases, the contribution of inductance to wire impedance increases linearly, while the contribution of resistance to wire impedance is constant for frequences not too high, as shown in Figure 2.5.

2.3.4 RC Model

On-chip VLSI interconnect is most often modeled by RC trees. The RC trees are a forest with three stipulations: (1) a set of resistors forms a forest of trees where each tree is connected and there is no loop; (2) each tree has at least a source node but no resistor is connected to ground; (3) capacitances are connected between ground and each node of the trees and between nodes of different trees. Figure 2.2 shows a typical RC tree. RC trees lend themselves to efficient analysis and synthesis. These analysis and synthesis techniques will be covered later in this book.

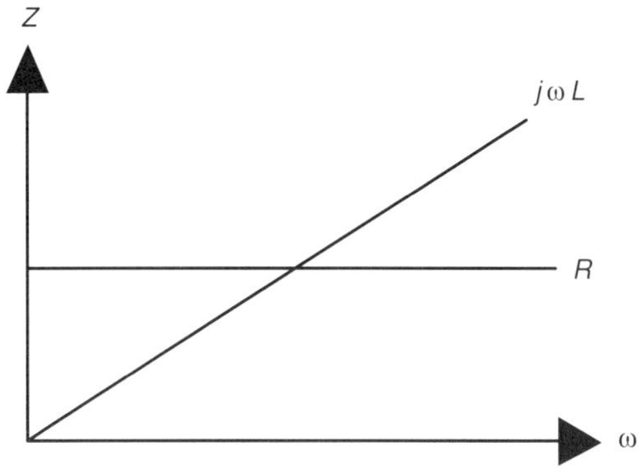

FIGURE 2.5 *Contributions to wire impedance of resistance (R) and inductance (L) as functions of frequency.*

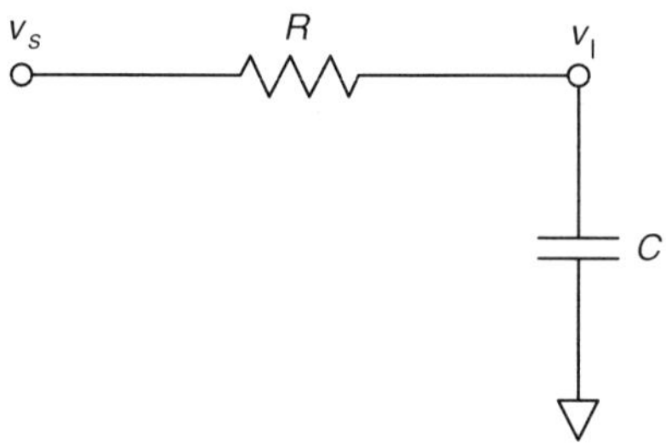

FIGURE 2.6 *The simple RC model.*

The simplest model for a VLSI interconnect is the RC model. In the RC model, each wire is modeled by a resistor equal to the resistance per unit length of the wire, as shown in Figure 2.6.

In Figure 2.6, R is the total wire resistance, that is, $R = R_{unit}l$, where R_{unit} is the resistance per unit length and l the wire length, and C is the load capacitance. From Kirchhoff's current law we have

$$\frac{v_s - v_l}{R} = C\frac{dv_l}{dt} \tag{2.11}$$

Suppose $v_s(S)$ is a step function with magnitude v_{dd} and assume $v_l(0) = 0$; we can solve the above differential equation for v_l and get

$$v_l = v_{dd}\left(1 - e^{(-t/RC)}\right) \tag{2.12}$$

The waveform for v_l is shown in Figure 2.7.

Alternatively, we can obtain equation (2.12) from frequency domain analysis using the Laplace transform:

$$v_l(s) = \frac{1/sC}{R + 1/sC}v_s(s) = \frac{v_s(s)}{sRC + 1} \tag{2.13}$$

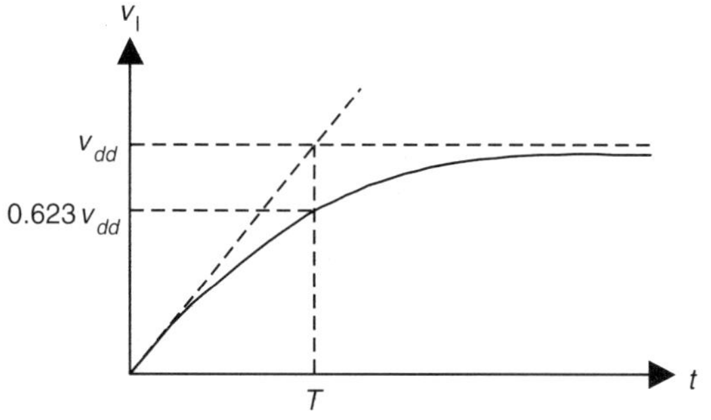

FIGURE 2.7 *Waveform for v_1 in Figure 2.6.*

where $v_s(s) = \mathcal{L}\{v_s(t)\}$ and $v_1(s) = \mathcal{L}\{v_1(t)\}$. Since the input is a step function, we have $v_s(s) = v_{dd}/s$. Thus,

$$v_1(s) = \frac{v_{dd}}{s(sRC + 1)} \tag{2.14}$$

$$= v_{dd}\left(\frac{1}{s} - \frac{RC}{sRC + 1}\right) \tag{2.15}$$

$$= v_{dd}\left(\frac{1}{s} - \frac{1}{s + 1/RC}\right) \tag{2.16}$$

Applying the inverse Laplace transform to equation (2.16), we can obtain the same result as in equation (2.12).

The output expression in time domain (2.12) concludes that the output delay is proportional to the RC constant. The exact portion is determined by the switch voltage. Suppose we assume that the switch voltage is 50% of the voltage source; then the output delay can be calculated from the following equation:

$$1 - e^{(-t_{0.5}/RC)} = 0.5 \tag{2.17}$$

Thus, the output delay is

$$t_{0.5} = \ln 2 * RC = 0.69RC \tag{2.18}$$

Transfer Function of RC Network The transfer function is the ratio between output and input voltages in the s domain:

$$H(s) = v_1(s)/v_s(s) = \frac{1/RC}{s + 1/RC} \tag{2.19}$$

If we view the RC network as a filter, as the frequency range $s = j\omega$ increases, the absolute value of the above equation decreases. Thus, the RC network can be viewed as a *lows pass filter*.

DEFINITION 2.3.1

Impulse Response: The impulse response $H(t)$ is the inverse Laplace transform of the transfer function $H(s)$:

$$H(t) = \frac{1}{RC}e^{-(t/RC)} \tag{2.20}$$

The impulse response can be viewed as the output response when the input is an impulse at time $t = 0$. The impulse wave is an ideal wave with extremely high magnitude and very short interval. The area of the wave is unity. Thus, when the interval approaches zero, the magnitude advances toward infinity. The convolution of the impulse response with input waveform gives the output result. We use the impulse response to characterize the properties of the output versus input relation in time domain.

Taylor's Expansion of Transfer Function We can also characterize the circuit using the Taylor's expansion of the transfer function (2.19) in the s domain. We expand the function about $s = 0$.

$$H(s) = 1 + s(-RC) + s^2(-RC)^2 + \cdots \tag{2.21}$$

From the first few coefficients of the series, we can derive the characteristics of the waveform of the impulse response.

1. The area of the RC impulse response waveform is equal to the first coefficient in the series.

$$\int_0^{\infty} H(t)\,dt = \int_0^{\infty} H(t)e^{-st}\,dt\bigg|_{s=0} = H(s)|_{s=0} = 1 \tag{2.22}$$

2. The *mean* response time of the RC impulse waveform is equal to the coefficient of the first power of s with a sign change.

$$\int_0^{\infty} H(t)t\,dt = \int_0^{\infty} H(t)\frac{-de^{-st}}{ds}\,dt\bigg|_{s=0} = -\frac{dH(s)}{ds}\bigg|_{s=0} = RC \tag{2.23}$$

The mean response time (2.23) indicates that the output delay is proportional to RC. This result is consistent with equation (2.18).

Note that by using Taylor's expansion, we assume that frequency is not too high, ie sRC $\ll$ 1. In other words the angular frequency $\omega \ll (1/\mathrm{RC})$.

2.3.5 RLC Model

When inductance (L) is considered, a VLSI interconnect can be modeled as shown in Figure 2.8.

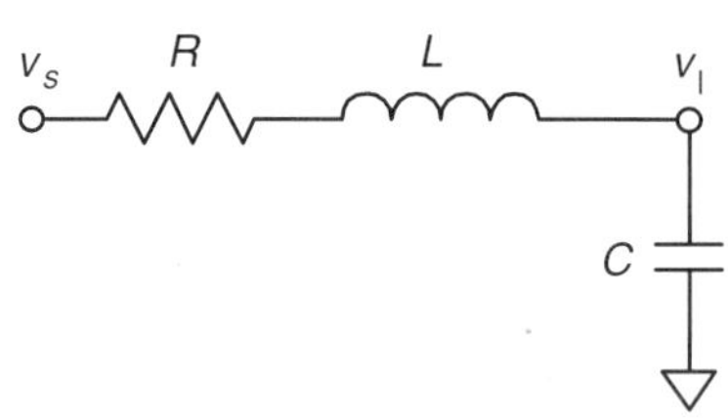

FIGURE 2.8 *The RLC model.*

For the circuit in Figure 2.8, we have from Kirchhoff's voltage law:

$$v_1 + L\frac{di}{dt} + Ri = v_s \tag{2.24}$$

and from Kirchhoff's current law:

$$C\frac{dv_1}{dt} = i \tag{2.25}$$

Substituting equation (2.25) into equation (2.24), we have

$$LC\frac{d^2 v_1}{dt^2} + RC\frac{dv_1}{dt} + v_1 = v_s \tag{2.26}$$

Using Laplace transform, we derive the equation in the s domain.

$$s^2 LC v_1(s) + sRC v_1(s) + v_1(s) = v_s(s) \tag{2.27}$$

Let the input be a step function, that is $v_s(s) = v_{dd}/s$. From frequency domain analysis, we have

$$v_1(s) = \frac{1/sC}{R + Ls + 1/sC} v_s(s) = \frac{1/LC}{s(s^2 + sR/L + 1/LC)} v_{dd} \tag{2.28}$$

We use the following notation to simplify the above equation.

$$\alpha = \frac{R}{2L} \tag{2.29}$$

$$\omega_0 = \frac{1}{\sqrt{LC}} \tag{2.30}$$

Thus, the equation can be expressed as

$$v_1(s) = \frac{\omega^2}{s(s^2 + 2\alpha s + \omega^2)} v_{dd} \tag{2.31}$$

Then, we decompose the output function $v_1(s)$ with partial fraction expressions.

$$v_1(s) = v_{dd} \left\{ \frac{1}{s} - \left(\frac{1}{2} - \frac{\alpha}{2} \frac{1}{\sqrt{\alpha^2 - \omega_0^2}} \right) \frac{1}{s + \alpha + \sqrt{\alpha^2 - \omega_0^2}} \right.$$
$$\left. - \left(\frac{1}{2} + \frac{\alpha}{2} \frac{1}{\sqrt{\alpha^2 - \omega_0^2}} \right) \frac{1}{s + \alpha - \sqrt{\alpha^2 - \omega_0^2}} \right\} \tag{2.32}$$

We define the *quality factor* of the RLC circuit in Figure 2.8 as

$$Q = \frac{\omega_0}{2\alpha} = \frac{1}{R}\sqrt{\frac{L}{C}} \tag{2.33}$$

According to the value of the quality factor, we have three cases:

1. Overdamped, $Q\langle 1/2, \alpha\rangle\omega_0$, or $L < R^2C/4$: From frequency domain analysis, we have equation (2.32)
   We can calculate the inverse Laplace transform to yield

$$v_1 = v_{dd}\left[1 - \left(\frac{1}{2} - \frac{\alpha}{2}\frac{1}{\sqrt{\alpha^2-\omega_0^2}}\right)e^{\left(-\alpha-\sqrt{\alpha^2-\omega_0^2}\right)t} - \left(\frac{1}{2} + \frac{\alpha}{2}\frac{1}{\sqrt{\alpha^2-\omega_0^2}}\right)e^{\left(-\alpha+\sqrt{\alpha^2-\omega_0^2}\right)t}\right] \tag{2.34}$$

   The response for an overdamped RLC circuit is depicted in Figure 2.9.
2. Critically damped, $Q = 1/2$, $\alpha = \omega_0$, or $L = R^2C/4$: From frequency domain analysis, we have

$$v_1(s) = v_{dd}\left[\frac{1}{s} - \frac{\omega_0}{(s+\omega_0)^2} - \frac{1}{s+\omega_0}\right] \tag{2.35}$$

   which can be inverse Laplace transformed to yield

$$v_1 = v_{dd}[1 - (\omega_0 t + 1)e^{-\omega_0 t}] \tag{2.36}$$

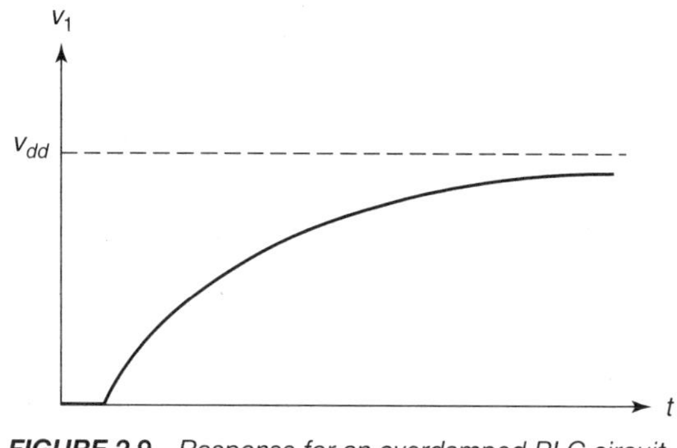

FIGURE 2.9 *Response for an overdamped RLC circuit.*

3. Underdamped, $Q > 1/2$, $\alpha < \omega_0$, or $L > R^2C/4$: From frequency domain analysis, we have

$$v_1(s) = v_{dd}\left(\frac{1}{s} + \frac{\alpha - j\omega_d}{2j\omega_d}\frac{1}{s+\alpha+j\omega_d} - \frac{\alpha + j\omega_d}{2j\omega_d}\frac{1}{s+\alpha-j\omega_d}\right) \tag{2.37}$$

where

$$\omega_d = \sqrt{\omega_0^2 - \alpha^2} \tag{2.38}$$

Applying the inverse Laplace transform to equation (2.37), we have

$$v_1 = v_{dd}\left(1 + \frac{\alpha - j\omega_d}{2j\omega_d}e^{(-\alpha-j\omega_d)t} - \frac{\alpha + j\omega_d}{2j\omega_d}e^{(-\alpha+j\omega_d)t}\right) \tag{2.39}$$

The sum of the two exponential terms can be bounded by the sum of the absolute values of their coefficients.

$$v_{dd}\frac{(\alpha^2 + \omega_d^2)^{1/2}}{\omega_d} = v_{dd}\frac{\omega_0}{\omega_d} \tag{2.40}$$

The response of an underdamped RLC circuit is depicted in Figure 2.10. The waveform is bounded by the first rising edge and the envelope between curves $v_{dd}(1 - (\omega_0/\omega_d)e^{-\alpha t})$ and $v_{dd}(1 + (\omega_0/\omega_d)e^{-\alpha t})$.

Given constant R, C values, the following lemma states that for any inductance L the RLC network output voltage-time area is a constant.

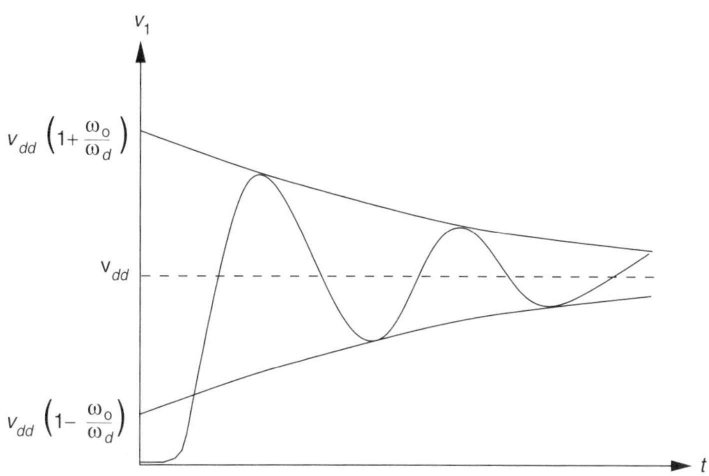

FIGURE 2.10 *The underdamped RLC circuit.*

Lemma 2.3.1 *Let $v_{RLC}(t)$ and $v_{RC}(t)$ be the time-domain waveforms of an RC and an RCL networks with equal R,C values. Suppose that the input in s domain, $v_s(s)$, can be expressed as a rational function $P(s)/Q(s)$ and the denominator $Q(s)$ has at most one pole at $s = 0$. Then the voltage-time areas of the two outputs are equal, that is,*

$$\int_0^\infty (v_{RLC}(t) - v_{RC}(t))\,dt = 0 \tag{2.41}$$

Proof: Let $v_{RLC}(s)$ and $v_{RC}(s)$ be the Laplace transforms of $v_{RLC}(t)$ and $v_{RC}(t)$. The equality can be proved as follows.

$$\int_0^\infty (v_{RLC}(t) - v_{RC}(t))\,dt = \int_0^\infty (v_{RLC}(t) - v_{RC}(t))e^{-st}\,dt\Big|_{s=0} \tag{2.42}$$

$$= (v_{RLC}(s) - v_{RC}(s))|_{s=0} \tag{2.43}$$

$$= \left(\frac{1/sC}{R + Ls + 1/sC} v_s(s) - \frac{1/sC}{R + 1/sC} v_s(s)\right)\Bigg|_{s=0} \tag{2.44}$$

$$= \frac{-LCs^2}{(LCs^2 + RCs + 1)(RCs + 1)} v_s(s)\Bigg|_{s=0} \tag{2.45}$$

$$= 0 \quad \square \tag{2.46}$$

Exercise Assuming the input is a step function, use the waveforms in time domain to derive that the voltage-time area of two outputs, $v_{RLC}(t)$ and $v_{RC}(t)$ are equal.

Remark The above lemma does not hold if input v_s has two or more poles at $s = 0$. For example, if the input is a ramp function, $v_s = 1/s^2$, then equation (2.45) becomes negative:

$$\int_0^\infty (v_{RLC}(t) - v_{RC}(t))\,dt = \frac{-LCs^2}{(LCs^2 + RCs + 1)(RCs + 1)} \frac{1}{s^2}\Bigg|_{s=0} \tag{2.47}$$

$$= -LC \tag{2.48}$$

The above equation indicates that the voltage-time area of RLC network output is smaller. Conceptually, RLC network output becomes slower in response to ramp input. However, in practice, the input signal is not a ramp function. The input is clamped at v_{dd}. It may behave like a ramp function when the signal transition is slow.

Observations

1. At time zero, the slope of the rising edge of an RLC circuit is zero, $dv_{RLC}/dt|_{t=0} = 0$, while the slope of the rising edge of an RC circuit is inversely proportional to the RC product, $dv_{RC}/dt|_{t=0} = v_{dd}/RC$.

2. The rising slope of an RLC circuit increases with time, while the rising slope of an RC circuit decreases with time as $dv_{RC}/dt = v_{dd}/RCe^{-(t/RC)}$.
3. Initially, the waveform of the RLC circuit rises more slowly than that of the RC circuit; however, the RLC waveform will then rise faster and saturate at the voltage source level v_{dd} sooner.
4. For fixed resistances and capacitances, the slew rate and rise time of the response improves with the amount of inductance in the circuit (i.e., as there is more and more inductance, the rising edge gets sharper).
5. If the switch voltage is low, $\ll 50\%\ v_{dd}$, then the response of the RC network is faster than the RLC network. RCL network responds faster when the switch voltage is high.
6. A critically damped RLC circuit will become underdamped once enough inductance is added in the circuit. Then, the waveform will exhibit the behavior of ringing.

Transfer Function of RLC Network The transfer function of the RLC network is

$$H(s) = v_1(s)/v_s(s) = \frac{1/LC}{s^2 + sR/L + 1/LC} \tag{2.49}$$

If we view the RLC network as a filter, as the frequency range $s = jw$ increases, the absolute value of the above equation decreases. Hence, this RLC network can also be viewed as a *low-pass filter*.

Impulse Response of RLC Network Let us use the underdamped case as an example to demonstrate the impulse response of the RLC network. Replacing the coefficients in the transfer function with the notations α, ω_0, and ω_d, equations (2.30), (2.38), we have

$$H(s) = \frac{\omega_0^2}{s^2 + 2\alpha s + \omega_0^2} = \frac{\omega_0^2}{2\omega_d}\left(\frac{-1}{s + \alpha + j\omega_d} + \frac{1}{s + \alpha - j\omega_d}\right) \tag{2.50}$$

We derive the response (Fig 2.11) via inverse Laplace transform.

$$H(t) = \frac{\omega_0^2}{\omega_d} e^{-\alpha t} \sin \omega_d t \tag{2.51}$$

Taylor's Expansion The Taylor's series of the transfer function is derived by recursive division of the rational function (2.49) from small to high order of s.

$$H(s) = 1 + s(-RC) + s^2(R^2C^2 - LC) + \cdots \tag{2.52}$$

Note that the first two terms of the above expression are the same as the transfer function (2.21) of RC network. Thus, the characteristics of the waveform of the impulse response are identical to RC network in terms of area and mean response

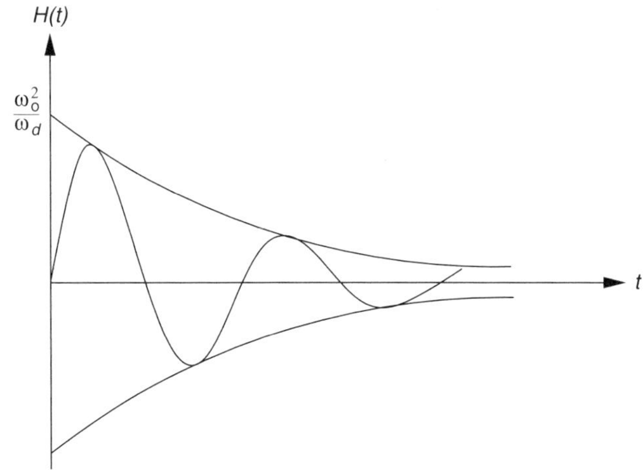

FIGURE 2.11 *Impulse response of an RLC network.*

time. This is consistent with the previous derivation that the output voltage-time area of RLC network is the same as the area of RC network.

The third term starts to differentiate between the two series: the RLC network is smaller by LC.

2.4 THE EFFECT OF CAPACITIVE COUPLING

In this section, we discuss the effect of capacitive coupling. We use a two-line circuit (Fig. 2.12) to illustrate the coupling effect. We derive the exact coupling voltage, its peak, and span in response to a step input voltage source. A simplified symmetric bus line is explored. We also derive the output response of a ramp input.

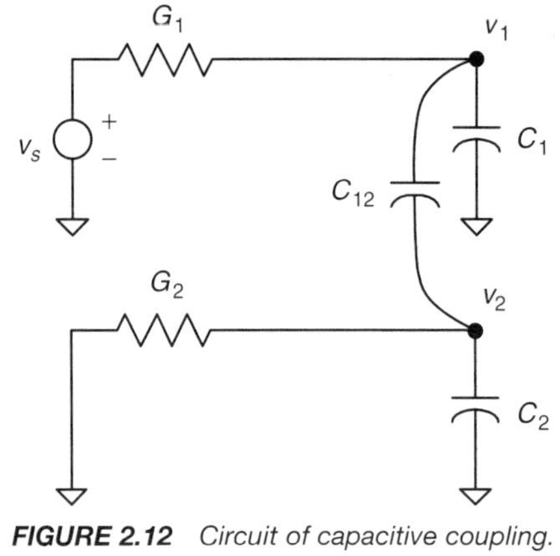

FIGURE 2.12 *Circuit of capacitive coupling.*

Let us derive the capacitive coupling voltage on Line 2. Kirchhoff's current law on the two lines can be expressed as

$$\begin{bmatrix} C_1 + C_{12} & -C_{12} \\ -C_{12} & C_2 + C_{12} \end{bmatrix} \begin{bmatrix} \dfrac{dv_1}{dt} \\ \dfrac{dv_2}{dt} \end{bmatrix} = -\begin{bmatrix} G_1 & 0 \\ 0 & G_2 \end{bmatrix} \begin{bmatrix} v_1 \\ v_2 \end{bmatrix} + \begin{bmatrix} G_1 v_s \\ 0 \end{bmatrix} \tag{2.53}$$

where $G_1 = 1/R_1$ and $G_2 = 1/R_2$. Since we are interested in the coupling voltage v_2, the output relation is

$$v_2 = [0 \quad 1] \begin{bmatrix} v_1 \\ v_2 \end{bmatrix} \tag{2.54}$$

Let us simplify the derivation by assuming that initially there is no voltage across the capacitors, that is, $v_1(t)|_{t=0} = v_2(t)|_{t=0} = 0$. Applying Laplace transform to equation (2.53), we have

$$\left(s \begin{bmatrix} C_1 + C_{12} & -C_{12} \\ -C_{12} & C_2 + C_{12} \end{bmatrix} + \begin{bmatrix} G_1 & 0 \\ 0 & G_2 \end{bmatrix} \right) \begin{bmatrix} v_1(s) \\ v_2(s) \end{bmatrix} = \begin{bmatrix} G_1 v_s(s) \\ 0 \end{bmatrix} \tag{2.55}$$

where $v_1(s) = \mathcal{L}(v_1(t))$, $v_2(s) = \mathcal{L}(v_2(t))$, and $v_s(s) = \mathcal{L}(v_s(t))$ are the Laplace transforms of $v_1(t)$, $v_2(t)$, and $v_s(t)$, respectively. Solving equation (2.55) for $v_1(s)$ and $v_2(s)$, we have

$$\begin{bmatrix} v_1(s) \\ v_2(s) \end{bmatrix} = \left(s \begin{bmatrix} C_1 + C_{12} & -C_{12} \\ -C_{12} & C_2 + C_{12} \end{bmatrix} + \begin{bmatrix} G_1 & 0 \\ 0 & G_2 \end{bmatrix} \right)^{-1} \begin{bmatrix} G_1 v_s(s) \\ 0 \end{bmatrix} \tag{2.56}$$

Combining equation (2.54) and equation (2.56) and rearranging terms, we have the output function.

$$v_2(s) = [0 \quad 1] \begin{bmatrix} G_1 + s(C_1 + C_{12}) & -sC_{12} \\ -sC_{12} & G_2 + s(C_2 + C_{12}) \end{bmatrix}^{-1} \begin{bmatrix} G_1 v_s(s) \\ 0 \end{bmatrix} \tag{2.57}$$

$$= \frac{sC_{12} G_1 v_s(s)}{[G_1 + s(C_1 + C_{12})][G_2 + s(C_2 + C_{12})] - s^2 C_{12}^2} \tag{2.58}$$

$$= \frac{sC_{12} G_1 v_s(s)}{s^2(C_1C_2 + C_1C_{12} + C_2C_{12}) + s(G_1(C_2 + C_{12}) + G_2(C_1 + C_{12})) + G_1G_2} \tag{2.59}$$

2.4.1 Output Response of Step Input

Suppose $v_s(t)$ is a unit step rising at time $t = 0$ so that the s domain expression $v_s(s) = \frac{1}{s}$, and let us use a, b, and d to replace the coefficients of the denominator of $v_2(s)$ as a polynomial expression of s.

$$\begin{aligned} a &= C_1C_2 + C_1C_{12} + C_2C_{12} \\ b &= G_1(C_2 + C_{12}) + G_2(C_1 + C_{12}) \\ d &= G_1G_2 \end{aligned} \tag{2.60}$$

We can simplify the coupling effect in the following rational function.

$$v_2(s) = \frac{C_{12}G_1}{as^2 + bs + d} \tag{2.61}$$

Rearranging terms and doing partial fraction decomposition, we have

$$v_2(s) = \frac{C_{12}G_1}{a} \frac{1}{s^2 + \frac{b}{a}s + \frac{d}{a}} \tag{2.62}$$

$$= \frac{C_{12}G_1}{a} \frac{1}{\left(s + \frac{b + \sqrt{b^2 - 4ad}}{2a}\right)\left(s + \frac{b - \sqrt{b^2 - 4ad}}{2a}\right)} \tag{2.63}$$

$$= \frac{-\frac{C_{12}G_1}{\sqrt{b^2 - 4ad}}}{s + \frac{b + \sqrt{b^2 - 4ad}}{2a}} + \frac{\frac{C_{12}G_1}{\sqrt{b^2 - 4ad}}}{s + \frac{b - \sqrt{b^2 - 4ad}}{2a}} \tag{2.64}$$

Note that $b^2 - 4ad \geq 0$ in the above expression. Applying inverse Laplace transform to equation (2.64), we get the time-domain expression:

$$v_2 = -\frac{C_{12}G_1}{\sqrt{b^2 - 4ad}} e^{-(b+\sqrt{b^2-4ad}/2a)t} + \frac{C_{12}G_1}{\sqrt{b^2 - 4ad}} e^{-(b-\sqrt{b^2-4ad}/2a)t} \tag{2.65}$$

Peak of Capacitive Coupling Voltage To find the maximum value of v_2 from equation (2.65), we extract the format of the expression with a function y and obtain the solution with the following lemma.

Lemma 2.4.1 *Given a function $y(t)$ in the format of:*

$$y(t) = -e^{(-f-h)t} + e^{(-f+h)t} \tag{2.66}$$

where f and h are positive constants independent of t, the maximum of $y(t)$ is equal to

$$y_{max} = 2h(f + h)^{-(1/2)-(f/2h)}(f - h)^{-(1/2)+(f/2h)} \tag{2.67}$$

Proof: The maximum of $y(t)$ occurs at time $t = t_{max}$ when its derivative with respect to time is zero:

$$\left.\frac{dy(t)}{dt}\right|_{t=t_{max}} = (f+h)e^{(-f-h)t_{max}} + (-f+h)e^{(-f+h)t_{max}} = 0 \tag{2.68}$$

Solving equation (2.68), we can obtain the value of t_{max}:

$$t_{max} = \frac{1}{2h}\ln\frac{f+h}{f-h} \tag{2.69}$$

Substituting t_{max} into equation (2.66), we can obtain y_{max}, the maximum value of $y(t)$:

$$y_{max} - -e^{(-f-h)1/2h\ln(f+h/f-h)} + e^{(-f+h)1/2h\ln(f+h/f-h)} \tag{2.70}$$

$$= e^{-f1/2h\ln(f+h/f-h)}\left(-e^{-h1/2h\ln(f+h/f-h)} + e^{h1/2h\ln(f+h/f-h)}\right) \tag{2.71}$$

$$= \left(\frac{f+h}{f-h}\right)^{-(f/2h)}\left(-e^{-(1/2)\ln(f+h/f-h)} + e^{1/2\ln(f+h/f-h)}\right) \tag{2.72}$$

$$= \left(\frac{f+h}{f-h}\right)^{-(f/2h)} e^{1/2\ln(f+h/f-h)}\left(-e^{-\ln(f+h/f-h)} + 1\right) \tag{2.73}$$

$$= \left(\frac{f+h}{f-h}\right)^{-(f/2h)}\left(\frac{f+h}{f-h}\right)^{1/2}\left(1 - \left(\frac{f+h}{f-h}\right)^{-1}\right) \tag{2.74}$$

$$= \frac{2h}{f+h}\frac{(f+h)^{-(f/2h)+(1/2)}}{(f-h)^{-(f/2h)+(1/2)}} \tag{2.75}$$

$$= 2h(f+h)^{-(1/2)-(f/2h)}(f-h)^{-(1/2)+(f/2h)} \quad \square \tag{2.76}$$

Since the format of $y(t)$ matches that of v_2 in equation (2.65) with $f = b/2a$ and $h = \sqrt{(b^2 - 4ad/2a)}$, we can obtain $v_{2,max}$, the maximum value of v_2:

$$v_{2,max} = \frac{2hC_{12}G_1}{\sqrt{b^2-4ad}}\frac{(f+h)^{-(1/2)-(f/2h)}}{(f-h)^{(1/2)-(f/2h)}} \tag{2.78}$$

$$= \frac{2hC_{12}G_1}{\sqrt{b^2-4ad}}\frac{1}{(f-h)^{(1/2)-(f/2h)}(f+h)^{(1/2)+(f/2h)}} \tag{2.79}$$

Rearranging the terms in equation (2.79), we can express $v_{2,max}$ in the notations of a, b, and d. Thus, the *peak capacitive coupling voltage* is

$$v_{2,max} = \frac{C_{12}G_1}{\sqrt{ad}}\left(\frac{b-\sqrt{b^2-4ad}}{b+\sqrt{b^2-4ad}}\right)^{b/2\sqrt{b^2-4ad}} \tag{2.80}$$

Let us set $x = 4ad/b^2$ and expand a and d; we rewrite the above equation in the following format:

$$v_{2,max} = \sqrt{\frac{G_1}{G_2}} \frac{C_{12}}{\sqrt{C_1C_2 + C_1C_{12} + C_2C_{12}}} \left(\frac{1-\sqrt{1-x}}{1+\sqrt{1-x}}\right)^{1/2\sqrt{1-x}} \tag{2.81}$$

Figure 2.13 shows the relation of the last term $g(x) = \left(1 - \sqrt{1-x}/1+\sqrt{1-x}\right)^{1/2\sqrt{1-x}}$ versus x.

As x increases from 0 to 0.3, the last term rises sharply from 0 to 0.235. When x approaches one, the last term saturates in the range of 0.368. If we assume that x and the last term do not vary much, the peak of the coupling voltage is proportional to the first two terms: $v_{2,max} \propto \sqrt{(G_1/G_2)}\,(C_{12}/\sqrt{C_1C_2 + C_1C_{12} + C_2C_{12}})$.

Exercise Replace the parameters with a few sets of numbers to compare the peak of the coupling voltage with a simplified bound set by Vittal and Marek-Sadowska [6]:

$$v_{2,max} \leq \frac{C_{12}}{C_2 + C_{12} + \dfrac{G_2}{G_1}(C_1 + C_{12})}. \tag{2.82}$$

Symmetric Bus Lines To simplify the expression of the worst-case coupling, we assume that the two coupled lines have identical conductance, resistance, and self-capacitance. In other words, we assume that they are symmetric. Symmetric lines are often observed in bus lines and differential nets. Hence, the simplified results are applicable to those cases.

Let us denote

$$C = C_1 = C_2 \tag{2.83}$$

$$G = G_1 = G_2 \tag{2.84}$$

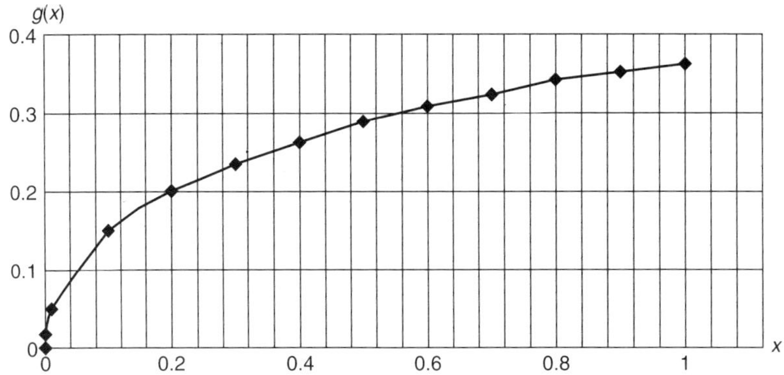

FIGURE 2.13 *g(x) versus x.*

We can, therefore, simplify the expression of a, b, and d.

$$a = C^2 + 2CC_{12} \tag{2.85}$$

$$b = 2G(C + C_{12}) \tag{2.86}$$

$$d = G^2 \tag{2.87}$$

Consequently, the terms in the worst-case coupling expression (2.80) can be expressed in the resistor and capacitor values.

$$\frac{b + \sqrt{b^2 - 4ad}}{2a} = \frac{G}{C} \tag{2.88}$$

$$\frac{b - \sqrt{b^2 - 4ad}}{2a} = \frac{G}{C + 2C_{12}} \tag{2.89}$$

$$\frac{b}{2\sqrt{b^2 - 4ad}} = \frac{C + C_{12}}{2C_{12}} \tag{2.90}$$

Replacing the terms in the worst-case coupling expression (2.80) with the above equations we derive the solution of $v_{2,max}$.

$$v_{2,max} = \frac{C^{(C/2C_{12})}C_{12}}{(C + 2C_{12})^{(C/2C_{12})}(C + 2C_{12})} \tag{2.91}$$

Let us denote x the ratio of self capacitance C and mutual capacitance C_{12} (i.e., $x = C/C_{12}$). We rewrite $v_{2,max}$ as a function of x.

$$v_{2,max} = \frac{x^{(x/2)}}{(2 + x)^{(1+x/2)}} \tag{2.92}$$

In Figure 2.14, we list the relation of peak coupling voltage $v_{2,max}$ versus coupling capacitance ratio C/C_{12}.

1. We can observe that for a pair of symmetric signal nets, the peak of the capacitive coupled voltage is independent of the conductance. The peak value is a function of C/C_{12}.
2. When the ratio C/C_{12} approaches zero, mutual capacitance dominates the capacitance of the wire. The peak coupling voltage progresses toward 0.5.
3. As the ratio C/C_{12} increases, mutual capacitance becomes a smaller portion of the net capacitance. In the case that the self-capacitance equals the mutual capacitance, $C/C_{12} = 1$, the peak value drops to 0.19. The peak value drops sharply and reduces to zero.

Span of Capacitive Coupling Voltage The span of the voltage is also an important factor to measure the coupling. It is dominated by the interconnect

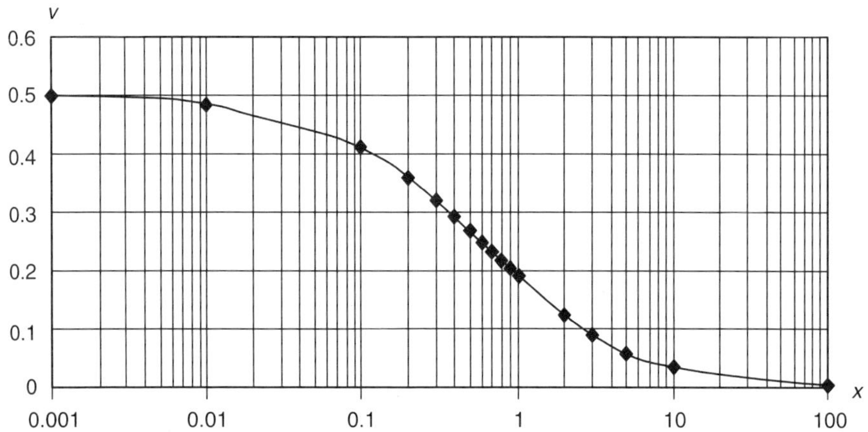

FIGURE 2.14 *Peak capacitive coupling voltage $v_{2,max}$ versus $x = C/C_{12}$ [equation (2.92)].*

resistance and coupling capacitance. This relation can be derived by the voltage-time area of the coupling.

$$\int_0^\infty v_2(t)\,dt = v_2(s)|_{s=0} = \frac{C_{12}G_1}{d} = \frac{C_{12}G_1}{G_1G_2} = R_2C_{12} \tag{2.93}$$

The above equation shows that the coupling voltage-time area is proportional to the net resistance. We can reduce the span of the coupling voltage by decreasing the resistance.

Transfer Function of Capacitive Coupling The transfer function of the capacitive coupling can be derived from equation (2.59) and the notations a, b, and d in (2.60).

$$H(s) = \frac{C_{12}G_1 s}{as^2 + bs + d} \tag{2.94}$$

Applying recursive division from low to high order of s, we can derive Taylor's expansion as

$$H(s) = \frac{C_{12}G_1}{d}s - \frac{C_{12}G_1 b}{d^2}s^2 + \cdots \tag{2.95}$$

Since the first term starts with first order of s, we can conclude that the areas of the impulse response above and below the time t axis are balanced, that is,

$$\int_0^\infty H(t)\,dt = 0 \tag{2.96}$$

Replacing d with G_1G_2, equation (2.60), we have the mean time of the response:

$$\int_0^\infty H(t)t\,dt = \frac{C_{12}G_1}{d} = \frac{C_{12}G_1}{G_1G_2} = \frac{C_{12}}{G_2} = R_2C_{12} \tag{2.97}$$

which is equal to the span of the coupling voltage, equation (2.93).

2.4.2 Output Response of Ramp Input

We can derive another upper bound on the coupled voltage output using a ramp input. For a ramp input, the slope of the transition has strong effect on the coupling noise. Thus, we simplify the case by assuming that the ramp input has no clamp on the voltage range, that is, $v_s(t) = \alpha t$, to derive the worst-case behavior. This model is pertinent when the input voltage has a slow-rising slope.

For a ramp input, its expression in s domain is

$$v_s(s) = \alpha\frac{1}{s^2} \tag{2.98}$$

The output response can be derived from equations (2.59) and (2.60) in s domain.

$$v_2(s) = H(s)v_s(s) = \frac{\alpha C_{12}G_1}{(as^2 + bs + d)s} \tag{2.99}$$

We decompose the above function in a partial fraction expression.

$$\begin{aligned} v_2(s) = {} & \frac{\alpha C_{12}G_1}{ds} + \frac{\alpha C_12G_1}{2d}\frac{\left(-1 + \dfrac{b}{\sqrt{b^2 - 4ad}}\right)}{s + \dfrac{b + \sqrt{b^2 - 4ad}}{2a}} \\ & + \frac{\alpha C_12G_1}{2d}\frac{\left(-1 - \dfrac{b}{\sqrt{b^2 - 4ad}}\right)}{s + \dfrac{b - \sqrt{b^2 - 4ad}}{2a}} \end{aligned} \tag{2.100}$$

Suppose we transform the above equation into time domain; the first term can be described as a step function. The second and third terms are exponential functions. Because the third term is more negative and it attenuates more slowly, we can derive that the net value of these two terms is negative and approaches zero as time progresses toward infinity. Thus, we conclude that the maximum value of the output voltage $v_{2,max}$ is bounded by the first term. Thus, voltage $v_{2,max}$ is no larger than the coefficient of the step function, that is,

$$v_{2,max}(t) \leq \frac{\alpha C_{12}G_1}{d} = \frac{\alpha C_{12}G_1}{G_1G_2} = \frac{\alpha C_{12}}{G_2} = \alpha R_2C_{12}. \tag{2.101}$$

Note that in the above expression α multiplies the term R_2C_{12} where R_2C_{12} is equal to the coefficient of the first power s of the transfer function $H(s)$.

2.5 THE EFFECT OF INDUCTIVE COUPLING

We present the effect of inductive coupling using a two-line circuit (Fig. 2.15). We assume that the lines are symmetric to simplify the notations in the derivation. The Taylor's expansion of the transfer function is discussed.

We derive the circuit analysis with Kirchhoff's voltage and current laws. The following simultaneous equation is Kirchhoff's current law for currents injecting into the capacitors.

$$\begin{bmatrix} C_1 & 0 \\ 0 & C_2 \end{bmatrix} \begin{bmatrix} \dfrac{dv_1}{dt} \\ \dfrac{dv_2}{dt} \end{bmatrix} = \begin{bmatrix} i_1 \\ i_2 \end{bmatrix} \tag{2.102}$$

Kirchhoff's voltage law is for the voltages across the inductors.

$$\begin{bmatrix} L_1 & L_m \\ L_m & L_2 \end{bmatrix} \begin{bmatrix} \dfrac{di_1}{dt} \\ \dfrac{di_2}{dt} \end{bmatrix} = - \begin{bmatrix} v_1 \\ v_2 \end{bmatrix} - \begin{bmatrix} G_1 & 0 \\ 0 & G_2 \end{bmatrix} \begin{bmatrix} i_1 \\ i_2 \end{bmatrix} + \begin{bmatrix} v_s \\ 0 \end{bmatrix} \tag{2.103}$$

Combining the above two equations, we can derive a second-order differential equation.

$$\begin{bmatrix} L_1 & L_m \\ L_m & L_2 \end{bmatrix} \begin{bmatrix} C_1 & 0 \\ 0 & C_2 \end{bmatrix} \begin{bmatrix} \dfrac{d^2v_1}{dt^2} \\ \dfrac{d^2v_2}{dt^2} \end{bmatrix} = - \begin{bmatrix} v_1 \\ v_2 \end{bmatrix} - \begin{bmatrix} G_1 & 0 \\ 0 & G_2 \end{bmatrix} \times \begin{bmatrix} C_1 & 0 \\ 0 & C_2 \end{bmatrix} \begin{bmatrix} \dfrac{dv_1}{dt} \\ \dfrac{dv_2}{dt} \end{bmatrix} + \begin{bmatrix} v_s \\ 0 \end{bmatrix} \tag{2.104}$$

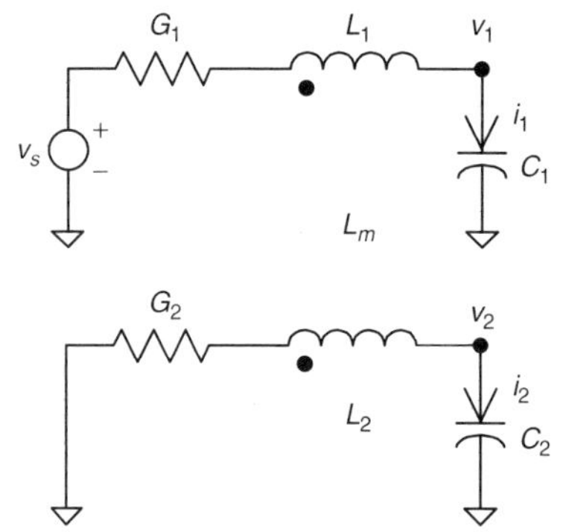

FIGURE 2.15 Circuit of Inductive Coupling.

Using the Laplace transform, we can express the above differential equation in the s domain.

$$\left(s^2\begin{bmatrix} L_1 & L_m \\ L_m & L_2 \end{bmatrix}\begin{bmatrix} C_1 & 0 \\ 0 & C_2 \end{bmatrix} + s\begin{bmatrix} G_1 & 0 \\ 0 & G_2 \end{bmatrix}\begin{bmatrix} C_1 & 0 \\ 0 & C_2 \end{bmatrix} + \begin{bmatrix} 1 & 0 \\ 0 & 1 \end{bmatrix}\right) \times \begin{bmatrix} v_1(s) \\ v_2(s) \end{bmatrix} = \begin{bmatrix} v_s(s) \\ 0 \end{bmatrix} \tag{2.105}$$

Again, we assume that the two nets are symmetric to simplify the derivation, that is,

$$L = L_1 = L_2 \tag{2.106}$$

$$C = C_1 = C_2 \tag{2.107}$$

$$G = G_1 = G_2 \tag{2.108}$$

Thus, we can express the system equation as

$$\left(s^2\begin{bmatrix} L & L_m \\ L_m & L \end{bmatrix}\begin{bmatrix} C & 0 \\ 0 & C \end{bmatrix} + s\begin{bmatrix} G & 0 \\ 0 & G \end{bmatrix}\begin{bmatrix} C & 0 \\ 0 & C \end{bmatrix} + \begin{bmatrix} 1 & 0 \\ 0 & 1 \end{bmatrix}\right) \times \begin{bmatrix} v_1(s) \\ v_2(s) \end{bmatrix} = \begin{bmatrix} v_s(s) \\ 0 \end{bmatrix} \tag{2.109}$$

Since capacitance C is attached to either L or G in all its occurrence, we use the following replacement to simply the notations.

$$l = CL, l_m = CL_m, \quad g = CG \tag{2.110}$$

Thus, we have

$$\left(s^2\begin{bmatrix} l & l_m \\ l_m & l \end{bmatrix} + s\begin{bmatrix} g & 0 \\ 0 & g \end{bmatrix} + \begin{bmatrix} 1 & 0 \\ 0 & 1 \end{bmatrix}\right)\begin{bmatrix} v_1(s) \\ v_2(s) \end{bmatrix} = \begin{bmatrix} Gv_s(s) \\ 0 \end{bmatrix} \tag{2.111}$$

$$\begin{bmatrix} s^2 l + sg + 1 & s^2 l_m \\ s^2 l_m & s^2 l + sg + 1 \end{bmatrix}\begin{bmatrix} v_1(s) \\ v_2(s) \end{bmatrix} = \begin{bmatrix} Gv_s(s) \\ 0 \end{bmatrix} \tag{2.112}$$

The voltage vector can be derived by inverting the first matrix.

$$\begin{bmatrix} v_1(s) \\ v_2(s) \end{bmatrix} = \frac{1}{(s^2 l + sg + 1)^2 - s^4 l_m^2} \times \begin{bmatrix} s^2 l + sg + 1 & -s^2 l_m \\ -s^2 l_m & s^2 l + sg + 1 \end{bmatrix}\begin{bmatrix} v_s(s) \\ 0 \end{bmatrix} \tag{2.113}$$

Therefore, we can express the output function as

$$v_2(s) = \frac{-s^2 l_m v_s(s)}{(s^2 l + sg + 1)^2 - s^4 l_m^2} \tag{2.114}$$

$$= \frac{-s^2 l_m v_s(s)}{(s^2(l + l_m) + sg + 1)(s^2(l - l_m) + sg + 1)} \tag{2.115}$$

Let us assume that input $v_s(s)$ is a step function, that is, $v_s(s) = v_{dd}/s$; we have

$$v_2(s) = -\frac{1}{2}\left(\frac{-1}{s^2(l + l_m) + sg + 1} + \frac{1}{s^2(l - l_m) + sg + 1}\right)\frac{v_{dd}}{s} \tag{2.116}$$

$$= v_{dd}\left(\frac{A}{s + p_1} + \frac{B}{s + p_2} + \frac{C}{s + p_3} + \frac{D}{s + p_4} + \frac{E}{s}\right) \tag{2.117}$$

The output response in time domain can be expressed as

$$v_2 = v_{dd}\left(Ae^{-p_1 t} + Be^{-p_2 t} + Ce^{-p_3 t} + De^{-p_4 t} + E\right). \tag{2.118}$$

where

$$p_1 = \frac{g + \sqrt{g^2 - 4(l + l_m)}}{2(l + l_m)} \qquad p_2 = \frac{g - \sqrt{g^2 - 4(l + l_m)}}{2(l + l_m)} \tag{2.119}$$

$$p_3 = \frac{g + \sqrt{g^2 - 4(l - l_m)}}{2(l - l_m)} \qquad p_4 = \frac{g - \sqrt{g^2 - 4(l - l_m)}}{2(l - l_m)} \tag{2.120}$$

$$A = \frac{l}{4(l + l_m)^2}\left(-1 + \frac{g}{\sqrt{g^2 - 4(l + l_m)}}\right) \tag{2.121}$$

$$B = \frac{l}{4(l + l_m)^2}\left(-1 - \frac{g}{\sqrt{g^2 - 4(l + l_m)}}\right) \tag{2.122}$$

$$C = \frac{l}{4(l - l_m)^2}\left(-1 + \frac{g}{\sqrt{g^2 - 4(l - l_m)}}\right) \tag{2.123}$$

$$D = \frac{l}{4(l - l_m)^2}\left(-1 - \frac{g}{\sqrt{g^2 - 4(l . l_m)}}\right) \tag{2.124}$$

$$E = \frac{2 l l_m}{(l + l_m)^2 (l - l_m)^2} \tag{2.125}$$

Transfer Function and Taylor's Expansion The transfer function of the network can be derived from equation (2.115).

$$H(s) = \frac{v_2(s)}{v_s(s)} = \frac{-s^2 l_m}{(s^2(l + l_m) + sg + 1)(s^2(l - l_m) + sg + 1)} \tag{2.126}$$

The Taylor's expansion of the transfer function (2.126) can be written as follows.

$$H(s) = -l_m s^2 + 2g l_m s^3 + \cdots \tag{2.127}$$

Since the first term starts with second order of s, we can conclude that areas of the impulse response above and below time-t axis are balanced, that is,

$$\int_0^\infty H(t)\, dt = 0 \tag{2.128}$$

The first-order moment is also zero.

$$\int_0^\infty H(t) t\, dt = 0 \tag{2.129}$$

2.6 TRANSMISSION LINE MODEL

Transmission line properties of on-chip wiring need to be taken into account due to the long lengths and fast rise times encountered. Below is a description of the speed of light in air and in a media with relative permittivity of 4.

$$\begin{aligned} \text{Air} \quad & v = 3 \times 10^{10}\ \text{cm/s} \Rightarrow v = 3\ \text{cm/0.1ns} \\ k = 4 \quad & v = 1.5 \times 10^{10}\ \text{cm/s} \Rightarrow v = 1.5\ \text{cm/0.1ns} \end{aligned}$$

We can check that very soon the distance for a signal to travel in the period of a 0–1 transition will be comparable to the dimension of the die. The traditional lumped-circuit representation is no longer adequate, since it results in substantial underestimation of both crosstalk and delay. However, on-chip interconnects have unique characteristics, namely high capacitive and inductive coupling, nonnegligible resistive losses, and nonuniform transmission-line structure.

As a rule of thumb, transmission line behavior becomes significant when the signal rise time is smaller than 2.5 times the time of flight. The line acts as a lumped element circuit when the rise time is larger than 5 times the time of flight. In between there is a gray area where either approximation might work depending on the accuracy required.

The effective resistance of the transmission line can increase significantly with frequency, while the inductance decreases. This is due to the inductance and resistance trade-off of the return paths. At higher frequencies, the return current concentrates in the low inductance paths, which may contain high resistance. Therefore, the inductance drops while the resistance increases.

2.7 POWER DISSIPATION

The *dynamic power dissipation* of a VLSI interconnect with a capacitive load can be calculated by

$$P_{dynamic} = f \int v_{dd} i \, dt = f v_{dd} \int C \frac{dv}{dt} dt = f C v_{dd}^2 \tag{2.130}$$

where f is the frequency of the wire. Note that in an RC net model, the dynamic power dissipation is independent of the resistance. The resistance plays a role in determining the signal delay. Once the delay constraint is satisfied, the power dissipation is determined by the capacitance.

In addition to dynamic power dissipation, power is also dissipated by short circuit current and leakage current. Thus, the total power dissipation is

$$P_{total} = P_{dynamic} + P_{short} + P_{leakage} = f C v_{dd}^2 + f v_{dd} i_{short} + v_{dd} i_{leakage} \tag{2.131}$$

where i_{short} is the short circuit current and $i_{leakage}$ is the leakage current.

2.8 INTERCONNECT RELIABILITY

There are various issues associated with interconnect reliability [7]. The reliability behavior is very much dependent upon the materials of the interconnect [9]. Four sources of failure have been identified in aluminum and aluminum-based alloy films: (1) electromigration, (2) formation of stress-induced voids, (3) formation of thermal hillocks, and (4) grain collapse. Among the four sources, electromigration is one dominant factor of failure when current density is high. In this section, we consider the design issues associated with electromigration.

Electromigration occurs when the metal ions in a wire are repeatedly impacted by electrons and thus transported. This transport of metal ions may cause open circuits or short circuits in the wire and may result in reliability problems.

For electromigration analysis, we decompose the current into alternating current and direct current two components.

One view contributes major electromigration to direct current. Alternating current results in a much lower percentage of electromigration damage [8] because the bidirectionanity of the current tends to cancel the sweeping effect, and tends to keep the metal ions more or less in their original positions. For example, in [10] Rodbell et al. identify in their experiments that when direct and alternating currents are superposed, the electromigration lifetimes are inversely proportional to the magnitude of the direct current applied. Yet in a conservative view, various formulas have also been used to model alternating current effect. In [11], an average current model is used for both alternating and direct currents.

One simple measure of electromigration is the current density in the direct current component. If the current density of a wire exceeds a certain threshold, electromigration is likely to occur. This poses a particular threat for deep-submicron technologies: in deep-submicron technologies, wire widths are shrunk due to the increasing integration density, while current drive is increased to enhance speed. Thus, deep-submicron technologies are particularly susceptible to electromigration. A conservative rule of thumb is that to avoid electromigration, current density should be kept below 1 mA/micron^2. However, other factors also affect electromigration, such as temperature and manufacturing process variations. Because of this quasi-random nature, electromigration usually occurs after the chip has been in operation for a while, and is thus hard to catch.

Major direct current component occurs in power ground distribution. Thus, one issue in the design of the power ground distribution is to keep the current density below the electromigration threshold under all possible usages.

In CMOS technology, the signal networks are loaded with capacitance. For a single-sourced interconnect tree, the portion of direct current component is zero. The signal is purely alternating current as described by equation (2.5). However, when the network is not a tree, but a mesh, we can anticipate direct current component. In case of a mesh, the current flow path is not unique. The "zero" and "one" signals can have different preference of the paths. The difference can be caused by the impedance variation due to different frequency ingredients of the "zero" and "one" waveforms. Consequently, for each path, the currents in two directions are different. We can then observe the direct current component flowing through the wire segments.

Other possible places of direct current distribution are multiple-sourced buses and active cells. For example, in a multiple-sourced bus, if the "one" signal is always driven by gate i whereas the "zero" signal is always driven by gate j, then wire segment (i,j) linking the outputs of gates i and j will experience a direct current from node i to node j.

A similar situation happens in active cells. Figure 2.16 shows an example of direct current inside a folded NOR cell. The two inputs a and b are tri-folded. The output z is connected to the left via a horizontal metal wire. In the top row, we have two versions of the cell layout. We assume that signals a and b are independent and each has 50% probability to be one or zero. The middle row shows the expected current flow of the horizontal metal wire. Five vertical arrows branch at the locations of five vertical wire segments in the layout. The labels Pi, Ni show the percentage of current flow through the p-channel and n-channel, respectively. The bottom row displays the magnitude of direct current. Thus, the cell in figure 2.16(b) has less direct current magnitude and span.

Note that although the direct current appears inside the cell, the magnitude of the current is proportional to the capacitance of the whole signal net driven by the cell. Thus, we need either to increase the wire width of segment (i,j) or to limit the load of the net.

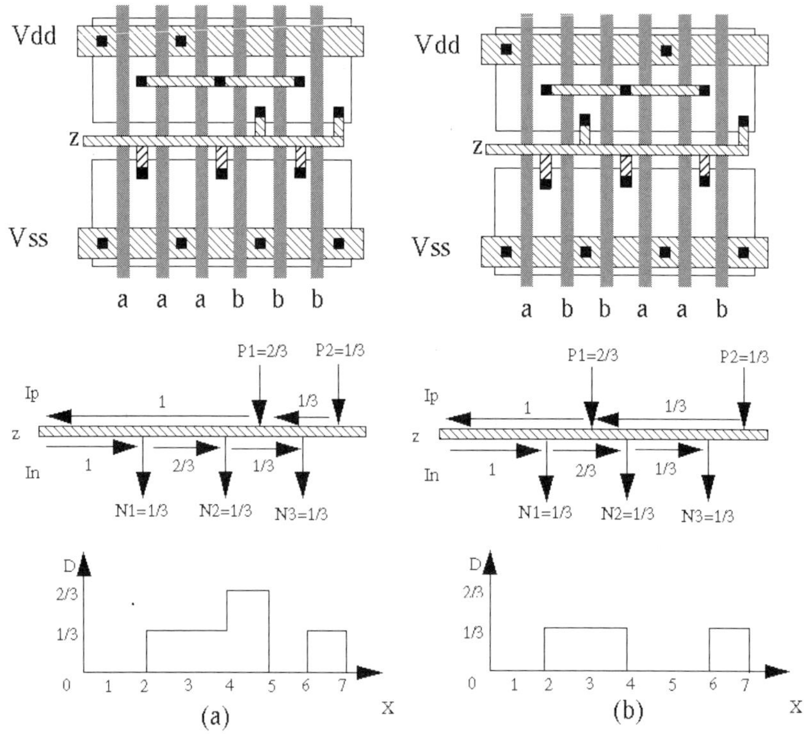

FIGURE 2.16 *Two layouts of a tri-folded NOR cell. The middle row shows the current distribution over the horizontal output segment. The bottom row shows the direct current component.*

BIBLIOGRAPHY

[1] The national technology roadmap for semiconductor, SIA, Semiconductor Industry Association, 1997.

[2] N. H. E. Weste, and K. Eshraghian, *Principles of CMOS VLSI Design*, Addison-Wesley, 1993.

[3] H. B. Bakoglu, *Circuits, Interconnections, and Packaging for VLSI*, Addison Wesley, 1990.

[4] A. Deutsch, Electrical characteristics of interconnections for high-performance systems, *IEEE Proceedings*, 86(2): 315–355, Feb. 1998.

[5] L. O. Chua, C. A. Desoer, and E. S. Kuh, *Linear and Nonlinear Circuits*, McGraw-Hill, 1987.

[6] A. Vittal, and M. Marek-Sadowska, Crosstalk reduction for VLSI, *IEEE Trans. on Computer-Aided Design*, 16(3): 290–298, March 1997.

[7] C. K. Hu, K. P. Rodbell, T. D. Sullivan, K. Y. Lee, and D. P. Bouldin, Electromigration and stress induced voiding in fine Al and Al-alloy thin film lines, *IBM J. Res. Develop.*, 39(4): 465–497, July 1995.

[8] L. M. Ting, J. S. May, W. R. Hunter, and J. W. McPherson, AC electromigration characterization and modeling of multilayered interconnects, *IEEE International Reliability Physics Symp.*, pp. 311–316, Atlanta, GA, 1993.

[9] C. K. Hu, and J. M. E. Harper, Copper interconnections and reliability, *Materials Chemistry and Physics* 52, 5–16, 1998.

[10] K. P. Rodbell, A. J. Castellano, and R. I. Kaufman, AC electromigration in Al metallization, *Int. Workshop on Stress Induced Phenomena in Metallization*, pp. 212–223, 1998.

[11] J. A. Maiz, Characterization of electromigration under bidirectional and pulsed unidirectional currents, IEEE Reliability Physics Symposium, pp. 220–228, 1989.

3

Device Models

3.1 INTRODUCTION

The delay of a device is strongly influenced by the interconnect. Likewise, the interconnect delay is also dependent upon the driving capability of the device. In this chapter, we first introduce the device I-V characteristics. For deep-submicron technologies, because of the short channel effects, the I-V relation becomes more complicated. We adopt a general format of device model using the input transition time and output load as two parameters. According to the format, we present the explicit expression and the table-lookup characterization. Due to the shielding effect of wire resistance, the lumped capacitance of the wires and the fanouts does not reflect the exact output load viewed from the device. We describe the method to find the effective capacitance so that the general format of the device model can be used.

3.2 DEVICE *I-V* CHARACTERISTICS

Figure 3.1 shows the top view of a typical MOSFET. S, G, and D represent the source, gate, and drain, respectively. W and L are the width and length of the MOSFET, respectively.

Given the three nodes, S, G, and D, we use node S as the reference. The voltages and currents of the two other nodes are denoted as v_{GS}, v_{DS}, i_{GS}, and i_{DS}. Among the four variables, the static current i_{GS} is zero in CMOS technology. We thus need only characterize the relation among v_{GS}, v_{DS}, and i_{DS} [1, 2, 3, 4].

Let us use an n-channel transistor as an example. The operation of a MOSFET can be separated into two regions [1]:

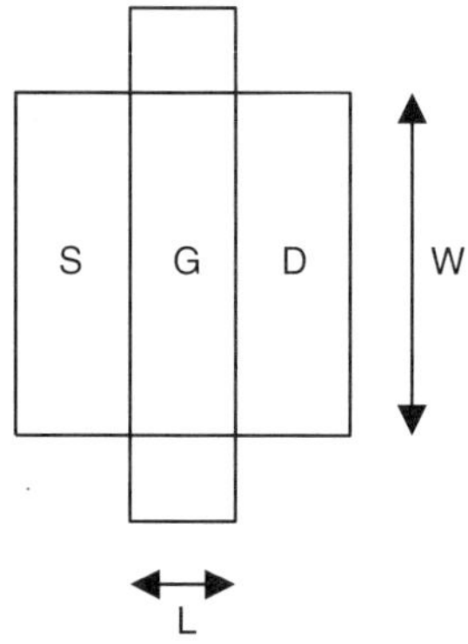

FIGURE 3.1 *Top view of a MOSFET.*

1. *Linear Region* ($v_{DS} \leq v_{GS} - v_T$): In this region, the drain-source current i_{DS} is proportional to the drain-source voltage v_{DS}, and we have

$$i_{DS} = \frac{\mu\varepsilon}{t}\frac{W}{L}\left[(v_{GS} - v_T)v_{DS} - \frac{v_{DS}^2}{2}\right] \tag{3.1}$$

where t is the oxide thickness, μ is the charge mobility, ε the oxide permittivity, and v_T is the threshold voltage.

2. *Saturation Region* ($v_{DS} > v_{GS} - v_T$): In this region, i_{DS} is independent of v_{DS}, and we have

$$i_{DS} = \frac{\mu\varepsilon}{t}\frac{W}{L}\frac{1}{2}(v_{GS} - v_T)^2 \tag{3.2}$$

The *I-V characteristics* of a MOSFET are shown in Figure 3.2.

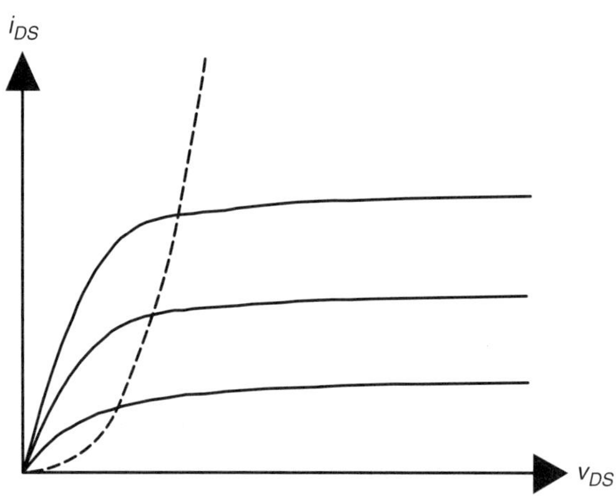

FIGURE 3.2 *I-V characteristics of a MOSFET.*

REMARKS

1. Suppose gate voltage v_{GS} is fixed; current i_{DS} becomes a function of v_{DS}. In the linear region, current i_{DS} increases as voltage v_{DS} increases. Thus, we can view the device between nodes D and S as a positive resistor. In the saturation region, current i_{DS} is independent of the voltage v_{DS}. Thus, we can view the device between nodes D and S as a constant current source.
2. i_{DS} starts to increases from zero when v_{GS} becomes larger than the threshold voltage v_T. Given an inverter, for the region where the input voltage is larger than the threshold voltage of the pull-down transistors and smaller than the difference between supply voltage and the pull-up threshold voltage, there is a short current from power to ground.
3. Because of the short channel effect in deep-submicron technologies, the I-V characteristics of a MOSFET do not follow equations (3.1), (3.2) as an ideal transistor. The I-V curves depend on more parameters.
4. In deep-submicron technologies, short channel effects encourage sophisticated single gates instead of cascaded multiple-level simple gates. The device delay model depends on different cases of the input/output combinations and becomes more involved.

3.3 GENERAL FORMAT OF DEVICE MODELS

When the load is purely capacitive, we can completely precharacterize a gate's delay and output signal behavior as a function of the input signal slew and of the load capacitance (Fig. 3.3). Let us assume that the output is a pure capacitive load with capacitance $C_L = F_{out} + W_L$ where F_{out} is the capacitance due to fanout load and W_L is the capacitance due to interconnect wire segments. Let us

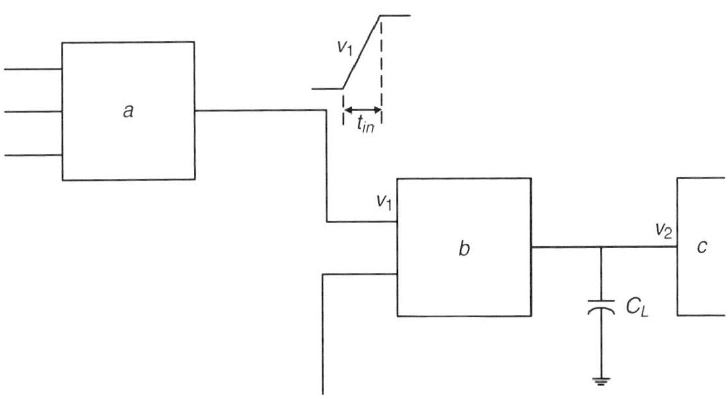

FIGURE 3.3 *Gate b has input v_1 and output v_2. Input v_1 has transition time t_{in}. The output is modeled by a capacitive load C_L.*

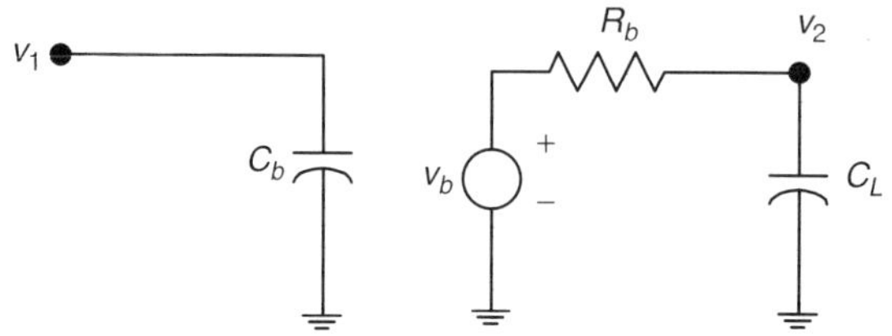

FIGURE 3.4 *The input of gate b is replaced by input capacitance C_b. The output is replaced by voltage source v_b in series with driving resistance R_b.*

describe the slew of the input with a transition time t_{in}. The output delay and the output slew are two functions of the input slew and output load, that is, delay $D(t_{in}, C_L)$, output transition time $t_r(t_{in}, C_L)$. Similarly, we can also characterize the output resistance as function $R_s(t_{in}, C_L)$ (Fig. 3.4).

3.4 DEVICE MODELS IN EXPLICIT EXPRESSION

Simplified RC model of a MOSFET To derive the optimal size of the transistors for the best circuit performance, one simple RC model can be used to characterize the drain to source and gate capacitance as the following functions of channel width and length:

$$R_s = R_0 \frac{L}{W} \tag{3.3}$$

$$C_{in} = C_0 LW \tag{3.4}$$

The model extracts the general trend according to which the driving resistance is proportional to the channel width and inversely proportional to the length of the transistor. The gate capacitance is in proportion to the channel area (i.e., LW). Due to its simplicity, this model is popular for interconnect synthesis operation such as single-level inverter insertion, input gate sizing, and output driver sizing.

Gate Delay Model with Explicit Expression An extended gate delay model describes the delay with two terms: *intrinsic delay* and *propagation delay*. The intrinsic delay is independent of the output load. Thus, it is derived with an open circuit at output. The propagation delay is the portion contributed by the load. It is formulated as the difference between the delay with output load and the delay without the load, the intrinsic delay.

For each different-sized gate, the timing is characterized with different set of constants. Usually, the transition to "one" signal has a different delay from the transition to "zero" signal. We use subscript lh and hl to denote the low-to-high and high-to-low transitions, respectively. Let us assume that input and output signals are inverted. We adopt a second-order nonlinear model to approximate the rising output delay D_{lh}, and falling output delay D_{hl}:

$$D_{lh}(t_{inhl}, C_L) = k_{1up} + k_{2up}C_L + k_{3up}t_{inhl} + k_{4up}t_{inhl}C_L \tag{3.5}$$

$$D_{hl}(t_{inlh}, C_L) = k_{1dn} + k_{2dn}C_L + k_{3dn}t_{inlh} + k_{4dn}t_{inlh}C_L \tag{3.6}$$

where k_{1up} and k_{1dn} can be interpreted as the intrinsic gate delays for up and down transitions, respectively; k_{2up} and k_{2dn} are the two driving resistances for the two transitions; k_{3up} and k_{3dn} are the linear factor of delay contributed by input transition time t_{in}; k_{4up} and k_{4dn} are the nonlinear factor of delay contributed by the product of rise time t_{in} and output load C_L. All constants are precharacterized and attached with the device as part of the library data.

The expression of output transition time t_r is similar to the delay function. A simplified version has t_r be a function of the load, that is,

$$t_{rlh} = T_{tup} + k_{tup}C_L \tag{3.7}$$

$$t_{rhl} = T_{tdn} + k_{tdn}C_L \tag{3.8}$$

where T_{tup}, T_{tdn}, k_{tup}, and k_{tdn} are constants.

For multiple inputs, the worst delay is derived from the worst case among all inputs. For components synchronized by clocks, such as registers and latches, a similar formula is used to characterize the *setup* and *hold* times.

3.5 DEVICE MODEL USING A TABLE-LOOKUP MODEL

Due to short channel effect, the gate delay models become complicated nonlinear functions. Furthermore, a lot of data are collected experimentally at discrete input configurations. Therefore, a table-lookup model is used to characterize the device.

When each component is built, the tables are constructed and attached with the design as a timing description in the library. For a given input slew and output load, we derive its delay via interpolation if the input slew and load fall inside the range of the index of the table. Since the table is used for each instance of the device, the issue of the table model is the efficiency in getting access to the table and deriving the solution.

One-Dimensional Table If we simplify the model of output transition time t_r to a function of single variable C_L, we can use a one-dimensional table to describe the model. Given a gate and its attached slew table, the table describes the output transition time with respect to discrete loads as indexes. The load is sorted by its value. For a load C_L, we can search two adjacent indexes such that the corresponding capacitances C_1 and C_2 bound load C_L from both sides (i.e., $C_1 \leq C_L \leq C_2$). Let t_{r1} and t_{r2} be the output transition time listed in the table with respect to the two indexes, that is,

$$\begin{aligned} t_r(C_1) &= t_{r1} \\ t_r(C_2) &= t_{r2} \end{aligned} \tag{3.9}$$

We use a linear expression to interpolate the output slew;

$$t_r(C) = a_1 C + a_2 \tag{3.10}$$

where a_1 and a_2 are the coefficients. Applying formula (3.10) to equations (3.9) and solving for coefficients a_1 and a_2, we have

$$\begin{aligned} a_1 &= \frac{t_{r2} - t_{r1}}{C_2 - C_1} \\ a_2 &= \frac{t_{r1}C_2 - t_{r2}C_1}{C_2 - C_1} \end{aligned} \tag{3.11}$$

Substituting equations (3.11) into formula (3.10), we can therefore derive the output transition time with respect to the given load C_L.

$$t_r(C_L) = \frac{t_{r2} - t_{r1}}{C_2 - C_1} C_L + \frac{t_{r1}C_2 - t_{r2}C_1}{C_2 - C_1} \tag{3.12}$$

Two-Dimensional Table We use a two-dimensional table to approximate the delay as a function of two variables. Given a gate, its delay table is described by a set of discrete capacitances as the index of the first dimension and a set of input slews as the index of the second dimension. Again the capacitances and slews are sorted by their values. For a pair of load C_L and input transition time t_{in}, we find the adjacent pair of indexes with corresponding capacitances C_1, C_2, and slews t_1, t_2 that bound C_L and t_{in} from both sides, that is, $C_1 \leq C_L \leq C_2$ and $t_1 \leq t_{in} \leq t_2$ (Fig. 3.5).

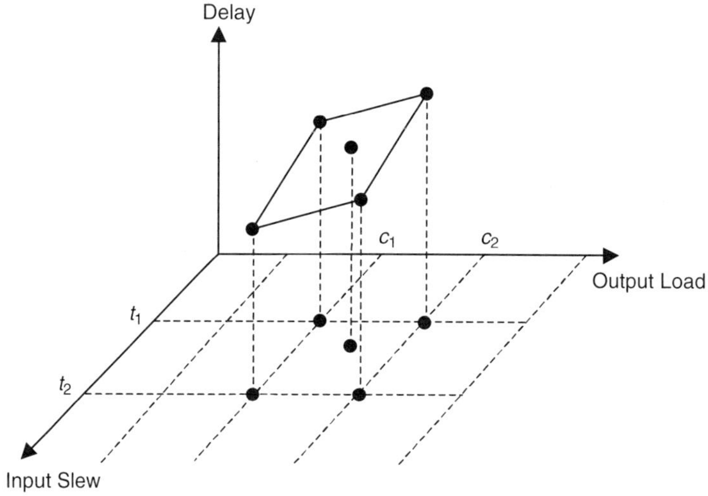

FIGURE 3.5 *Two-dimensional model of delay as a function of input skew and output load.*

Let D_i $(1 \leq i \leq 4)$ be the delays at the four points, that is,

$$\begin{aligned} D_1 &= D(C_1, t_1) \\ D_2 &= D(C_2, t_1) \\ D_3 &= D(C_1, t_2) \\ D_4 &= D(C_2, t_2). \end{aligned} \tag{3.13}$$

We approximate delay $D(C_L, t_{in})$ by an expression of four coefficients:

$$D(C_L, t_{in}) = k_1 + k_2 C_L + k_3 t_{in} + k_4 C_L t_{in} \tag{3.14}$$

Applying format (3.14) to equations (3.13), we have a simultaneous equation of the coefficients k_i.

$$\begin{bmatrix} 1 & C_1 & t_1 & C_1 t_1 \\ 1 & C_2 & t_1 & C_2 t_1 \\ 1 & C_1 & t_2 & C_1 t_2 \\ 1 & C_2 & t_2 & C_2 t_2 \end{bmatrix} \begin{bmatrix} k_1 \\ k_2 \\ k_3 \\ k_4 \end{bmatrix} = \begin{bmatrix} D_1 \\ D_2 \\ D_3 \\ D_4 \end{bmatrix} \tag{3.15}$$

Utilizing the structure of the matrix, we solve the simultaneous equation to obtain the expression of the four coefficients:

$$\begin{bmatrix} k_1 \\ k_2 \\ k_3 \\ k_4 \end{bmatrix} = \begin{bmatrix} (D_4 C_1 t_1 - D_3 C_2 t_1 - D_2 C_1 t_2 + D_1 C_2 t_2)/W \\ (D_3 t_1 - D_4 t_1 - D_1 t_2 + D_2 t_2)/W \\ (D_2 C_1 - D_4 C_1 - D_1 C_2 + D_3 C_2)/W \\ (D_1 - D_2 - D_3 + D_4)/W \end{bmatrix} \tag{3.16}$$

where $W = (C_1 - C_2)\ (t_1 - t_2)$. We can thus obtain the delay from formula (3.14).

3.6 EFFECTIVE CAPACITANCE MODEL

As the minimum feature sizes scale downward, the resistive component of the *RC* interconnect load is comparable to the gate output impedance. Due to the shielding effect of wire resistance, a capacitive load model of total wire and fanout capacitor does not reflect the exact output load viewed from the device. On the other hand, a load model more complicated than a capacitive load such as a second-order π load would be expensive in terms of the volume of the data.

In the *effective capacitance* model [5], we derive the output load different from the sum of wire and fanout capacitance. Actually, the output load is derived from iterations of process and is termed as effective capacitance C_{eff}. The gate is modeled by a driving resistance R_s with a voltage source (Fig 3.6). The voltage source is specified by an intrinsic delay T and a rise time t_r (Fig. 3.7).

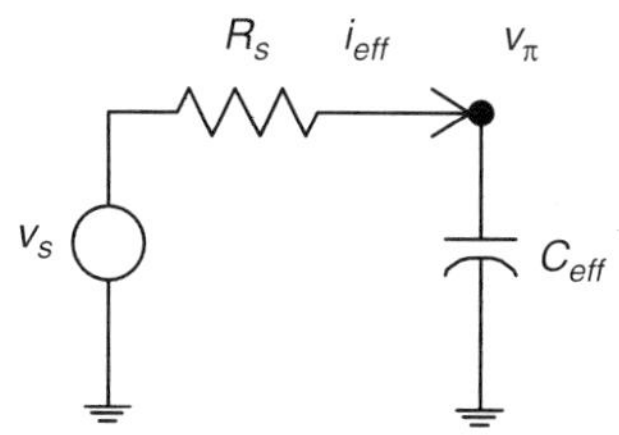

FIGURE 3.6 *The C_{eff} output model.*

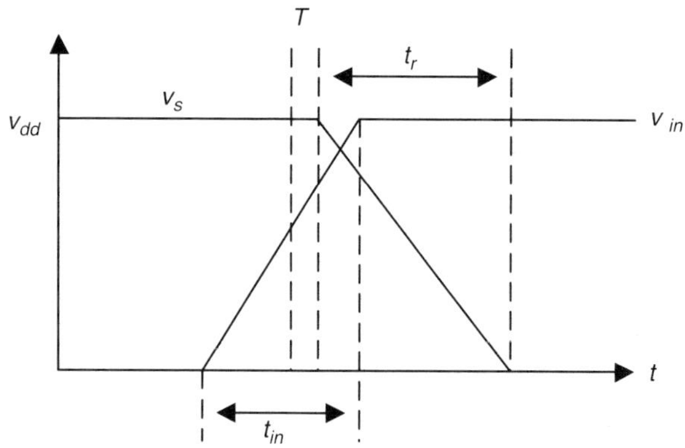

FIGURE 3.7 *The input and voltage source model.*

Following the general format of device models, the values of the above three parameters are characterized as functions of input transition time and output load. However, the output load is replaced by effective capacitance C_{eff}. Consequently, the precharacterized functions are written as $R_s(t_{in}, C_{eff})$, $T(t_{in}, C_{eff})$, and $t_r(t_{in}, C_{eff})$.

Given an output load of RC tree (Fig. 3.8), we first reduce the tree to a π model as shown in Figure 3.9. We then combine the π model with the gate model (Fig. 3.10) to derive the effective capacitance. To simplify the description, we assume that the input transition time t_{in} is given from the previous delay calculation of the input net. The π model generation and the transition time calculation of an interconnect tree will be discussed in the next chapter.

Given a gate with its input transition time t_{in} and output network as a π model (Fig. 3.10), the effective capacitance C_{eff} is set so that the voltage-time area over transition time t_r is equal to the voltage-time area of π model:

$$\int_0^{t_r(t_{in}, C_{eff})} v_\pi\left(T(t_{in}, C_{eff}), t_r(t_{in}, C_{eff}), R_s(t_{in}, C_{eff})\right) dt$$

$$= \int_0^{t_r(t_{in}, C_{eff})} v_{C_{eff}}\left(T(t_{in}, C_{eff}), t_r(t_{in}, C_{eff}), R_s(t_{in}, C_{eff}), C_{eff}\right) dt \quad (3.17)$$

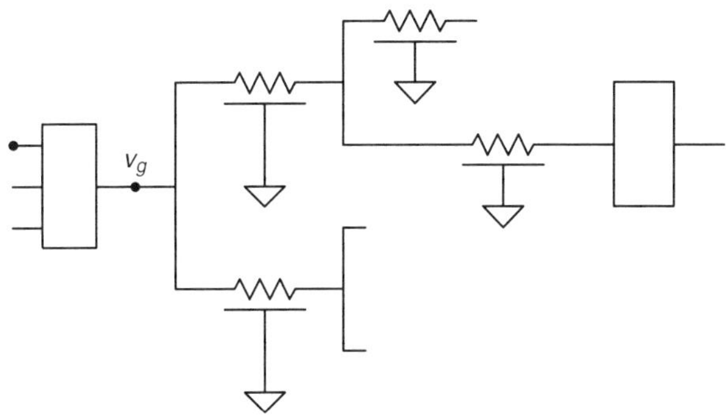

FIGURE 3.8 *RC tree model for output.*

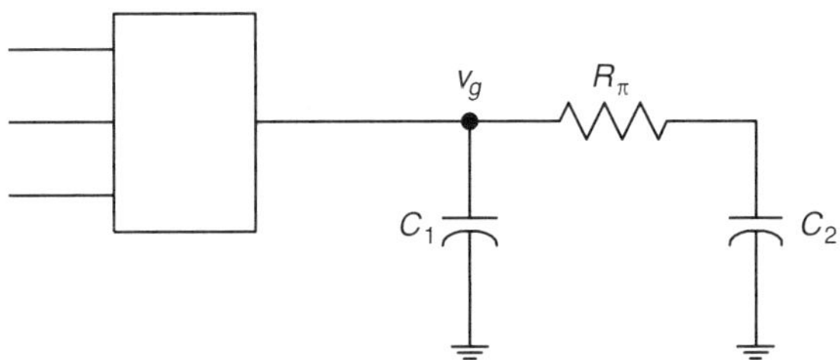

FIGURE 3.9 *A π model to replace an RC tree.*

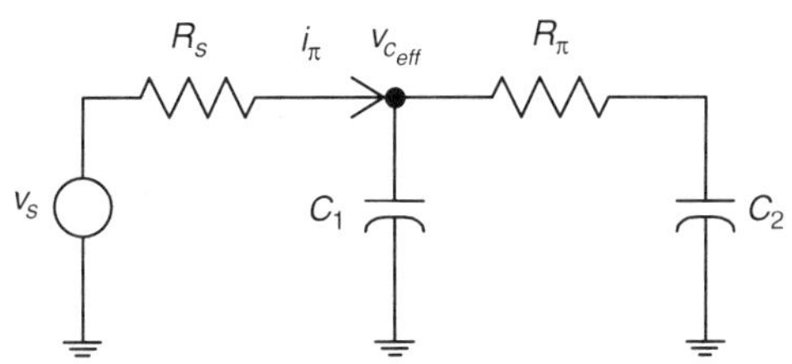

FIGURE 3.10 *The π model.*

where v_π is the voltage in π model and $v_{C_{eff}}$ is the voltage in C_{eff} model. The values of T, t_r, and R_s are functions of t_{in} and C_{eff}. Because the value of variable t_{in} is given, the above equation has only one variable C_{eff}.

Since the two models use the same voltage source v_s and driving resistance R_s, the equality of voltages can be derived from the equality of currents. Let i_π and $i_{C_{eff}}$ be the output currents of π model and C_{eff} model, respectively. We have

$$\int_0^{t_r(t_{in},C_{eff})} i_\pi\left(T(t_{in},C_{eff}),t_r(t_{in},C_{eff}),R_s(t_{in},C_{eff})\right)dt$$
$$=\int_0^{t_r(t_{in},C_{eff})} i_{C_{eff}}\left(T(t_{in},C_{eff}),t_r(t_{in},C_{eff}),R_s(t_{in},C_{eff}),C_{eff}\right)dt \quad (3.18)$$

We derive the integrated current of C_{eff} model:

$$\int_0^{t_r} i_{C_{eff}}(T, t_r, R_s)dt = \int_0^{t_r} C_{eff} \frac{v_{dd}}{t_r}(1 - e^{-(t/C_{eff}R_s)})dt$$
$$= C_{eff} \frac{v_{dd}}{t_r}(t_r + C_{eff}R_s(e^{-(t_r/C_{eff}R_s)} - 1)) \qquad (3.19)$$

Similarly, we can derive the integrated current of the π model:

$$\int_0^{t_r} i_\pi(T, t_r, R_s)dt = i_\pi(t_r) = \left[At_r + \frac{B}{p_1}(1 - e^{-p_1 t_r}) + \frac{D}{p_2}(1 - e^{-p_2 t_r})\right] \frac{v_{dd}}{R_s t_r} \qquad (3.20)$$

where

$$A = \frac{-z}{p_1 p_2} \qquad B = \frac{z - p_1}{p_1(p_2 - p_1)} \qquad D = \frac{z - p_2}{p_2(p_1 - p_2)} \qquad (3.21)$$

and z, p_1, and p_2 are the zero and the two poles of the transfer function of the current in π model.

$$\frac{i_\pi}{v_s} = \frac{1}{R_s + \frac{1}{sC_1 + \frac{1}{R_\pi + \frac{1}{sC_2}}}} \qquad (3.22)$$

$$= \frac{s^2 C_1 C_2 R_\pi + s(C_1 + C_2)}{s^2 C_1 C_2 R_s R_\pi + s(R_s C_2 + R_s C_1 + R_\pi C_2) + 1} \qquad (3.23)$$

Therefore, equation (3.18) can be expressed as the following equation:

$$At_r + \frac{B}{p_1}(1 - e^{-p_1 t_r}) + \frac{D}{p_2}(1 - e^{-p_2 t_r}) - R_s C_{eff} t_r$$
$$+ (R_s C_{eff})^2 (1 - e^{-(t_r/R_s C_{eff})}) = 0. \qquad (3.24)$$

Since effective capacitance C_{eff} is the only variable in the above equation, we can derive its solution using Newton-Raphson iterations. We use $C_{eff} = C_1 + C_2$ as an initial value for the iterations.

BIBLIOGRAPHY

[1] N. H. E. Weste, and K. Eshranghian, *Principles of CMOS VLSI Design*, Addison-Wesley, 1993.

[2] H. W. Johnson, and M. Graham, *High Speed Digital Design*, PTR Prentice Hall, 1993.

[3] J. M. Rabaey, *Digital Integrated Circuits*, Prentice Hall, 1996.

[4] M. Smith, *Application-Specific Integrated Circuits*, Addison-Wesley, 1997.

[5] F. Dartu, N. Menezes, and L. T. Pileggi, Performance computation for precharacterized CMOS gates with RC loads, *IEEE Trans. Computer-Aided Design*, pp. 544–553, May 1996.

4

Interconnect Analysis

4.1 INTRODUCTION

In the synchronous system, the delay of the combinational logic plus setup time cannot be larger than the system clock period T. On the other hand, the combinational logic delay cannot be shorter than the hold time of the registers.

$$D_{comb} + T_{setup} \leq T \tag{4.1}$$

$$D_{comb} \geq T_{hold} \tag{4.2}$$

Thus, the delay is bounded on both sides with an upper bound $T - T_{setup}$ and a lower bound T_{hold}. The two-sided constraints complicate the design issues. Furthermore, the delay is strongly affected by the coupling of other signals. Thus, we need to derive a precise analysis of the signal propagation delay to ensure correct timing behavior.

In deep-submicron technology, each net needs to be modeled with many segments of lumped R, L, C elements for a precise estimation. Thus, the emphasis of the simulation is to reduce the model order of signal nets so that we can estimate the delay of a large network efficiently and accurately.

In this chapter, we assume that the circuit has perfect source supply and ground return. Under this assumption, we can use single nodes to represent the voltage source and the ground. However, in cases that the impedance of power and ground cannot be ignored, the derived delay tends to be optimistic [1].

In the following, we describe the time domain analysis and s domain analysis [2,3]. The relation between the two domains are discussed. In order to reduce the complexity of the circuit analysis, we introduce two methods: matrix reduction

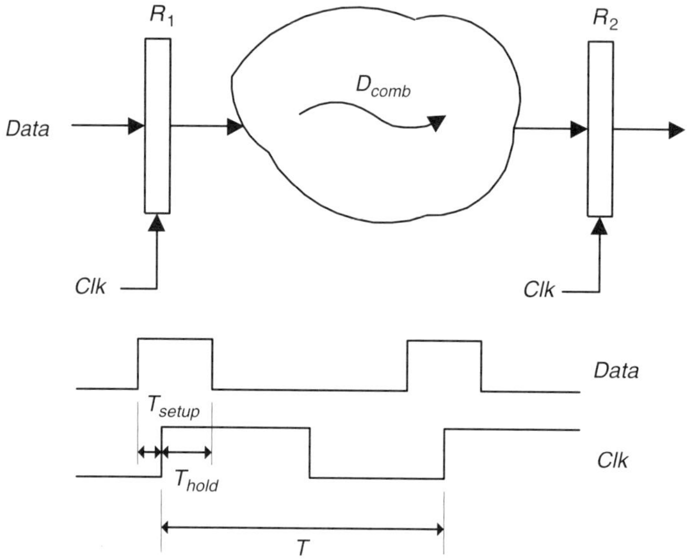

FIGURE 4.1 System with setup and hold time constraints. Data has to arrive at the registers before the rising edge of the clock by T_{setup} margin and keeps steady for T_{hold} interval after the rising edge of the clock.

and moment matching. The connection of the two approaches is discussed. The concept of moments has been exploited in many applications. We present the usage of moments as the Elmore model, a layout-driven analysis procedure, and a π model. At the end of the chapter, we touch on the transmission line models.

4.2 TIME DOMAIN ANALYSIS

In this section, we explain the RLC network analysis formulation. We use a mixed cut-set and mesh analysis to express the linear state equation. RC network is treated as a special case of RLC network. Using the formulation, we derive the transient response of the network.

4.2.1 RLC Network Analysis

In Figure 4.2, we can use the three segments of capacitors, C_1, C_2 and C_3 and three segments of resistors R_{14}, R_{25}, and $R_s = 1/G_s$ to construct the tree structure of the network (Fig. 4.3). The inductors L_{24}, L_{35} and voltage source v_s constitute the co-tree of the network. For each trunk on the tree, there is a unique set of branches on the co-tree that form a cut-set. Similarly, for each branch on the co-tree, there is a unique set of trunks on the tree that form a closed loop or mesh. The tree and co-tree structure supplies the framework for a mixed mode of cut-set and mesh analysis [2,3].

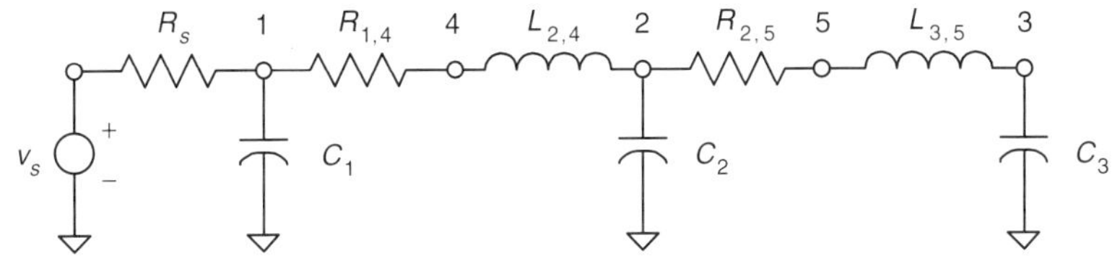

FIGURE 4.2 *One RLC line.*

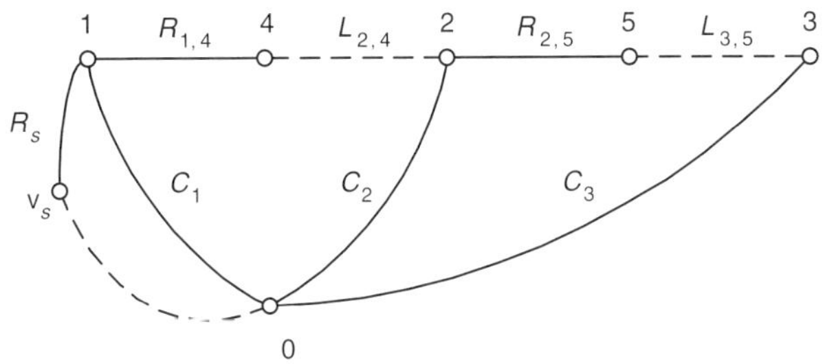

FIGURE 4.3 *Tree (solid lines) and co-tree (dashed lines) of the circuit in Figure 4.2.*

Let us use the voltages across the capacitors and the currents through the inductors as the state variables. We can write a mixed cut-set and mesh analysis with the linear state equation.

$$\left[\begin{array}{ccc|cc} C_1 & & & 0 & 0 \\ & C_2 & & 0 & 0 \\ & & C_3 & 0 & 0 \\ \hline 0 & 0 & 0 & L_{24} & \\ 0 & 0 & 0 & & L_{35} \end{array}\right] \begin{bmatrix} \frac{dv_1}{dt} \\ \frac{dv_2}{dt} \\ \frac{dv_3}{dt} \\ \frac{di_{42}}{dt} \\ \frac{di_{53}}{dt} \end{bmatrix} = - \left[\begin{array}{ccc|cc} G_{s1} & 0 & 0 & 1 & 0 \\ 0 & 0 & 0 & -1 & 1 \\ 0 & 0 & 0 & 0 & -1 \\ \hline -1 & 1 & 0 & R_{14} & 0 \\ 0 & -1 & 1 & 0 & R_{25} \end{array}\right]$$

$$\times \begin{bmatrix} v_1 \\ v_2 \\ v_3 \\ i_{42} \\ i_{53} \end{bmatrix} + \begin{bmatrix} G_{s1} v_s \\ 0 \\ 0 \\ 0 \\ 0 \end{bmatrix} \tag{4.3}$$

Let v_3 be the output voltage; we have

$$v_3 = [0 \quad 0 \quad 1 \quad 0 \quad 0] \begin{bmatrix} v_1 \\ v_2 \\ v_3 \\ i_{42} \\ i_{53} \end{bmatrix} \tag{4.4}$$

In equation (4.3), the first three rows are derived from Kirchhoff's current law, which describes that the total current flowing through each cut-set is zero. The cut-set is formed by a capacitive trunk on the tree structure and the corresponding branches on the co-tree. In other words, the equations describe that the current injecting into the capacitor is equal to the current flowing in from the branches of resistors and conductors.

The last two rows are derived from Kirchhoff's voltage law, which describes that the total voltage along a mesh is zero. The mesh is formed by an inductor in the co-tree and the corresponding resistors and capacitors along the tree structure. In other words, the equations describe that the voltage across the inductor is equal to the sum of the voltages along the mesh formed by the corresponding tree structure.

In general, we can use V_c to denote a vector of voltages across the capacitors and I_L to denote a vector of currents through the inductors. Hence, the linear state equation can be expressed as:

$$\mathbf{M}\frac{d\mathbf{x}(t)}{dt} = -\mathbf{G}\mathbf{x}(t) + \mathbf{P}\mathbf{u}(t) \tag{4.5}$$

$$\mathbf{y}(t) = \mathbf{Q}\mathbf{x}(t) \tag{4.6}$$

where

$$\mathbf{x}(t) \equiv \begin{bmatrix} \mathbf{V}_c(t) \\ \mathbf{I}_L(t) \end{bmatrix} \qquad \mathbf{M} \equiv \begin{bmatrix} \mathbf{C} & 0 \\ 0 & \mathbf{L} \end{bmatrix} \qquad \mathbf{G} \equiv \begin{bmatrix} \mathbf{Y} & \mathbf{E} \\ -\mathbf{E}^T & \mathbf{R} \end{bmatrix} \tag{4.7}$$

where $\mathbf{C}$ and $\mathbf{L}$ are the capacitance and inductance matrices of the network, respectively. Matrices $\mathbf{Y}$ and $\mathbf{R}$ are the admittance matrix and impedance matrix of cut-set and mesh [2,3]. In matrix $\mathbf{G}$, the two off-diagonal submatrices are represented by matrices $\mathbf{E}$ and $-\mathbf{E}^T$. Matrix $\mathbf{E}$ covers the co-tree branches in the cut-set analysis and matrix $-\mathbf{E}^T$ covers the tree trunks in the mesh analysis. For example, in Figures 4.2 and 4.3, inductor L_{24} is covered by the cut-set of tree trunk capacitor C_1 in a reverse direction and by the cut-set of tree trunk capacitor C_2 in a positive direction. Thus, the first column of matrix E in equation (4.3) is written as

$$[1, -1, 0]^T \tag{4.8}$$

Correspondingly, in the mesh analysis, tree trunks C_1 and C_2 are included in the mesh of co-tree branch L_{24} but in an opposite direction. Consequently, in equation (4.3), the first row of matrix $-E^T$ is written as

$$[-1, 1, 0] \tag{4.9}$$

which is the transpose of the vector in (4.8) with a multiplication of a negative sign.

4.2.2 RC Network Analysis

The RC network is a special case of RCL network. In RC network, the capacitors form the tree structure. The co-tree is constructed by the resistors. There is no inductor, and thus there are no state variables associated with the co-tree (Fig. 4.3). We can use the equations in RLC network with inductance $L = 0$. Or, we can use the cut-set analysis only to express in a linear state equation.

$$\begin{bmatrix} C_1 & & & \\ & C_2 & & \\ & & C_3 & \\ & & & C_4 \end{bmatrix} \begin{bmatrix} \frac{dv_1}{dt} \\ \frac{dv_2}{dt} \\ \frac{dv_3}{dt} \\ \frac{dv_4}{dt} \end{bmatrix} = - \begin{bmatrix} G_3 + G_{1,2} & -G_{1,2} & & \\ -G_{1,2} & G_{1,2} + G_{2,3} & -G_{2,3} & \\ & -G_{2,3} & G_{2,3} + G_{3,4} & -G_{3,4} \\ & & -G_{3,4} & G_{3,4} \end{bmatrix} \times \begin{bmatrix} v_1 \\ v_2 \\ v_3 \\ v_4 \end{bmatrix} + \begin{bmatrix} v_s G_s \\ 0 \\ 0 \\ 0 \end{bmatrix} \tag{4.10}$$

Let voltage v_4 be the output, we have

$$v_4 = [0 \quad 0 \quad 0 \quad 1] \begin{bmatrix} v_1 \\ v_2 \\ v_3 \\ v_4 \end{bmatrix} \tag{4.11}$$

Consequently, the RC network can also be expressed in the formats of equations (4.5) and (4.6).

4.2.3 Properties of the Matrices

From their construction, we state the properties of the matrices.

Lemma 4.2.1 *Matrices* **M**, **C**, **L**, **G**, **Y**, *and* **R** *are symmetric and diagonally dominant, that is, the diagonal entry* $e_{ii} \geq \sum_j |e_{ij}|$.

From the above lemma, we can deduce the nonnegative definite property.

Lemma 4.2.2 *Matrices* **M**, **C**, **L**, **G**, **Y**, *and* **R** *are nonnegative definite.*

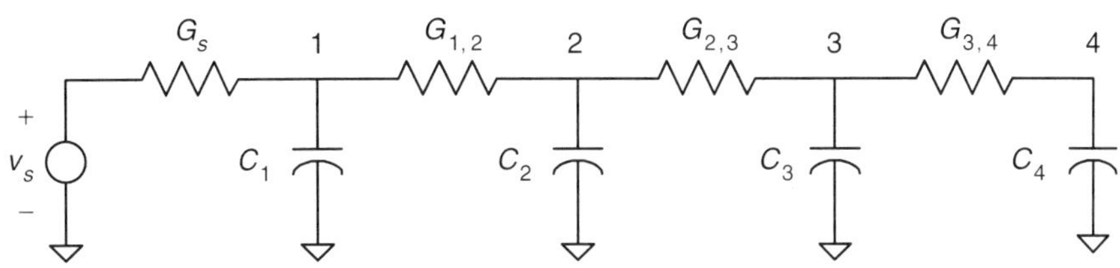

FIGURE 4.4 *An RC network example.*

4.2.4 Responses in Time Domain

The responses in time domain can be derived from the state equation. Let us shift matrix **M** to the right side of state equation (4.5); we have

$$\frac{d\mathbf{x}(t)}{dt} = -\mathbf{M}^{-1}\mathbf{G}\mathbf{x}(t) + \mathbf{M}^{-1}\mathbf{P}\mathbf{u}(t) \tag{4.12}$$

The solution of the above differential equation is the waveform of state variables in time domain:

$$\mathbf{x}(t) = e^{-\mathbf{M}^{-1}\mathbf{G}t}\mathbf{x}_0 + \int_0^t e^{-\mathbf{M}^{-1}\mathbf{G}\cdot(t-\tau)}\mathbf{M}^{-1}\mathbf{P}\mathbf{u}(\tau)\,d\tau \tag{4.13}$$

where $\mathbf{x}_0$ is the initial condition of the state variable, $\mathbf{x}_0 = \mathbf{x}(t)|_{t=0}$.

The output waveform in time domain is derived by the multiplication of matrix **Q** from the left:

$$\mathbf{y}(t) = \mathbf{Q}\mathbf{x}(t) = \mathbf{Q}e^{-\mathbf{M}^{-1}\mathbf{G}t}\mathbf{x}_0 + \mathbf{Q}\int_0^t e^{-\mathbf{M}^{-1}\mathbf{G}\cdot(t-\tau)}\mathbf{M}^{-1}\mathbf{P}\mathbf{u}(\tau)\,d\tau \tag{4.14}$$

The first term is the output waveform due to initial condition. The second term is the convolution of the impulse response and the input waveform. Note that in the power of the exponent, the matrices $\mathbf{M}^{-1}$ and **G** can be large in dimensions if the number of segments is large.

Taylor's Expansion in Time Domain For the case that time t approaches zero, we can express the exponential term in Taylor's series:

$$e^{-\mathbf{M}^{-1}\mathbf{G}t} = \mathbf{1} - \mathbf{M}^{-1}\mathbf{G}t + \frac{1}{2!}(\mathbf{M}^{-1}\mathbf{G})^2t^2 - \frac{1}{3!}(\mathbf{M}^{-1}\mathbf{G}t)^3t^3\ldots \tag{4.15}$$

The above expression is another view to interpretate the exponent of matrix.

4.3 S DOMAIN ANALYSIS

We can use Laplace transform to represent the state equations (4.5) and (4.6) in the s domain.

$$s\mathbf{M}\mathbf{x}(s) - \mathbf{M}\mathbf{x}_0 = -\mathbf{G}\mathbf{x}(s) + \mathbf{P}\mathbf{u}(s) \tag{4.16}$$

$$\mathbf{y}(s) = \mathbf{Q}\mathbf{x}(s) \tag{4.17}$$

Recall that $\mathbf{x}_0$ denotes the initial condition of the time domain state vector $\mathbf{x}(t)$ at time $t = 0$, that is, $\mathbf{x}(t)|_{t=0}$. We multiply both sides of equation (4.16) by matrix $\mathbf{G}^{-1}$ and obtain:

$$s\mathbf{G}^{-1}\mathbf{M}\mathbf{x}(s) - \mathbf{G}^{-1}\mathbf{M}\mathbf{x}_0 = -\mathbf{x}(s) + \mathbf{G}^{-1}\mathbf{P}\mathbf{u}(s) \tag{4.18}$$

Let us denote:

$$\mathbf{A} \equiv \mathbf{G}^{-1}\mathbf{M} \tag{4.19}$$

$$\mathbf{B} \equiv \mathbf{G}^{-1}\mathbf{P} \tag{4.20}$$

We substitute matrices **A** and **B** into system equation (4.18) and reduce the expression:

$$s\mathbf{A}\mathbf{x}(s) - \mathbf{A}\mathbf{x_0} = -\mathbf{x}(s) + \mathbf{B}\mathbf{u}(s) \tag{4.21}$$

Therefore, we can derive the state solution in s domain as:

$$\mathbf{x}(s) = (\mathbf{1} + s\mathbf{A})^{-1}(\mathbf{A}\mathbf{x_0} + \mathbf{B}\mathbf{u}(s)) \tag{4.22}$$

$$= (\mathbf{1} + s\mathbf{A})^{-1}\mathbf{A}\mathbf{x_0} + (\mathbf{1} + s\mathbf{A})^{-1}\mathbf{B}\mathbf{u}(s) \tag{4.23}$$

The output response in s domain is derived by multiplying the state solution with matrix **Q** from the left.

$$\mathbf{y}(s) = \mathbf{Q}\mathbf{x}(s) = \mathbf{Q}(\mathbf{1} + s\mathbf{A})^{-1}\mathbf{A}\mathbf{x_0} + \mathbf{Q}(\mathbf{1} + s\mathbf{A})^{-1}\mathbf{B}\mathbf{u}(s) \tag{4.24}$$

The transfer function derived from equation (4.24) setting $\mathbf{x}_0 = 0$ can be

$$\mathbf{H}(s) = \mathbf{Q}(\mathbf{1} + s\mathbf{A})^{-1}\mathbf{B} = \mathbf{Q}(\mathbf{1} + s\mathbf{G}^{-1}\mathbf{M})^{-1}\mathbf{G}^{-1}\mathbf{P} \tag{4.25}$$

The Correspondence between Output response in Time Domain and in S Domain To match the format of the output in time domain and in s domain, we expand matrix $\mathbf{B} = \mathbf{G}^{-1}\mathbf{P}$ and replace $\mathbf{G}^{-1}\mathbf{P}$ with $\mathbf{A}\mathbf{M}^{-1}\mathbf{P}$ in the last term of equation (4.24).

$$\mathbf{y}(s) = \mathbf{Q}(\mathbf{1} + s\mathbf{A})^{-1}\mathbf{A}\mathbf{x_0} + \mathbf{Q}(\mathbf{1} + s\mathbf{A})^{-1}\mathbf{G}^{-1}\mathbf{P}\mathbf{u}(s) \tag{4.26}$$

$$= \mathbf{Q}(\mathbf{1} + s\mathbf{A})^{-1}\mathbf{A}\mathbf{x_0} + \mathbf{Q}(\mathbf{1} + s\mathbf{A})^{-1}\mathbf{A}\mathbf{M}^{-1}\mathbf{P}\mathbf{u}(s) \tag{4.27}$$

Since $(\mathbf{1} + s\mathbf{A})^{-1}\mathbf{A}$ is equal to $(s\mathbf{1} + \mathbf{A}^{-1})^{-1}$, the output response can be expressed as follows.

$$\mathbf{y}(s) = \mathbf{Q}(s\mathbf{1} + \mathbf{A}^{-1})^{-1}\mathbf{x_0} + \mathbf{Q}(s\mathbf{1} + \mathbf{A}^{-1})^{-1}\mathbf{M}^{-1}\mathbf{P}\mathbf{u}(s) \tag{4.28}$$

To compare the format of the output response in time domain, we substitute matrix **A** and rewrite equation (4.14).

$$\mathbf{y}(t) = \mathbf{Q}\mathbf{x}(t) = \mathbf{Q}e^{-\mathbf{A}^{-1}t}\mathbf{x_0} + \mathbf{Q}\int_0^t e^{-\mathbf{A}^{-1}(t-\tau)}\mathbf{M}^{-1}\mathbf{P}\mathbf{u}(\tau)\,d\tau \tag{4.29}$$

Comparing the two equations above, we can conclude that the matrix inversion $(s\mathbf{1} + \mathbf{A}^{-1})^{-1}$ corresponds to the exponential function $e^{-\mathbf{A}^{-1}t}$. The multiplication with input $\mathbf{u}(s)$ in the last term of equation (4.28) corresponds to the convolution in time domain.

Expression in Rational Function Suppose that matrix $\mathbf{A}$ is diagonalizable (i.e., there is a matrix $\mathbf{S}$ and its inverse $\mathbf{S}^{-1}$ such that $\mathbf{S}^{-1}\mathbf{A}\mathbf{S} = \Lambda$, where Λ is a diagonal matrix). We can express the transfer function as

$$\mathbf{H}(s) = \mathbf{QS}(\mathbf{1} + s\mathbf{\Lambda})^{-1}\mathbf{S}^{-1}\mathbf{B} \tag{4.30}$$

The entry of the transfer function can be expressed as

$$\mathbf{H}[i, j](s) = \frac{f[i, 1]g[1, j]}{1 + s\lambda_1} + \frac{f[i, 2]g[2, j]}{1 + s\lambda_2} + \cdots + \frac{f[i, n]g[n, j]}{1 + s\lambda_n} \tag{4.31}$$

where $f[i, k]$ is the $[i, k]$ entry of matrix $\mathbf{QS}$, $g[k, j]$ is the $[k, j]$ entry of matrix $\mathbf{S}^{-1}\mathbf{B}$, and λ_k is the $[k, k]$ entry of the diagonal matrix Λ.

Suppose the system has a single input and a single output, the derivation of equation (4.31) suggests that the transfer function (4.25) can be represented in the format of a rational expression:

$$H(s) = \frac{v_{out}(s)}{v_s(s)} = \frac{P(s)}{Q(s)} \tag{4.32}$$

$$= \frac{a_0 + sa_1 + s^2a_2 + \cdots + s^m a_m}{b_0 + sb_1 + s^2b_2 + \cdots + s^n b_n} \tag{4.33}$$

$$= \frac{a_m(s - z_1)(s - z_2)\cdots(s - z_m)}{b_n(s - p_1)(s - p_2)\cdots(s - p_n)} \tag{4.34}$$

where $P(s)$ and $Q(s)$ are polynomial expressions of s in an order of m and n, respectively, a_j and b_j are the coefficients of the polynomial expressions, and z_j and p_j are the zeros and poles of the function, respectively. In general, the order, n, of the denominator is equal to the dimension of matrix $\mathbf{A}$. Number n can be large if we view a net as many pieces of wire segments with each segment modeled by lumped R, L, C elements.

Suppose $v_{in}(s) = (1/s - p_0)$; we can also expand the output $v_{out}(s)$ in the format of partial fraction decomposition.

$$v_{out}(s) = H(s)v_{in}(s) = \frac{P(s)}{Q(s)}\frac{1}{s - p_0} \tag{4.35}$$

$$= \frac{k_0}{s - p_0} + \frac{k_1}{s - p_1} + \frac{k_2}{s - p_2} + \cdots + \frac{k_n}{s - p_n} \tag{4.36}$$

From the expression in partial fraction decomposition, we can derive the time domain response via inverse Laplace transform.

$$v_{out}(t) = k_0e^{p_0t} + k_1e^{p_1t} + k_2e^{p_2t} \cdots + k_ne^{p_nt} \tag{4.37}$$

In equation (4.31), the eigenvalues λ_i of matrix $\mathbf{A}$ cannot be negative. In other words, the poles p_i cannot be positive. Otherwise, the waveform in time domain will expand as time t increases. Thus, we can conclude the nonnegative definite property of matrix $\mathbf{A}$.

Lemma 4.3.1 *Matrix* $\mathbf{A}$ *is nonnegative definite.*

***Serial Expansion of Matrix Inversion* $(\mathbf{1}+s\mathbf{A})^{-1}$** There are two extreme directions to expand the matrix inversion, $(\mathbf{1}+s\mathbf{A})^{-1}$. For the case that s approaches zero, assuming the initial condition $x_0 = 0$, we can express the state response function as:

$$\mathbf{x}(s) = (\mathbf{1} - s\mathbf{A} + s^2\mathbf{A}^2 - s^3\mathbf{A}^3 + \cdots)\mathbf{B}\mathbf{u}(s) \tag{4.38}$$

The transfer function is written as:

$$\mathbf{H}(s) = \mathbf{Q}(\mathbf{1} - s\mathbf{A} + s^2\mathbf{A}^2 - s^3\mathbf{A}^3 + \cdots)\mathbf{B} \tag{4.39}$$

For RLC networks with capacitive loads, the interconnect is usually viewed as a low-pass filter. Hence, the adoption of the expansion based on s approaching zero is a reasonable choice.

On the other hand, when s approaches infinity, we can expand the state response in the power of s^{-1}.

$$\mathbf{x}(s) = (s^{-1}\mathbf{A}^{-1} - s^{-2}\mathbf{A}^{-2} + s^{-3}\mathbf{A}^{-3} + \cdots)\mathbf{B}\mathbf{u}(s) \tag{4.40}$$

The transfer function is written as:

$$\mathbf{H}(s) = \mathbf{Q}(s^{-1}\mathbf{A}^{-1} - s^{-2}\mathbf{A}^{-2} + s^{-3}\mathbf{A}^{-3} + \cdots)\mathbf{B} \tag{4.41}$$

The expansion based on s approaching infinity yields mixed results on the accuracy of the simulation.

4.4 CIRCUIT REDUCTION VIA MATRIX APPROXIMATION

We reduce the size of the system equation using matrix approximation [4]. The process reduces the dimension of the matrix to be inverted. For example, in equation (4.24), we need to invert term $(\mathbf{1}+s\mathbf{A})$. However, this term contains a variable s. For such a matrix inversion, we can perform the eigenvalue eigenvector decomposition, do inversion symbolically, or process the inversion with a set of values for s. Nonetheless, the operation is much more involved than the inversion of a matrix with known entries. Furthermore, the complexity increases with the dimension of the matrix. Thus, this inversion is a major concern for large network analysis.

The reduction is performed by a multiplication of two matrices from the left and from the right of the term to be inverted. Let us denote $\mathbf{W}_{n,q}$ to be the matrix to multiply from the left and $\mathbf{X}_{n,q}$ to be the matrix to multiply from the right. The subscripts denote the dimensions. Thus, matrices $\mathbf{W}_{n,q}$ and $\mathbf{X}_{n,q}$ are of dimensions n by q. Matrices $\mathbf{X}_{n,q}$ and $\mathbf{W}_{n,q}$ are chosen so that they are bi-orthogonal, that is,

$$\mathbf{W}^T_{n,q}\mathbf{X}_{n,q} = \mathbf{1}_{q,q} \tag{4.42}$$

Note that n is the dimension of the state vector $\mathbf{x}(s)$ and q is the dimension of the reduced system. For an efficient interconnect simulation the reduced dimension q is much smaller than n. The increase of q improves the precision of the simulation, but prolong the simulation process.

The multiplication of matrix $\mathbf{X}_{n,q}$ is carried out by state variable approximation. We approximate the vector of n state variables with vector $\mathbf{z}_q(s)$ of reduced size q.

$$\mathbf{x}_n(s) = \mathbf{X}_{n,q}\mathbf{z}_q(s) \tag{4.43}$$

Assuming the initial condition $\mathbf{x_0} = 0$, the system equations (4.16), (4.17) can be approximated as:

$$s\mathbf{M}\mathbf{X}_{n,q}\mathbf{z}_q(s) = -\mathbf{G}\mathbf{X}_{n,q}\mathbf{z}_q(s) + \mathbf{P}\mathbf{u}(s) \tag{4.44}$$

$$\mathbf{y}(s) = \mathbf{Q}\mathbf{X}_{n,q}\mathbf{z}_q(s) \tag{4.45}$$

Again, we can multiply the system equation (4.44) with matrix $\mathbf{G}^{-1}$ from the left and obtain the following expression.

$$s\mathbf{G}^{-1}\mathbf{M}\mathbf{X}_{n,q}\mathbf{z}_q(s) = -\mathbf{X}_{n,q}\mathbf{z}_q(s) + \mathbf{G}^{-1}\mathbf{P}\mathbf{u}(s) \tag{4.46}$$

We simplify the above equation by substituting $\mathbf{A}$ and $\mathbf{B}$ (equations (4.19) and (4.20), into the expression.

$$s\mathbf{A}\mathbf{X}_{n,q}\mathbf{z}_q(s) = -\mathbf{X}_{n,q}\mathbf{z}_q(s) + \mathbf{B}\mathbf{u}(s) \tag{4.47}$$

There are two versions of operation to multiply matrix $\mathbf{W}_{n,q}$. The main difference is the stage of the operation. In version 1, we multiply matrix $\mathbf{W}_{n,q}$ after $\mathbf{G}^{-1}$ is applied to the system equation, that is, equation (4.47), while in version 2, we multiply matrix $\mathbf{W}^T_{n,q}$ directly to equation (4.44). The purpose of the multiplication is to derive the solution of $\mathbf{z}_q(s)$. Once $\mathbf{z}_q(s)$ is known, we can derive the transfer function. Version 1 is adopted by the approaches using Padé Via Lanczos (PVL) [5] and Arnoldi processes [6]. Version 2 is proposed by the approaches using congruence transformation [7,8].

Version 1: PVL and Arnoldi Approximation We multiply the system equation (4.47) with $\mathbf{W}^T_{n,q}$ from the left.

$$s\mathbf{W}^T_{n,q}\mathbf{A}\mathbf{X}_{n,q}\mathbf{z}_q = -\mathbf{W}^T_{n,q}\mathbf{X}_{n,q}\mathbf{z}_q + \mathbf{W}^T_{n,q}\mathbf{B}\mathbf{u}(s) \tag{4.48}$$

Note that since $\mathbf{W}_{n,q}^T \mathbf{X}_{n,q} = \mathbf{1}_{q,q}$ by construction, we can write the system equation as:

$$s\mathbf{W}_{n,q}^T \mathbf{A}\mathbf{X}_{n,q}\mathbf{z}_q = -\mathbf{z}_q + \mathbf{W}_{n,q}^T \mathbf{B}\mathbf{u}(s) \quad (4.49)$$

Let us denote

$$\begin{aligned} \mathbf{A}_{q,q} &\equiv \mathbf{W}_{n,q}^T \mathbf{A}\mathbf{X}_{n,q} \\ \mathbf{B}_{q,u} &\equiv \mathbf{W}_{n,q}^T \mathbf{B} \\ \mathbf{Q}_{y,q} &\equiv \mathbf{Q}\mathbf{X}_{n,q} \end{aligned} \quad (4.50)$$

where subscript u is the dimension of the input vector $\mathbf{u}(s)$ and subscript y is the dimension of output vector $\mathbf{y}(s)$.

Using the notations in equations (4.50), we rewrite system equation (4.48) of the reduced state variables.

$$s\mathbf{A}_{q,q}\mathbf{z}_q = -\mathbf{z}_q + \mathbf{B}_{q,u}\mathbf{u}(s) \quad (4.51)$$

The multiplication approximates matrix $\mathbf{A}$ to a reduced square matrix of size q by q. The approximated system equation (4.51) can be solved as

$$\mathbf{z_q} = (\mathbf{1}_{q,q} + s\mathbf{A}_{q,q})^{-1}\mathbf{B}_{q,u}\mathbf{u}(s) \quad (4.52)$$

Consequently, the transfer function can be expressed by

$$\mathbf{y}(s) = \mathbf{Q}_{y,q}(\mathbf{1}_{q,q} + s\mathbf{A}_{q,q})^{-1}\mathbf{B}_{q,u}\mathbf{u}(s) \quad (4.53)$$

Note that the dimensions of input $\mathbf{u}(s)$ and output $\mathbf{y}(s)$ have not changed. However, in the transfer function, the size of the term to be inversed is reduced to q by q. The reduction is made by the approximation of the state vector.

Construction of $X_{n,q}$ Matrix $\mathbf{X}_{n,q}$ is constructed with the structure of transfer function. Since the inverse term $(\mathbf{1} + s\mathbf{A})^{-1}$ needs to multiply matrix $\mathbf{B}$ at right, we can construct $\mathbf{X}_{n,q}$ using the columns spanned by the Krylov space $[\mathbf{B}, \mathbf{AB}, \mathbf{A}^2\mathbf{B}, \ldots, \mathbf{A}^{q/u}\mathbf{B}]$. Note that if matrix $\mathbf{B}$ has u columns, we expand the Krylov space into q/u powers of $\mathbf{A}$.

Construction of $W_{n,q}$ There are two choices of matrices $\mathbf{W}_{n,q}$: PVL and Arnoldi processes. The former derives accurate results while the later guarantees the reduced system will be stable.

I. THE PVL PROCESS Since the inverse term $(\mathbf{1} + s\mathbf{A})^{-1}$ needs to multiply $\mathbf{Q}$ at left, PVL process constructs $\mathbf{W}_{n,q}$ using the columns spanned by the Krylov space $[\mathbf{Q}, \mathbf{A}^T\mathbf{Q}, (\mathbf{A}^T)^2\mathbf{Q}, \ldots, (\mathbf{A}^T)^{q/y}\mathbf{Q}]$, where y is the number of rows of $\mathbf{Q}$. This process utilizes the input and output structure of the system. Thus, PVL derives very good precision. However, the matrix transformation is not symmetric, that is, $\mathbf{W}_{n,q} \neq \mathbf{X}_{n,q}$, thus the process is not stable in the sense that we may find positive poles in the reduced transfer function that are not in the original transfer function.

II. THE ARNOLDI PROCESS For Arnoldi process, matrix $\mathbf{W}_{n,q}$ is set to be equal to $\mathbf{X}_{n,q}$. In this case, matrix $\mathbf{X}_{n,q}$ is constructed to be orthogonal, $\mathbf{X}_{n,q}^T \mathbf{X}_{n,q} = \mathbf{1}_{q,q}$.

Since the output structure is ignored, the Arnoldi is less accurate than PVL process. However, the symmetry of the matrix transformation, $\mathbf{W}_{n,q} = \mathbf{X}_{n,q}$, preserves the nonnegative definite property of matrix $\mathbf{A}$ as stated in the following lemma.

Lemma 4.4.1 *Given a nonnegative definite matrix* $\mathbf{A}$, $\mathbf{A}_{q,q}$ *generated by the Arnoldi process (version 1.ii) remains nonnegative definite.*

Proof: In the Arnoldi process, we multiply matrix $\mathbf{A}$ with $\mathbf{X}_{n,q}$ and its transpose from two sides,

$$\mathbf{A}_{q,q} = \mathbf{X}_{n,q}^T \mathbf{A} \mathbf{X}_{n,q} \tag{4.54}$$

This is a congruence operation. Thus, the nonnegative definite property of matrix $\mathbf{A}$ is preserved. □

Note that in version 1 we multiply matrix $\mathbf{G}^{-1}$ before the transformation. This multiplication destroys the symmetric properties of the original matrices. In the next subsection, we will show that version 2 operation keeps the symmetric properties of the matrices. Thus, we can later prove the passivity of the reduced system.

Version 2: Congruence Transformation For version 2, we formulate the expression so that the transpose of output matrix $\mathbf{Q}$ in system equation (4.16) is equal to input matrix $\mathbf{P}$, that is, $\mathbf{Q}^T = \mathbf{P}$. This is accomplished by setting both the input and output as the external ports of the network. The input vector $\mathbf{u}(s)$ contains the voltage and current sources of the external ports. The output vector $\mathbf{y}(s)$ contains the currents for voltage sources and voltages for current sources.

Example We use the circuit in Figure 4.2 as an example to write the system equation. We set two ports: a port across v_s and a port across node 3 and ground. It is natural to set a port across v_s as a port driven by voltage source. Since we are interested in voltage v_3 as an output, we set node 3 and ground as a port driven by current source i_{3s}. Hence, in the system equation, the vector of driving sources is $[\, v_s \quad i_{3s} \,]^T$. The vector of states is $[\, v_1 \quad v_2 \quad v_3 \quad i_s \quad i_{42} \quad i_{53} \,]^T$ and the output vector is $[\, i_s \quad v_3 \,]^T$. According to the proposed setting of the vectors, we write the system equation as follow.

$$
\left[\begin{array}{ccc|ccc}
C_1 & & & 0 & 0 & 0 \\
 & C_2 & & 0 & 0 & 0 \\
 & & C_3 & 0 & 0 & 0 \\
\hline
0 & 0 & 0 & 0 & & \\
0 & 0 & 0 & & L_{24} & \\
0 & 0 & 0 & & & L_{35}
\end{array}\right]
\begin{bmatrix} \frac{dv_1}{dt} \\ \frac{dv_2}{dt} \\ \frac{dv_3}{dt} \\ \frac{di_s}{dt} \\ \frac{di_{42}}{dt} \\ \frac{di_{53}}{dt} \end{bmatrix}
=
\left[\begin{array}{ccc|ccc}
0 & 0 & 0 & -1 & 0 & 0 \\
0 & 0 & 0 & 0 & -1 & 0 \\
0 & 0 & 0 & 0 & 0 & -1 \\
\hline
1 & 0 & 0 & R_s & 0 & 0 \\
-1 & 1 & 0 & 0 & R_{14} & 0 \\
0 & -1 & 1 & 0 & 0 & R_{25}
\end{array}\right]
$$

$$
\times \begin{bmatrix} v_1 \\ v_2 \\ v_3 \\ i_s \\ i_{42} \\ i_{53} \end{bmatrix} + \begin{bmatrix} 0 & 0 \\ 0 & 0 \\ 0 & 1 \\ 1 & 0 \\ 0 & 0 \\ 0 & 0 \end{bmatrix} \begin{bmatrix} v_s \\ i_{3s} \end{bmatrix} \tag{4.55}
$$

$$
\begin{bmatrix} i_s \\ v_3 \end{bmatrix} = \begin{bmatrix} 0 & 0 & 0 & 1 & 0 & 0 \\ 0 & 0 & 1 & 0 & 0 & 0 \end{bmatrix} \begin{bmatrix} v_1 \\ v_2 \\ v_3 \\ i_s \\ i_{42} \\ i_{53} \end{bmatrix} \tag{4.56}
$$

Note that due to this multiport formulation, matrix $\mathbf{P}$ is equal to matrix $\mathbf{Q}^T$.

Exercise Explain that the equality $\mathbf{Q}^T = \mathbf{P}$ in the above example is not a coincidence.

We set matrix $\mathbf{W}_{n,q} = \mathbf{X}_{n,q}$. Equation (4.44) is multiplied with matrix $\mathbf{X}_{n,q}^T$ from the left.

$$
s\mathbf{X}_{\mathbf{n,q}}^{\mathbf{T}}\mathbf{M}\mathbf{X}_{\mathbf{n,q}}\mathbf{z_q} = -\mathbf{X}_{\mathbf{n,q}}^{\mathbf{T}}\mathbf{G}\mathbf{X}_{\mathbf{n,q}}\mathbf{z_q} + \mathbf{X}_{\mathbf{n,q}}^{\mathbf{T}}\mathbf{P}\mathbf{u}(\mathbf{s}) \tag{4.57}
$$

Let us denote

$$
\begin{aligned}
\mathbf{M}_{q,q} &\equiv \mathbf{X}_{n,q}^T\mathbf{M}\mathbf{X}_{n,q} \\
\mathbf{G}_{q,q} &\equiv \mathbf{X}_{n,q}^T\mathbf{G}\mathbf{X}_{n,q} \\
\mathbf{P}_{q,u} &\equiv \mathbf{X}_{n,q}^T\mathbf{P}
\end{aligned} \tag{4.58}
$$

Substituting the notations (4.58) into equation (4.57), we simplify the system equation as

$$
s\mathbf{M}_{q,q}\mathbf{z_q} = -\mathbf{G}_{q,q}\mathbf{z_q} + \mathbf{P}_{q,u}\mathbf{u}(\mathbf{s}) \tag{4.59}
$$

Therefore, we have the solution of the reduced state variable vector $\mathbf{z_q}$.

$$\mathbf{z_q} = (s\mathbf{M}_{q,q} + \mathbf{G}_{q,q})^{-1}\mathbf{P}_{q,u}\mathbf{u}(\mathbf{s}) \tag{4.60}$$

Consequently, the transfer function can be expressed as

$$\mathbf{y}(\mathbf{s}) = \mathbf{P}_{q,u}^T(s\mathbf{M}_{q,q} + \mathbf{G}_{q,q})^{-1}\mathbf{P}_{q,u}\mathbf{u}(\mathbf{s}) \tag{4.61}$$

The inversion of the reduced matrix in equation (4.61) can be performed via eigenvalue eigenvector decomposition.

The choice of matrix $\mathbf{X}_{n,q}$ follows the block Arnoldi process. We construct $\mathbf{X}_{n,q}$ using the columns spanned by the Krylov space $[\mathbf{B}, \mathbf{AB}, \mathbf{A}^2\mathbf{B}, \ldots, \mathbf{A}^{q/u}\mathbf{B}]$.

The congruence transformation keeps the structure of the original matrices. For example, the transformed matrix $\mathbf{M}_{q,q}$ remains symmetric and nonnegative definite.

Lemma 4.4.2 *Given a symmetric and nonnegative definite matrix* $\mathbf{M}$, *matrix* $\mathbf{M}_{q,q}$ *remains symmetric and nonnegative definite.*

Proof: Since

$$\mathbf{M}_{q,q} = \mathbf{X}_{n,q}^T\mathbf{M}\mathbf{X}_{n,q} \tag{4.62}$$

is a congruence transformation, matrix $\mathbf{M}_{q,q}$ remains symmetric and nonnegative definite. □

Matrix $\mathbf{G}$ is composed of submatrices $\mathbf{Y}$ and $\mathbf{R}$ in equation (4.7). Since it was shown that submatrices $\mathbf{Y}$ and $\mathbf{R}$ in equation (4.7) are symmetric and nonnegative definite, we can prove that the sum of matrix $\mathbf{G}_{q,q}$ and its transpose, $\mathbf{G}_{q,q} + \mathbf{G}_{q,q}^T$ is nonnegative definite.

Lemma 4.4.3 *Let matrix* $\mathbf{G}_{q,q}$ *be derived from matrix* $\mathbf{G}$ *by congruence transformation (version 2). Then, matrix* $\mathbf{G}_{q,q} + \mathbf{G}_{q,q}^T$ *is symmetric and nonnegative definite.*

Proof: Since

$$\mathbf{G}_{q,q} = \mathbf{X}_{n,q}^T\mathbf{G}\mathbf{X}_{n,q} \tag{4.63}$$

we have

$$\mathbf{G}_{q,q} + \mathbf{G}_{q,q}^T = \mathbf{X}_{n,q}^T(\mathbf{G} + \mathbf{G}^T)\mathbf{X}_{n,q} \tag{4.64}$$

After expanding matrix $\mathbf{G}$ by equation (4.7), we have

$$\mathbf{G}_{q,q} + \mathbf{G}_{q,q}^T = \mathbf{X}_{n,q}^T\left(\begin{bmatrix} \mathbf{Y} & \mathbf{E} \\ -\mathbf{E}^T & \mathbf{R} \end{bmatrix} + \begin{bmatrix} \mathbf{Y}^T & -\mathbf{E} \\ \mathbf{E}^T & \mathbf{R}^T \end{bmatrix}\right)\mathbf{X}_{n,q} \tag{4.65}$$

Because matrices **Y** and **R** are symmetric, we have

$$\mathbf{G}_{q,q} + \mathbf{G}_{q,q}^T = \mathbf{X}_{n,q}^T \begin{bmatrix} 2\mathbf{Y} & 0 \\ 0 & 2\mathbf{R} \end{bmatrix} \mathbf{X}_{n,q} \tag{4.66}$$

Because matrices **Y** and **R** are nonnegative definite, we conclude that after the transformation $\mathbf{G}_{q,q} + \mathbf{G}_{q,q}^T$ is nonnegative definite. □

4.4.1 Stability and Passivity

There are no active components in the interconnect segments. Thus, the network is supposed to be stable and passive [9]. Passivity is an important property to satisfy because stable but not passive macromodels can become unstable systems after they are connected to other stable loads [13]. In the following, we will give definitions of stability and passivity and discuss the properties of the reduced network.

DEFINITION 4.4.1

Stability: A network is stable if, for all bounded inputs, the output is bounded. An unforced network is stable if, for all initial states at time t_0, the state trajectory is bounded in the interval $[t_0, \infty)$

From the above definition, we can deduce the property of the poles. Suppose that a network function is stable. Each of its pole p is negative, that is, the real component $Re(p) \leq 0$; and if $Re(p) = 0$, the pole is simple.

DEFINITION 4.4.2

Passivity of One Port: A one-port network is said to be passive if the total power dissipation is nonnegative for all initial time t_0, for all time $t > t_0$, and for all possible input waveforms, that is,

$$\int_{t_0}^{t} v(\tau)i(\tau)\,d\tau + E(t_0) \geq 0$$

where $E(t_0)$ is the energy stored at time t_0.

We can extend the concept to the whole network by the following definition.

DEFINITION 4.4.3

Passivity of a Network: A network is passive if all the elements of the network are passive.

Passivity in Complex Space Representation For a steady state response, we can describe the passivity in complex space representation. For a one-port network, we first write its voltage and current as sinusoidal waves with exponential component.

$$v(t) = v_0 e^{\alpha t} \cos(\omega t + \varphi_1) \tag{4.68}$$

$$i(t) = i_0 e^{\alpha t} \cos(\omega t + \varphi_2) \tag{4.69}$$

The energy injection is the product of voltage and current:

$$\begin{aligned} v(t)i(t) &= v_0 i_0 e^{2\alpha t} \cos(\omega t + \varphi_1) \cos(\omega t + \varphi 2) \\ &= \tfrac{1}{2} v_0 i_0 e^{2\alpha t} (\cos(2\omega t + \varphi_1 + \varphi_2) \\ &\quad + \cos(\varphi_1 - \varphi_2)) \end{aligned} \tag{4.70}$$

The average power injecting into the network is equal to:

$$\tfrac{1}{2} v_0 i_0 e^{2\alpha t} \cos(\varphi_1 - \varphi_2) \tag{4.71}$$

The above power expression can be derived using complex space representation. In complex space, the sinusoidal waves can be expressed as

$$i(t) = Re\{\tilde{i}(t)\}, \qquad \tilde{i}(t) = i_0 e^{\alpha t + j(\omega t + \varphi_1)} \tag{4.72}$$

$$v(t) = Re\{\tilde{v}(t)\}, \qquad \tilde{v}(t) = v_0 e^{\alpha t + j(\omega t + \varphi_2)} \tag{4.73}$$

The power is written as:

$$\tfrac{1}{2} \tilde{v}^*(t)\tilde{i}(t) = \tfrac{1}{2} v_0 i_0 e^{\alpha t - j(\omega t + \varphi_2)} e^{\alpha t + j(\omega t + \varphi_1)} = \tfrac{1}{2} v_0 i_0 e^{2\alpha t + j(\varphi_1 - \varphi_2)} \tag{4.74}$$

where the superscript * denotes the complex conjugate operation. The real component of expression (4.74) is equal to the average power, expression (4.71).

Let $Z(s)$ be the impedance of the port. The steady state voltage response is expressed as

$$\tilde{v}(t) = Z(\alpha + j\omega)\tilde{i}(t) \tag{4.75}$$

The power can be rewritten as:

$$\tfrac{1}{2} \tilde{v}^*(t)\tilde{i}(t) = \tfrac{1}{2} \hat{i}^*(\alpha + j\omega) Z^*(\alpha + j\omega) \hat{i}(\alpha + j\omega) \tag{4.76}$$

where $\tilde{i}(\alpha + j\omega)$ represents the magnitude and phase of current $i(t)$ at $s = \alpha + j\omega$, that is, $\hat{i}(\alpha + j\omega) = i_0 e^{j\varphi_1}$.

We extend the above concept of complex representation to multiple ports. Each port is either a voltage source or current source. For a voltage source port, the input is the voltage and the output is a current. For a current source port, the input is the current and the output is a voltage. Let u(t) be the input vector of all ports, and $\mathbf{H}^*(s)$ be the transfer function of the output vector versus the input vector. Matrix $\mathbf{H}^*(s)$ is a square matrix because the dimension of the input vector is equal to the dimension of the output vector. We express the power dissipation of multiple port as

$$\frac{1}{2}\hat{\mathbf{u}}^*(s_0)\mathbf{H}^*(s_0)\hat{\mathbf{u}}(s_0) \tag{4.77}$$

where $\hat{\mathbf{u}}(s_0)$ represents the magnitude and phase of input vector $\mathbf{u}(t)$ and matrix $\mathbf{H}(s_0)$ is the transfer function at $s = s_0$. Therefore, we deduce that the network is passive if the real component of expression (4.77) is positive for all s_0 in the right-half-plane of the complex space,

$$Re(\hat{\mathbf{u}}^*(s_0)\mathbf{H}^*(s_0)\hat{\mathbf{u}}(s_0)) \geq 0 \tag{4.78}$$

We can sum up power dissipation (4.77) and its own transpose to cancel the imaginary component. The above expression is equivalent to

$$\frac{1}{2}\hat{\mathbf{u}}^*(s_0)(\mathbf{H}^*(s_0) + \mathbf{H}(s_0))\hat{\mathbf{u}}(s_0) \geq 0 \tag{4.79}$$

Note that in expression (4.79), vector $\hat{\mathbf{u}}(s_0)$ is determined by the inputs. Thus, expression (4.79) needs to be true for all vectors $\hat{\mathbf{u}}(s_0)$ in complex space to ensure that the network is passive.

Stability and Passivity in System Matrix Expression For a stable network, transfer function $\mathbf{H}(s)$ equation (4.25), should have only negative poles p_j, that is, the real component $Re(p_j) \leq 0$. If a pole falls on the imaginary axis, it must be a simple pole. Otherwise, the output response will grow out of any bound as time t approaches infinity.

Stability of Arnoldi Transformation (version 1.ii) From lemma 4.4.2, we can derive that the reduced system equation by Arnoldi transformation in version 1.ii remains stable.

Theorem 4.4.1 *The network reduced by Arnoldi transformation in version 1.ii remains to be stable.*

Proof: From Arnoldi transformation version 1.ii, equation (4.53), we have the transfer function:

$$\mathbf{H}(s) = \mathbf{Q}_{y,q}(\mathbf{1}_{q,q} + s\mathbf{A}_{q,q})^{-1}\mathbf{B}_{q,u} \tag{4.80}$$

Since matrix $\mathbf{A}_{q,q}$ is nonnegative definite (lemma 4.4.1), the eigenvalues of $\mathbf{A}_{q,q}$ are nonnegative. Consequently, the poles of $(\mathbf{1}_{q,q} + s\mathbf{A}_{q,q})^{-1}$ are in the left plane of the complex space. Therefore, the reduced network is stable. □

Passivity of Congruence Transformation (version 2) In congruence transformation, we set output dimension equal to input dimension, $y = u$. Thus, matrix

$\mathbf{H}(s)$ is square. From lemmas 4.4.2 and 4.4.3, we can derive that the reduced system equation of congruence transformation in version 2 is passive.

Theorem 4.4.2 *The network reduced by congruence transformation in version 2 is passive.*

Proof: According to the definition of passivity, we need to prove that

$$\hat{\mathbf{u}}^*(s)(\mathbf{H}^*(s) + \mathbf{H}(s))\hat{\mathbf{u}}(s) \geq 0 \tag{4.81}$$

For congruence transformation, version 2, we have

$$\mathbf{H}(s) = \mathbf{P}_{q,u}^T(s\mathbf{M}_{q,q} + \mathbf{G}_{q,q})^{-1}\mathbf{P}_{q,u} \tag{4.82}$$

Thus, we can derive

$$\mathbf{H}^*(s) + \mathbf{H}(s) = \mathbf{P}_{q,u}^T((s\mathbf{M}_{q,q} + \mathbf{G}_{q,q})^{-1} + (s^*\mathbf{M}_{q,q} + \mathbf{G}_{q,q})^{-T})\mathbf{P}_{q,u} \tag{4.83}$$

$$= \mathbf{P}_{q,u}^T(s^*\mathbf{M}_{q,q} + \mathbf{G}_{q,q})^{-T}(s\mathbf{M}_{q,q} + \mathbf{G}_{q,q} + s^*\mathbf{M}_{q,q}^* + \mathbf{G}_{q,q}^*) \times (s\mathbf{M}_{q,q} + \mathbf{G}_{q,q})^{-1}\mathbf{P}_{q,u}) \tag{4.84}$$

$$= ((s\mathbf{M}_{q,q} + \mathbf{G}_{q,q})^{-1}\mathbf{P}_{q,u})^*(s\mathbf{M}_{q,q} + s^*\mathbf{M}_{q,q}^* + \mathbf{G}_{q,q} + \mathbf{G}_{q,q}^*) \times ((s\mathbf{M}_{q,q} + \mathbf{G}_{q,q})^{-1}\mathbf{P}_{q,u}) \tag{4.85}$$

From lemma 4.4.2, $\mathbf{M}_{q,q} = \mathbf{M}_{q,q}^T$, we can rewrite the middle term. The above expression becomes

$$\mathbf{H}^*(s) + \mathbf{H}(s) = ((s\mathbf{M}_{q,q} + \mathbf{G}_{q,q})^{-1}\mathbf{P}_{q,u})^*(2\,Re(s)\mathbf{M}_{q,q} + \mathbf{G}_{q,q} + \mathbf{G}_{q,q}^T) \times ((s\mathbf{M}_{q,q} + \mathbf{G}_{q,q})^{-1}\mathbf{P}_{q,u}) \tag{4.86}$$

From lemma 4.4.3, we can deduce that the middle term

$$2\,Re(s)\mathbf{M}_{q,q} + \mathbf{G}_{q,q} + \mathbf{G}_{q,q}^T \tag{4.87}$$

is positive definite. The left term of equation (4.85) is the conjugate transpose of the right term. Thus, we can conclude that $\mathbf{H}^*(s) + \mathbf{H}(s)$ is positive definite. Therefore, we have $\hat{\mathbf{u}}^*(s)(\mathbf{H}^*(s) + \mathbf{H}(s))\hat{\mathbf{u}}(s) \geq 0$. □

4.5 ANALYSIS USING MOMENT MATCHING

We use the matching of moments to approximate the derivation of signal delay [10,11]. In this section, we first introduce the definition and characteristics of moments. We touch on simple central moments to estimate delay. Given the moments, we describe the transfer function reduction using a Padé approximation [12]. We then present the calculation of moments utilizing the structure of the interconnect circuits. The connection between Padé approximation and PVL matrix approximation is discussed. Equipped with the above concepts, we investigate the relation between RLC and RC network, the crosstalk, the Elmore delay model (a first-order moment analysis of RC trees), and the π model.

4.5.1 Concept of Moments

We first assume that the system is single input and single output. Thus, the transfer function is written in scalar.

DEFINITION 4.5.1

Moments: The moments are the coefficients of the Taylor's expansion about $s = 0$, or *Maclaurin expansion*, of the transfer function in s domain, that is, moment $m^{(j)} = (1/j!)(d^jH/ds^j)|_{s=0}$.

Following this definition, we can represent the coefficients of the Maclaurin expansion with the notations of moments.

$$H(s) = H_0 + s\left.\frac{dH}{ds}\right|_{s=0} + s^2\frac{1}{2!}\left.\frac{d^2H}{ds^2}\right|_{s=0} + s^3\frac{1}{3!}\left.\frac{d^3H}{ds^3}\right|_{s=0} + \cdots$$

$$= \sum_{j=0}^{\infty} s^j \frac{1}{j!}\frac{d^jH}{ds^j} \tag{4.88}$$

$$= m^{(0)} + sm^{(1)} + s^2m^{(2)} + s^3m^{(3)} + \cdots = \sum_{j=0}^{\infty} s^j m^{(j)} \tag{4.89}$$

The definition of moments can be extended from scalar to vectors and matrices for systems of multiple input-outputs.

Derivation of Moments The exact Maclaurin expansion of transfer function was derived in the expansion of transfer function (4.39). We rewrite the equation here for reference.

$$\mathbf{H}(s) = \mathbf{Q}(\mathbf{1} + s\mathbf{A})^{-1}\mathbf{B} = \mathbf{Q}(\mathbf{1} - s\mathbf{A} + s^2\mathbf{A}^2 - s^3\mathbf{A}^3 + \cdots)\mathbf{B}$$

$$= \sum_{j=0}^{\infty} s^j(-1)^j\mathbf{Q}\mathbf{A}^j\mathbf{B} \tag{4.90}$$

Mapping the above equation with the definition of moments, we derive the expression of moments $\mathbf{m}^{(j)}$.

$$\mathbf{m}^{(j)} = (-1)^j\mathbf{Q}\mathbf{A}^j\mathbf{B} = (-1)^j\mathbf{Q}(\mathbf{G}^{-1}\mathbf{M})^j\mathbf{G}^{-1}\mathbf{P} \tag{4.91}$$

Characteristics of Moments The characteristics of the moments in terms of waveform can be derived from the definition of the Laplace transform. We use Taylor's series to expand the exponential function e^{-st}. Let $H(t)$ be the impulse response in time domain; we have

$$H(s) = \int_0^{\infty} H(t)e^{-st}\,dt = \int_0^{\infty} H(t)\left[1 - st + s^2\frac{1}{2}t^2 - s^3\frac{1}{3!}t^3 + \cdots\right] dt \tag{4.92}$$

$$= \sum_{j=0}^{\infty} s^j \frac{(-1)^j}{j!} \int_0^{\infty} t^j H(t)\,dt \tag{4.93}$$

Mapping the above equation with the definition that $H(s) = \sum_{j=0}^{\infty} s^j m^{(j)}$, we derive that moment $m^{(j)}$ is equal to the integration of the impulse response $H(t)$ and t^j with time t,

$$m^{(0)} = \int_0^{\infty} H(t)\,dt \tag{4.94}$$

$$m^{(1)} = -\int_0^{\infty} tH(t)\,dt \tag{4.95}$$

$$m^{(2)} = \frac{1}{2!}\int_0^{\infty} t^2 H(t)\,dt \tag{4.96}$$

$$m^{(3)} = -\frac{1}{3!}\int_0^{\infty} t^3 H(t)\,dt \tag{4.97}$$

$$\cdots = \cdots \tag{4.98}$$

$$m^{(j)} = \frac{(-1)^j}{j!}\int_0^{\infty} t^j H(t)\,dt \tag{4.99}$$

Remark: We can also use Taylor's series to expand the transfer function about $s = s_0$. In this case, we replace s with $s = s_0 + \sigma$.

4.5.2 Delay Estimation Using Central Moments

Moments can be used to approximate the curve of the impulse response. We choose from a family of functions to fit the first few moments. The function that matches the characteristics of the moments is used to derive the signal delay. The method is straightforward and stable. This approach works well for monotonic outputs such as RC networks and overdamped RLC networks [14].

The mean time of the impulse response is considered the center of the moments, that is, $\int_0^{\infty} tH(t)\,dt / \int_0^{\infty} H(t)\,dt$. We assume that the total area of impulse response is a unit, $\int_0^{\infty} H(t)\,dt = 1$. This is true in the case of capacitive load. Otherwise, a simple factor is applied. Therefore, the center of the mass is $\int_0^{\infty} tH(t)\,dt = -m^{(1)}$.

The central moments are measured using the mass center as the origin,

$$\mu^{(j)} = \int_0^{\infty} (t - (-m^{(1)}))^j H(t)\, dt \tag{4.100}$$

Therefore we have:

AREA UNDER THE CURVE $\mu^{(0)} = \int_0^{\infty} (t - (-m^{(1)}))^0 H(t)\, dt = 1$. The area under the curve is unity.

MASS CENTER $\mu^{(1)} = \int_0^{\infty} (t - (-m^{(1)}))H(t)\, dt = 0$. The mass center is set as the origin.

Note that we can also interpret the mass center as circuit delay from the viewpoint of step input response. Suppose the initial state is zero, that is, $x_0 = 0$. For step input $u(t)$, using equation (4.14), we express output $y(t)$ in time domain:

$$y(t) = \int_0^t H(t-\tau)u(\tau)\, d\tau = \int_0^t H(\tau)u(t-\tau)\, d\tau = \int_0^t H(\tau)\, d\tau \tag{4.101}$$

Differentiating the expression with time t, we have

$$\frac{dy(t)}{dt} = H(t) \tag{4.102}$$

Substituting equation (4.102) into moment one expression, we describe the mass center as:

$$-m^{(1)}) = \int_0^{\infty} t\frac{dy(t)}{dt}\, dt \tag{4.103}$$

This is the voltage-time area of step input response between output curve $(y(t))$ and vertical (voltage) axis. Suppose that output $y(t)$ is a step function. Then this voltage-time area is equal to its delay. In the case that switch voltage is half of the supply voltage level, if the output rises as a ramp function, the voltage-time area is also equal to the signal delay.

INERTIAL AROUND THE MASS CENTER $\mu^{(2)} = \int_0^{\infty} (t - (-m^{(1)}))^2 H(t)\, dt = 2m^{(2)} - (-m^{(1)})^2$. This is the variance of the distribution that spreads around the center.

SKEWNESS OF THE DISTRIBUTION $\mu^{(3)} = \int_0^{\infty} (t - (-m^{(1)}))^3 H(t)\, dt = 6(-m^{(3)}) - 6m^{(2)}(-m^{(1)}) + 2(-m^{(1)})^3$. For a unimodal function, the skewness of the distribution determines the relative position of the peak to the mass center. Its magnitude measures the distance between the peak and the center.

PEAKNESS OF THE DISTRIBUTION $\mu^{(4)} = \int_0^{\infty} (t - (-m^{(1)}))^4 H(t)\, dt = 24m^{(4)} - 24(-m^{(3)})(-m^{(1)}) + 12m^{(2)}(-m^{(1)})^2 - 3(-m^{(1)})^4$. For a unimodal function, its relation to the inertial $\mu^{(2)}$ reflects the portion of the area under the curve attributable to the tails of the curve.

Given the central moments, we can use a family of curves to approximate the impulse response. It is shown that the Gamma function is one reasonable representative of the impulse responses in RC trees since it provides a bell-shaped curve bounded on the left and decaying to the right exponentially.

The output waveform is derived from the convolution of the impulse response and the input waveform. The delay is thus calculated according to the specified switching voltage. For specified functions and input waveforms, we can precharacterize the function with a table to speed up the circuit analysis.

4.5.3 Padé Approximation

Once the moments are derived, Padé approximation finds a reduced rational function of a small order of s that exactly matches the first few moments. The inverse Laplace transform of the rational function approximates the impulse response. Padé approximation is efficient and effective for circuit analysis. The disadvantage of the approach is that the method is not stable. Once in a while, we may observe positive poles in the reduced rational function.

DEFINITION 4.5.2

Padé Approximation: Given two integers p, q, the type-(p/q) Padé approximation of the transfer function $H(s)$ is a rational function

$$H_{p,q}(s) = \frac{P_p(s)}{Q_q(s)} = \frac{a_0 + sa_1 + s^2a_2 + \cdots + s^pa_p}{1 + sb_1 + s^2b_2 + \cdots + s^qb_q} \tag{4.104}$$

where $P_p(s)$ and $Q_q(s)$ are polynomial functions of order p and q, respectively. The Maclaurin series of $H_{p,q}(s)$ agrees with the Maclaurin series of $H(s)$ in the first $p + q + 1$ terms, that is,

$$H(s) = H_{p,q}(s) + O(s^{p+q+1}) \tag{4.105}$$

Note that coefficient b_0 in equation (4.104) is set to one as the base of the expression. Otherwise, the rational expression has an arbitrary multiplicative factor.

Derivation of Padé Approximation A type-(p/q) Padé approximation has $p + q + 1$ coefficients of a_i and b_i. The coefficients are uniquely determined by the first $p + q + 1$ moments of the transfer function $H(s)$.

For the denominator $Q_q(s)$, we derives its coefficients b_i using the simultaneous equation in the following lemma.

Lemma 4.5.1 *For $H_{p,q}$, type-(p/q) Padé approximation, we have the following equality*

$$\begin{bmatrix} m^{(0)} & m^{(1)} & m^{(2)} & \dots & m^{(q-1)} \\ m^{(1)} & m^{(2)} & m^{(3)} & \dots & m^{(q)} \\ \dots & \dots & \dots & \dots & \dots \\ m^{(q-2)} & m^{(q-1)} & m^{(q)} & \dots & m^{(2q-3)} \\ m^{(q-1)} & m^{(q)} & m^{(q+1)} & \dots & m^{(2q-2)} \end{bmatrix} \begin{bmatrix} b_q \\ b_{q-1} \\ \vdots \\ b_2 \\ b_1 \end{bmatrix} = - \begin{bmatrix} m^{(q)} \\ m^{(q+1)} \\ \vdots \\ m^{(2q-2)} \\ m^{(2q-1)} \end{bmatrix} \tag{4.106}$$

Proof: The equality can be derived from the fact that the expression of $H_{p,q}(s) = P_p(s)/Q_q(s)$ needs to match the first $p+q+1$ moments of $H(s)$. We divide the numerator $P_p(s)$ by the denominator $Q_q(s)$ recursively from low to high order of s. By definition, the first $p+q+1$ terms of the quotient should match the first $p+q+1$ moments with the remainder in the format of $s^{p+q+1}r(s)$ where $r(s)$ is a polynomial function of s

$$\frac{P_p(s)}{Q_q(s)} = m^{(0)} + sm^{(1)} + s^2 m^{(2)} + \cdots + s^{p+q} m^{(p+q)} + \frac{s^{p+q+1} r(s)}{Q_q(s)} \tag{4.107}$$

Multiplying both sides by the denominator $Q_q(s)$, we have

$$\begin{aligned} a_0 + sa_1 + s^2 a_2 + \cdots + s^p a_p = {} & (1 + sb_1 + s^2 b_2 + \cdots + s^q b_q) \\ & \times (m^{(0)} + sm^{(1)} + s^2 m^{(2)} \\ & + \ldots + s^{p+q} m^{(p+q)}) + s^{p+q+1} r(s). \end{aligned} \tag{4.108}$$

We can then expand the right-hand-side terms by multiplication. The coefficients of powers s^i for $i \le p+q$ on both sides should be equal. Hence, we have equation (4.106) that matches the coefficients of terms s^i for $p < i \le p+q$. □

Once the coefficients b_i of the denominator are derived, we can calculate the coefficients a_i of the numerator from the equations in the following lemma.

Lemma 4.5.2 The coefficients a_i of numerator $Q_q(s)$ satisfy the equations:

$$\begin{aligned} a_0 &= m^{(0)} \\ a_1 &= m^{(1)} + m^{(0)} b_1 \\ a_2 &= m^{(2)} + m^{(1)} b_1 + m^{(0)} b_2 \\ a_3 &= m^{(3)} + m^{(2)} b_1 + m^{(1)} b_2 + m^{(0)} b_3 \\ \cdot &= \cdots \\ a_p &= m^{(p)} + \sum_{i=1}^{\min(p,q)} m^{(p-j)} b_j \end{aligned} \tag{4.109}$$

Proof: The equations are derived from matching the coefficients of s_i for $0 < i \leq p$ in equation (4.108). □

Partial Fraction Decomposition of Padé Approximation To derive the partial fraction decomposition, we first search the q roots of the denominator $Q_q(s)$;

$$Q_q(s) = \prod_{i=1}^{q}(s - p_i) \tag{4.110}$$

Assuming input $v_{in} = 1/(s - p_0)$, the partial fraction decomposition can be expressed in the following format.

$$v_{out}(s) = H_{p,q}v_{in}(s) = \frac{P_p(s)}{Q_q(s)}\frac{1}{s - p_0} = \frac{P_p(s)}{\prod_{i=0}^{q}(s - p_i)} \tag{4.111}$$

$$= \frac{\hat{k}_0}{s - p_0} + \frac{\hat{k}_1}{s - p_1} + \frac{\hat{k}_2}{s - p_2} + \cdots + \frac{\hat{k}_q}{s - p_q} = \sum_{i=0}^{q} \frac{\hat{k}_i}{s - p_i} \tag{4.112}$$

There are two versions to derive the residues $\hat{k}_j$. In version 1, we skip the derivation of the numerator $P_p(s)$. We find the residues of transfer function $H_{p,q}(s)$ from the moments directly. We then find output function $v_{out}(s)$ by multiplying the input to the decomposed transfer function. In version 2, we utilize rational function $(P_p(s)/Q_q(s))$. Residues k_j are derived by multiplying factor $s - p_j$ and replacing variable s with pole p_j.

Version 1 of Partial Fraction Expansion The derivation is obtained first from the partial fraction decomposition of the transfer function.

$$H_{p,q}(s) = \frac{k_1}{s - P_1} + \frac{k_2}{s - P_2} + \frac{k_3}{s - P_3} + \cdots + \frac{k_q}{s - P_q} \tag{4.113}$$

The residues k_j are obtained from the simultaneous equation stated in the following lemma.

Lemma 4.5.3 *The residues k_j of the partial fraction decomposition (4.113) satisfy the simultaneous equation:*

$$\begin{bmatrix} p_1^{-1} & p_2^{-1} & \cdots & p_q^{-1} \\ p_1^{-2} & p_2^{-2} & \cdots & p_q^{-2} \\ \vdots & \vdots & \vdots & \vdots \\ p_1^{-q} & p_2^{-q} & \cdots & p_q^{-q} \end{bmatrix} \begin{bmatrix} k_1 \\ k_2 \\ \vdots \\ k_q \end{bmatrix} = \begin{bmatrix} m^{(0)} \\ m^{(1)} \\ \vdots \\ m^{(q-1)} \end{bmatrix} \tag{4.114}$$

Proof: The above expression is derived from the terms in equation (4.89).

$$\left. \frac{1}{i!}\frac{d^i H_{p,q}}{ds^j} \right|_{s=0} = m^{(j)} \tag{4.115}$$

Replacing $H_{p,q}(s)$ in the above expression with equation (4.113) and expanding after the differential operation, we have

$$k_1 p_1^{-j} + k_2 p_2^{-j} + \cdots + k_q p_q^{-j} = m^{(j)} \tag{4.116}$$

which constitutes the simultaneous equation for residues k_j. □

Once the fraction decomposition of the transfer function (4.113) is derived, we can multiply the input and derive the output function.

$$\begin{aligned} v_{out}(s) &= H_{p,q} v_{in}(s) \\ &= \left(\frac{k_1}{s - p_1} + \frac{k_2}{s - p_2} + \frac{k_3}{s - p_3} + \cdots + \frac{k_q}{s - p_q} \right) \frac{1}{s - p_0} \end{aligned} \tag{4.117}$$

Decomposing each $(k_j/s - p_j)$ with $(1/s - p_0)$ pairwise, we obtain the decomposed expression.

$$v_{out}(s) = -\left(\sum_{j=1}^{q} \frac{k_j}{p_j - p_0} \right) \frac{1}{s - p_0} + \frac{k_1}{p_1 - p_0} \frac{1}{s - p_1} \tag{4.118}$$

$$+ \frac{k_2}{p_2 - p_0} \frac{1}{s - p_2} + \frac{k_3}{p_3 - p_0} \frac{1}{s - p_3}$$

$$+ \cdots + \frac{k_q}{p_q - p_0} \frac{1}{s - p_q} \tag{4.119}$$

Version 2 of Partial Fraction Expansion Version 2 utilizes the derived numerator $P_p(s)$ equation (4.109). We write the output function in the rational function and partial fraction decomposition forms:

$$v_{out}(s) = \frac{P_p(s)}{\prod_{i=0}^{q}(s - p_i)} = \sum_{i=0}^{q} \frac{\hat{k}_i}{s - p_i} \tag{4.120}$$

Multiplying factor $(s - p_j)$ to both sides, we obtain

$$\frac{P_p(s)}{\prod_{i=0, i \neq j}^{q}(s - p_i)} = \sum_{i=0, i \neq j}^{q} \frac{\hat{k}_i}{s - p_i}(s - p_j) + \hat{k}_j \tag{4.121}$$

We derive the residues $\hat{k}_j$ by replacing variable s with p_j,

$$\hat{k}_j = \frac{P_p(p_j)}{\prod_{i=0, i \neq j}^{q}(p_j - p_i)} \tag{4.122}$$

4.5.4 Moment Derivation Using Interconnect Tree

From the state response expression (4.38), we express the state vectors with moments, that is,

$$\begin{aligned}\mathbf{x}(s) &= \mathbf{m}^{(0)} + s\mathbf{m}^{(1)} + s^2\mathbf{m}^{(2)} + \cdots \\ &= (\mathbf{1} - s\mathbf{A} + s^2\mathbf{A}^2 - s^3\mathbf{A}^3 + \cdots)\mathbf{B} \qquad (4.123) \\ &= \mathbf{B} - s\mathbf{AB} + s^2\mathbf{A}^2\mathbf{B} - s^3\mathbf{A}^3\mathbf{B} + \cdots \qquad (4.124)\end{aligned}$$

where $\mathbf{m}^{(j)}$ is a vector of moment j of state variables. Since the coefficients of the same power of s should be equal, we can derive the moments of the state variables.

$$\mathbf{m}^{(0)} = \mathbf{B} \qquad (4.125)$$

$$\mathbf{m}^{(1)} = -\mathbf{AB} \qquad (4.126)$$

$$\mathbf{m}^{(2)} = \mathbf{A}^2\mathbf{B} \qquad (4.127)$$

$$\cdots = \cdots \qquad (4.128)$$

$$\mathbf{m}^{(j)} = (-1)^j\mathbf{A}^j\mathbf{B} \qquad (4.129)$$

The expression can be rewritten in a recursive form.

$$\mathbf{m}^{(0)} = \mathbf{B} \qquad (4.130)$$

$$\mathbf{m}^{(j)} = -\mathbf{A}\mathbf{m}^{(j-1)} \qquad (4.131)$$

Recall that matrix $\mathbf{B} = \mathbf{G}^{-1}\mathbf{P}$ and matrix $\mathbf{A} = \mathbf{G}^{-1}\mathbf{M}$. We can expand the expression and obtain:

$$\mathbf{m}^{(0)} = \mathbf{G}^{-1}\mathbf{P} \qquad (4.132)$$

$$\mathbf{m}^{(j)} = -\mathbf{G}^{-1}\mathbf{M}\mathbf{m}^{(j-1)} \qquad (4.133)$$

Shifting matrix $\mathbf{G}$ to the left side of the equality, we express the equation in the following format.

$$\mathbf{G}\mathbf{m}^{(0)} = \mathbf{P} \qquad (4.134)$$

$$\mathbf{G}\mathbf{m}^{(j)} = -\mathbf{M}\mathbf{m}^{(j-1)} \qquad (4.135)$$

Utilizing a circuit analogy, we view the above expression as a static circuit analysis expression. All the resistors are kept. The inductors and capacitors are replaced by voltage and current sources, respectively. Vector $\mathbf{m}^{(j)}$ in equation (4.135) is the vector of variables representing the trunk voltages of the tree structure and the branch currents of the co-tree. The equation describes Kirchhoff's voltage and current laws.

The vector on the right of equality $-\mathbf{M}\mathbf{m}^{(j-1)}$ describes the voltage and current sources. It includes the voltage sources that replace the inductors and the current sources that replace the capacitors.

Equation (4.135) enables us to derive the moments from the lower to higher order recursively. In general, the interconnect is composed of a tree of segments. This is a special case of circuit analysis. It takes operation in a linear complexity of the tree to calculate moment $\mathbf{m}^{(j)}$. For the case that the interconnect has meshes, the derivation of $\mathbf{m}^{(j)}$ is similar to general circuit analysis.

Example We use Figure 4.5 to illustrate the moment derivation. For moment zero, the equivalent circuit is derived by shorting the inductors and opening the capacitors. Thus $v_1^{(0)} = v_2^{(0)} = v_{sa}^{(0)}$ and $v_{(0)}^3 = v_4^{(0)} = v_{sb}^{(0)}$ (Fig. 4.6). We assume that the inputs are impulse signals. Hence, the voltage sources are shorted for higher-order moments, $v_{sa}^{(j)} = v_{sb}^{(j)} = 0$ for $j > 0$. The inductors are replaced by voltage sources. Thus, we have (Fig. 4.7):

$$v_{51}^{(j)} = L_{15} i_{51}^{(j-1)} + L_{m1} i_{73}^{(j-1)} \tag{4.136}$$

$$v_{73}^{(j)} = L_{37} i_{73}^{(j-1)} + L_{m1} i_{51}^{(j-i)} \tag{4.137}$$

$$v_{62}^{(j)} = L_{26} i_{62}^{(j-1)} + L_{m2} i_{84}^{(j-1)} \tag{4.138}$$

$$v_{84}^{(j)} = L_{48} i_{84}^{(j-1)} + L_{m2} i_{62}^{(j-1)} \tag{4.139}$$

The capacitors are replaced by current sources.

$$i_1^{(j)} = C_1 v_1^{(j-1)} \tag{4.140}$$

$$i_2^{(j)} = C_2 v_2^{(j-1)} \tag{4.141}$$

$$i_3^{(j)} = C_3 v_3^{(j-1)} \tag{4.142}$$

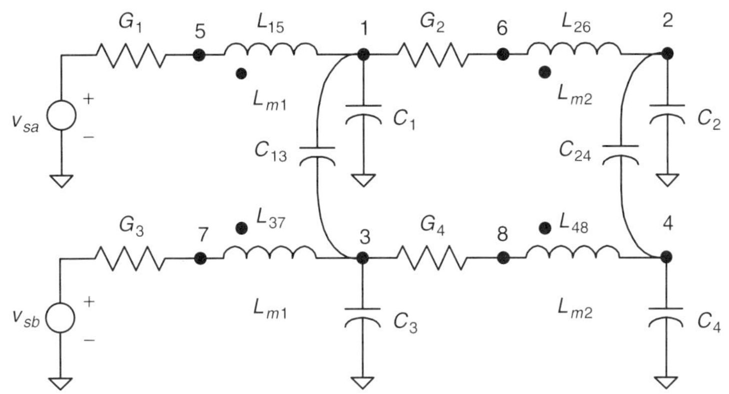

FIGURE 4.5 *Circuit to illustrate the moment derivation.*

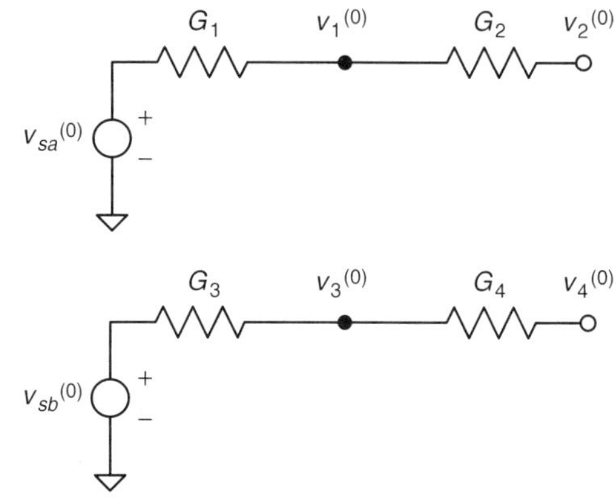

FIGURE 4.6 *Derivation of moment zero.*

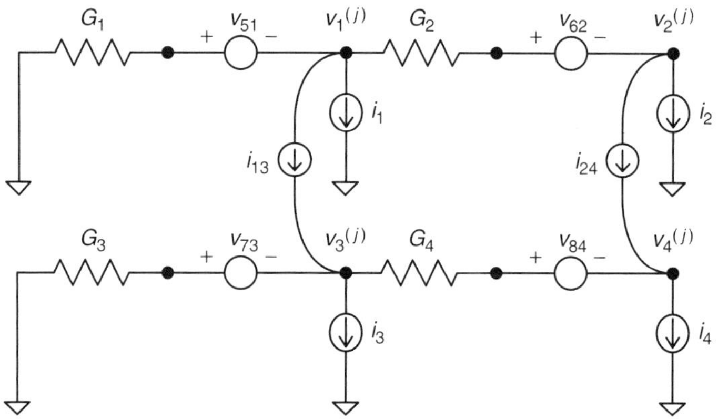

FIGURE 4.7 *Derivation of moment j.*

$$i_4^{(j)} = C_4 v_4^{j-1} \tag{4.143}$$

$$i_{13}^{(j)} = C_{1,3}(v_1^{(j-1)} - v_3^{(j-1)} \tag{4.144}$$

$$i_{24}^{(j)} = C_{2,4}(v_2^{(j-1)} - v_4^{(j-1)}) \tag{4.415}$$

4.5.5 Connection between Padé Approximation and PVL Matrix Approximation

Feldmann and Freund [5] showed that PVL matrix approximation is equivalent to Padé approximation. They derived that using PVL approximation the reduced transfer function (4.53) is equivalent to the original system in the sense that the first 2_q moments, $m^{(0)}, m^{(1)}, \ldots, m^{(2q-1)}$ remain the same. In other words, we have

$$\mathbf{Q}_{y,q}\mathbf{A}_{q,q}^{k}\mathbf{B}_{q,u} = \mathbf{Q}\mathbf{A}^{k}\mathbf{B} \; for \; 0 \leq k \leq 2q-1 \tag{4.146}$$

The reduced transfer function has a transfer function with $q-1$ zeros and q poles. Since the function matches the first $2q$ moments, it is exactly a type-$(q-1/q)$ Padé approximation. The advantage of PVL approximation is its numerical stability. However, it inherits the problem of Padé approximation that the reduction does not preserve the stability and passivity of the original network.

4.5.6 Voltage-Time Area of RLC and RC Network

We compare the voltage-time area of RLC and RC network. Given constant conductance G, resistance R, and capacitance C values, the following lemma states that for any inductances L, the RLC output voltage-time area is a constant. This is an extension of the derivation in lumped networks. Utilizing the derivation of moments, we show that the statement is true for distributed interconnect network.

Lemma 4.5.4 *Let $v_{RLC}(t)$ and $v_{RC}(t)$ be the time-domain waveforms of the RC and RLC network with equal R,C values. Suppose that the input in s domain, $v_s(s)$, can be expressed as a rational function $(P(s)/Q(s))$ and the denominator $Q(s)$ has at most one pole at $s=0$. Then the voltage-time area of the two outputs are equal,*

$$\int_0^{\infty} (v_{RLC}(t) - v_{RC}(t))\, dt = 0 \tag{4.147}$$

Proof: In RLC network, the state variable vector can be decomposed into a vector of capacitor voltage $\mathbf{v}_C$ and a vector of inductor current $\mathbf{i}_L$,

$$\mathbf{x} = \begin{bmatrix} \mathbf{v}_C \\ \mathbf{i}_L \end{bmatrix} \tag{4.148}$$

For the derivation of the moments, initially all capacitors are replaced by an open circuit and inductors are replaced by a short circuit. We solve the circuit analysis

$$\mathbf{Gm}^{(0)} = \mathbf{P} \tag{4.149}$$

Thus, moment zero of capacitor voltage $\mathbf{v}_C^{(0)}$ is independent of the values of inductors. As shown in Figure 4.6, moment zero of inductor currents is a vector of zeros, $\mathbf{i}_L^{(0)} = \mathbf{0}$.

Once the moment zero is derived, we solve the circuit analysis again to obtain moment one:

$$\mathbf{G}\begin{bmatrix} \mathbf{v}_C^{(1)} \\ \mathbf{i}_L^{(1)} \end{bmatrix} = -\begin{bmatrix} \mathbf{C} & 0 \\ 0 & \mathbf{L} \end{bmatrix}\begin{bmatrix} \mathbf{v}_C^{(0)} \\ \mathbf{i}_L^{(0)} \end{bmatrix} = -\begin{bmatrix} \mathbf{C}\mathbf{v}_C^{(0)} \\ \mathbf{L}\mathbf{i}_L^{(0)} \end{bmatrix} \tag{4.150}$$

Since $\mathbf{i}_L^{(0)} = \mathbf{0}$, moment one is also independent of the values of the inductors.

Because moment zero and moment one are independent of the values of the inductors, the difference happens only at moment two and thereafter. Hence, we can express the difference of the two output voltages $v_{RLC}(s) - v_{RC}(s)$ in the format $s^2 r(s)$, where $r(s)$ is a polynomial function of s.

Consequently, we can conclude that the voltage-time area is a constant, that is,

$$\begin{aligned}\int_0^{\infty} (v_{RLC}(t) - v_{RC}(t))\, dt &= \left.\int_0^{\infty} (v_{RLC}(t) - v_{RC}(t))e^{-st}\, dt\right|_{s=0} \\ &= (v_{RLC}(s) - v_{RC}(s))|_{s=0} \\ &= s^2 r(s) v_s|_{s=0} = 0 \end{aligned} \tag{4.151}$$

4.5.7 Capacitive Coupling Output with Respect to Ramp Input

Given a distributed RC network, we demonstrate that using moments we can derive a simple bound on the coupled voltage output with a ramp input. For a ramp input, the slope of the transition has a strong effect on the coupling noise. In this subsection, we simplify the case to assume that the ramp input has no clamp on the voltage range that is, $v_s(t) = \alpha t$, to derive the worst-case behavior. The model is pertinent when the signal transition is slow.

We denote the wire connected to the voltage source v_s as the aggressive line and the wires coupled to the aggressive line as victim lines. Let $v_2(t)$ be the output response on a victim line; its value reaches maximum as time t approaches infinity which can be derived in the s domain as:

$$\begin{aligned} lim_{t\to\infty} v_2(t) &= s v_2(s)|_{s=0} = sH_2(s)v_s(s)|_{s=0} \\ &= \left. sH_2(s)\frac{\alpha}{s^2}\right|_{s=0} = \left. H_2(s)\frac{\alpha}{s}\right|_{s=0} \end{aligned} \tag{4.152}$$

where $H_2(s)$ is the transfer function of output v_2 versus input v_s.

The following lemma states a bound on the peak of the coupling voltage.

Lemma 4.5.5 *The coupling output voltage of the victim line is bounded by*

$$v_2(t) \le v_2^{(1)} \alpha \tag{4.153}$$

where $v_2^{(1)}$ is moment one of the transfer function $H_2(s)$.

Proof: We prove the lemma by showing that moment zero of the victim line is zero. Consequently, moment one contributes to the solution of equation (4.152).

We decompose the state variable vector $\mathbf{x}$ into two subvectors: a vector of capacitor voltage $\mathbf{V}_{C1}$ of the nodes in the aggressive line and a vector of capacitor voltage $\mathbf{v}_{C2}$ of the nodes in the victim lines,

$$\mathbf{x} = \begin{bmatrix} \mathbf{v}_{C1} \\ \mathbf{v}_{C2} \end{bmatrix} \tag{4.154}$$

Moment zero is derived from the circuit analysis:

$$\mathbf{G}\mathbf{m}^{(0)} = \mathbf{P} \tag{4.155}$$

Since initially all capacitors are set to an open circuit, we derive that capacitor voltage on the aggressive line is unity and the capacitor voltage on the victim line is zero,

$$\mathbf{v}_{C1}^{(0)} = \mathbf{1} \tag{4.156}$$

$$\mathbf{v}_{C2}^{(0)} = \mathbf{0} \quad \square \tag{4.157}$$

4.5.8 Elmore Model (First-Order Moment) Analysis for Single Tree

Given a distributed RC tree, the Elmore model [15] uses moment one as a simplified delay model. Recall that moment one can be interpreted as the mass center of the voltage-time area.

$$m^{(1)} = \int_0^{\infty} tH(t)\,dt \tag{4.158}$$

Suppose the switch voltage of each gate is 0.5. For a step voltage input, from output response function (4.14), we can derive that the exact delay is equal to the median, t_{median}, of the voltage-time area,

$$\int_0^{t_{median}} H(t)\,dt = 0.5 \tag{4.159}$$

If the impulse response is symmetric, we have the mass center equal to the median, $-m^{(1)} = t_{median}$. Moment one delay model is accurate. However, in RC network, the impulse responses are exponential functions. The waveform is lopsided with a heavy front. The median is always smaller than the mass center, $t_{median} < -m^{(1)}$. Thus, the Elmore model gives an upper bound of the delay.

In general, we can multiply moment one by a ratio smaller than one to estimate the delay, for example, a ratio of 0.69 is used in equation (2.18). This strategy works when the tree is balanced. For a tree with uneven branches, the exact delay along short branches is usually smaller than the Elmore model delay. This is due to the shield effect of wire resistance. Higher-order moments are needed to derive an accurate delay.

Suppose the source voltage applies at node s of the tree. The whole tree is ordered from node s toward leaves. We are interested in the voltage response at a node t. Following the derivation of the previous subsection, we have moment zero equal to a unit, $m_{st}^{(0)} = 1$. Let us denote $m_{st}^{(1)}$ moment one of the transfer function from node s to node t. We use $-m_{st}^{(1)}$ as the Elmore delay from node s to node t. There are two ways to formulate moment one [16,17]: *resistance-oriented* and *capacitance-oriented* formulas. For the resistance-oriented formula, the number of terms is equal to the number of segments from node s to node t. Each segment resistance appears in each term. However, a node capacitance is included in many different terms. For the capacitance-oriented formula, each capacitance is included in at most one term, whereas the resistance is included in many different terms.

RESISTANCE-ORIENTED FORMULA In the resistance-oriented formula, the delay is derived segment by segment along the path from node s to node t. For each segment in the path, we multiply the segment resistance with lumped capacitance viewed at node v, yielding the following formula:

$$-m_{st}^{(1)} = \sum_{(u,v)\in P(s,t)} r_{uv} c(T_v) \tag{4.160}$$

where (u, v) is a segment on path $P(s, t)$ from node s to node t. r_{uv} is the segment resistance and $c(T_v)$ is the total capacitance in the subtree with root at node v. If node v is a leaf, the subtree capacitance is equal to the node capacitance, $c(T_v) = c_v$. Otherwise, we sum up the subtree capacitances of its direct descendants $DD(v)$ plus its own node capacitance, $c(T_v) = \Sigma_{k\in DD(v)} c(T_k) + c_v$.

We derive the moment with a bottom-up process to accumulate subtree capacitance $c(T_v)$ for each node u on the path $P(s, t)$. The term $r_{uv}c(T_v)$ is viewed as the delay contributed by segment (u, v). The sum of the segment delay along the path is the delay between the two ends.

DISTRIBUTED RC LINES For metal interconnect, it is natural to view wiring segments as distributed RC lines rather than as discrete R and C components. In this case the delay from s to t is given by

$$\sum_{(u,v)\in P(s,t)} r_{uv} \left(\frac{c_{uv}}{2} + c(T_v) \right) \tag{4.161}$$

where r_{uv} and c_{uv} are respectively the distributed resistance and capacitance of the segment connecting nodes u and v.

The correctness of this formula can be seen by uniformly dividing the segment (u, v) into a sequence of n L-model RC segments and applying (4.160) to obtain the following expression for the delay of segment (u, v):

$$r_{uv}c(T_v) + \frac{r_{uv}c_{uv}}{n^2} \left(\sum_{i=1}^{n} i \right) = \tag{4.162}$$

$$r_{uv}c(T_v) + \frac{r_{uv}c_{uv}}{n^2} \left(\frac{n(n+1)}{2} \right) \tag{4.163}$$

Taking the limit as n approaches infinity gives (4.161).

The net result is that we may replace distributed RC segments with the corresponding π-model and simply apply (4.160) or the capacitance-oriented formula presented below.

CAPACITANCE-ORIENTED FORMULA In the capacitance-oriented formula, for each node i along the path from node s to node t, we use an accumulated resistance from node s to node u and the capacitance related only to node i to derive each term. Let us use $n_{st}^{(1)}$ to denote the following expression:

$$n_{st}^{(1)} = \sum_{(u,v)\in P(s,t)} r_{su}(c(T_u) - c(T_v)) \tag{4.164}$$

where r_{su} is the sum of series resistance on path $P(s,i)$, that is, $r_{su} = \Sigma_{(k,l)\in P(s,u)} r_{kl}$. The second term $(c(T_u) - c(T_v))$ is the capacitance contributed to the delay at node u. The deduction of subtree capacitance $c(T_v)$ leaves its delay contribution to node v and nodes downstream.

We express the capacitance-oriented formula from $n_{st}^{(1)}$.

$$-m_{st}^{(1)} = n_{st}^{(1)} + r_{st}c(T_t) \tag{4.165}$$

The last term is the delay contributed by the lumped subtree capacitance at node t and the accumulated resistance from node s to node t.

Futhermore, we can separate the path at any segment $(u, v) \in P(s, t)$ and derive the Elmore delay with the two formats on the two sides of the segment.

$$-m_{st}^{(1)} = n_{su}^{(1)} + (-m_{vt}^{(1)}) + r_{sv}c(T_v) \tag{4.166}$$

The last term is the delay contributed by the accumulated resistance form node s to node j and the lumped subtree capacitance at node j. The above mixed formula can be used to calculate the interconnect delay from both ends and integrate the result in the middle.

Exercise Show that the Elmore delay expressions (4.160), (4.165), and (4.166) are equivalent.

4.5.9 Layout-Driven Analysis Procedure

In this subsection, we describe a layout-driven analysis procedure to analyze RC tree during routing. The procedure enables the bottom-up computation of interconnect tree moments, and can be easily integrated with a global router constructed by a bottom-up process. The purpose of the integration is to achieve incremental layout optimization (i.e., timing analysis and routing are tightly coupled, with feedback between them).

In the previous moment generation, we derive the moments with an assumption that the whole interconnect tree is given. For interconnect synthesis, we need to route first, then analyze the resulting topology. If it does conform to the electrical and timing requirements, we need to ripup and reroute, and so on. This not only limits the applicability of moments only to circuit analysis, it still does not break the traditional route-extraction-simulation loop, which is often the bottleneck in design cycle time.

The layout-driven analysis procedure extends the applicability of moments to the routing domain. That is, while routing is being performed, timing analysis is done at the same time, and subsequent routing is performed in accordance with the timing analysis results. Since routing is adjusted dynamically with the feedback from timing analysis, routing and timing analysis can be integrated and truly become one, achieving incremental layout optimization.

Let node i be a source and node t be a sink. Segment $(i, i+1)$ is in the path $P_{i,t}$ from i to t. Let us denote $m_{i,t}^{(k)}$ the kth moment of the transfer function form i to t. Moment $m_{i,t}^{(k)}$ can be derived from the information downstream:

$$m_{i,t}^{(k)} = m_{i,i+1}^{(k)} + m_{i+1,t}^{(k)} + Y_{i,t}^{(k)} \tag{4.167}$$

where $Y_{i,t}^{(k)}$ is the kth-order coupling between path $P_{i,i+1}$ and path $P_{i+1,t}$. For a bottom-up analysis procedure, we assume that moment $m_{i+1,t}^{(j)}$ has been calculated for $j \leq k$. Thus, we only need to derive the first term $m_{i,i+1}^{(k)}$ and third term $Y_{i,t}^{(k)}$ in order to obtain moment $m_{i,t}^{(k)}$.

Moment $m_{i,i+1}^{(k)}$ of path $P_{i,i+1}$ can be derived from

$$m_{i,i+1}^{(k)} = R_{i,i+1} X_{i,i+1}^{(k)} \tag{4.168}$$

where $R_{i,i+1}$ is the resistance of segment $(i, i+1)$, and $X_{i,i+1}^{(k)}$ is a tree recursive term:

$$X_{i,i+1}^{(k)} = \sum_{j \in DD(i+1)} X_{i+1,j}^{(k)} + m_{i,i+1}^{(k-1)} C_{T_{i+1}} + \sum_{t=1,\ldots,k-2} \left(m_{i,i+1}^{(1)} \sum_{j \in DD(i+1)} X_{i+1,j}^{(k-1)} \right) \tag{4.169}$$

where $DD(i+1)$ is the set of direct descendants of node $i+1$ and $C_{T_{i+1}}$ is the total lumped capacitance of the subtree rooted at node $i+1$. Thus, the tree recursive term $X_{i,i+1}^{(k)}$ of segment $(i, i+1)$ can be derived from the recursive terms of connected segments in downstream.

Once the moments of segment $(i, i+1)$ are obtained, we can compute the coupling term.

$$Y_{i,t}^{(k)} = \sum_{l=1,\ldots,k-1} m_{i,i+1}^{(l)} m_{i+1,t}^{(k-l)} \tag{4.170}$$

Exercise Verify that equation (4.167) derives the moments by induction.

4.5.10 Output π Model for RC Tree Reduction

In the preceding chapter, we use π model replacement (Fig 3.9) to demonstrate RC tree reduction at the output of a driving gate [16]. In this subsection, we show the derivation of π model using the moments. We assume that the output response with type-(2/1) Padé approximation is given.

$$\frac{i_g}{v_g} = \frac{P_2(s)}{Q_1(s)} = \frac{a_0 + sa_1 + s^2 a_2}{1 + sb_1} \tag{4.171}$$

Note that $a_0 = 0$ because of the capacitive loading.

For a π model (Fig 3.9), the output response is expressed as

$$\frac{i_g}{v_g} = sC_1 + \frac{1}{R_\pi + \frac{1}{sC_2}} = \frac{s(C_1 + C_2) + s^2 C_1 C_2 R_\pi}{1 + sC_2 R_\pi} \tag{4.172}$$

To set the equivalence of the above two equations (4.171), (4.172), we have the equality on the coefficients,

$$C_1 + C_2 = a_1 \tag{4.173}$$

$$C_1 C_2 R_\pi = a_2 \tag{4.174}$$

$$C_2 R_\pi = b_1 \tag{4.175}$$

In the above three equations, we have three unknowns, C_1, C_2, and R_π. The solution of the nonlinear equations is:

$$C_1 = \frac{a_2}{b_1} \tag{4.176}$$

$$C_2 = a_1 - \frac{a_2}{b_1} \tag{4.177}$$

$$R_\pi = \frac{b_1^2}{a_1 b_1 - a_2} \tag{4.178}$$

4.6 TRANSMISSION LINES

In this section, we first formulate the transmission line with telegrapher equations. In integrated circuit chips, the interconnects are R, L, C lossy lines. We describe the step input response of R, L, C transmission lines. We view the waveform as the composite of sinusoidal wave. This viewpoint allows us to derive the real and imaginary components of the exponential coefficient of the traveling wave. The real component is the attenuation and the imaginary component is the phase shift of the wave. The difference of the speed of wave and the phase velocity is explained. Finally, we derive the characteristic impedance of the transmission line. Using the characteristic impedance, we derive the wave transmission and reflection at the boundary of the line.

For transmission line analysis, the voltage and current are functions of its distance z along the line and time t [18,22,23]. Since SiO_2 is a good insulator, we can assume that there is no shunt conductance. Let us use R, L, and C to

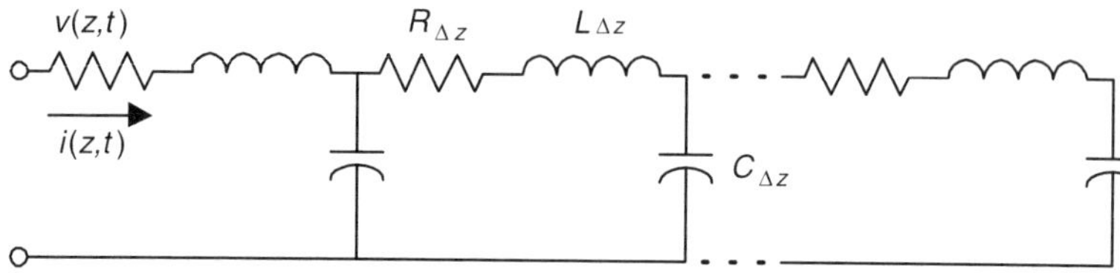

FIGURE 4.8 *Transmission line with RLC model.*

denote the series resistance, series inductance, and shunt capacitance per unit length, respectively. We can write the differential expressions of wave propagation, *telegrapher* equations:

$$\frac{dv(z,t)}{dz} = -Ri(z,t) - L\frac{di(z,t)}{dt} \tag{4.179}$$

$$\frac{di(z,t)}{dz} = -C\frac{dv(z,t)}{dt} \tag{4.180}$$

The first equation shows that the voltage drop along the line is caused by current flowing through the RL series. The second equation illustrates that some current is lost on the shunt capacitance. The above equations can be expressed in s domain after Laplace transform.

$$\frac{dv(z,s)}{dz} = -(R+sL)i(z,s) \tag{4.181}$$

$$\frac{di(z,s)}{dz} = -sCv(z,s) \tag{4.182}$$

Thus, we have two equations for two unknowns $v(z,s)$ and $i(z,s)$. We combine the above two equations and derive the expression for voltage $v(z,s)$ only.

$$\frac{d^2v(z,s)}{dz^2} = sC(R+sL)v(z,s) \tag{4.183}$$

From the above differential equation, we derive the solution of voltage $v(z,s)$.

$$v(z,s) = v_i e^{-\sqrt{sC(R+sL)}\cdot z} + v_r e^{\sqrt{sC(R+sL)}\cdot z} \tag{4.184}$$

where v_i is the magnitude of the voltage traveling in the forward direction, $+z$, and v_r is the magnitude of the voltage traveling backward, $-z$. Substituting the solution of voltage $v(z,s)$ to equation (4.181), we have the solution of current $i(z,s)$.

$$i(z,s) = i_i e^{-\sqrt{sC(R+sL)}\cdot z} - i_r e^{\sqrt{sC(R+sL)}\cdot z} \tag{4.185}$$

$$= v_i\sqrt{\frac{sC}{R+sL}}e^{-\sqrt{sC(R+sL)}\cdot z} - v_r\sqrt{\frac{sC}{R+sL}}e^{\sqrt{sC(R+sL)}\cdot z} \tag{4.186}$$

where i_i is the magnitude of the current traveling in the forward direction and i_r is the magnitude of the current traveling backward.

4.6.1 Step Input Response

The step input response can be derived using inverse Laplace transform. In transient response to the step voltage, a forward wave travels from the source to the sink. Let us express the input voltage as $v(0, s) = (v_{dd}/s)$. We derive forward voltage $v(z, s)$ from equation (4.184).

$$v(z, s) = \frac{v_{dd}}{s} e^{-\sqrt{sC(R+sL)} \cdot z} \tag{4.187}$$

Substituting the voltage expression to equation (4.181), we have the solution for forward current $i(z, s)$.

$$i(z, s) = \frac{v_{dd}}{s} \sqrt{\frac{sC}{R+sL}} e^{-\sqrt{sC(R+sL)} \cdot z} \tag{4.188}$$

We rewrite the current solution in the following format, which is useful for inverse Laplace transform.

$$i(z, s) = v_{dd} C \frac{1}{\sqrt{sC(R+sL)}} e^{-\sqrt{sC(R+sL)} \cdot z} \tag{4.189}$$

$$= v_{dd} \sqrt{\frac{C}{L}} \frac{e^{-\sqrt{s(s+(R/L)) \cdot (z/(1/\sqrt{LC}))}}}{\sqrt{s(s + \frac{R}{L})}} \tag{4.190}$$

The above expression fits the format of the *Bessel* function [19,20,21]. Thus, we derive the following lemma that describes the solution of current in time domain.

Lemma 4.6.1 *The forward current due to a step input can be expressed as:*

$$\begin{aligned} i(z, t) &= 0 \; if \; t < z\sqrt{LC} \\ &= v_{dd} \sqrt{\frac{C}{L}} e^{-(R/2L)t} I_0 \left(\frac{R}{2L} \sqrt{t^2 - z^2 LC} \right) \; if \; t \geq z\sqrt{LC} \end{aligned} \tag{4.191}$$

where $I_0()$ is a modified Bessel function of the first kind of zero order.

$$I_0(x) = 1 + \frac{x^2}{2^2} + \frac{x^4}{2^2 \cdot 4^2} + \frac{x^6}{2^2 \cdot 4^2 \cdot 6^2} + \cdots \tag{4.192}$$

Or, we can write $I_0(x)$ in a series formula:

$$I_0(x) = \sum_{j=0}^{\infty} u^{(j)}, \; where \; u^{(0)} = 1, u^{(j)} = u^{(j-1)} \frac{x^2}{(2j)^2} \tag{4.193}$$

Proof: The derivation follows the formula of the inverse Laplace transform of the Bessel function.

$$\mathcal{L}^{-1}\left(\frac{e^{-\sqrt{s(s+2\alpha)}\cdot(z/\omega_0)}}{\sqrt{s(s+2\alpha)}}\right) = e^{-\alpha t} I_0\left(\alpha\sqrt{t^2 - \frac{z^2}{\omega_0^2}}\right) \tag{4.194}$$

We map expression (4.190) to the above formula with the replacement of notations ω_0 and α:

$$\omega_0 = \frac{1}{\sqrt{LC}} \; and \; \alpha = \frac{R}{2L} \tag{4.195}$$

Hence, following the formula of the inverse Laplace transform, we prove the lemma. □

Speed of Wave The if condition, $t \geq z\sqrt{LC}$, in forward current wave expression (4.191) shows that the current wavefront moves at the speed of light [18],

$$\frac{z}{t} = \omega_0 = \frac{1}{\sqrt{LC}} \tag{4.196}$$

Since $I_0(0) = 1$, the height of the wavefront is equal to

$$v_{dd}\sqrt{\frac{C}{L}} e^{-(R/2L)z\sqrt{LC}} = v_{dd}\sqrt{\frac{C}{L}} e^{-(R/2)\sqrt{(C/L)}\cdot z} \tag{4.197}$$

The solution of the voltage in time domain is derived from the above lemma and equation (4.180). First, from telegrapher equation (4.180), we have

$$v(z,t) = \int_0^t \frac{-1}{C}\frac{di(z,\tau)}{dz}\,d\tau \tag{4.198}$$

To find the differential of current $i(z,\tau)$, we simplify the if conditions in equation (4.191) with a step function,

$$i(z,t) = \left\{v_{dd}\sqrt{\frac{C}{L}} e^{-(R/2L)t} I_0\left(\frac{R}{2L}\sqrt{t^2 - z^2 LC}\right)\right\} s\left(t - \frac{z}{\omega_0}\right) \tag{4.199}$$

where $s(x) = 0$ if $x < 0$, $s(x) = 1$ otherwise. Replacing $i(z,\tau)$ in equation (4.198) with equation (4.199), we derive the expression:

$$v(z,t) = \left\{\frac{-v_{dd}}{C}\sqrt{\frac{C}{L}}\left[\frac{-1}{\omega_0} e^{-\alpha(z/\omega_0)} + \int_{(z/\omega_0)}^t e^{-\alpha\tau}\frac{dI_0\left(\alpha\sqrt{\tau^2 - \frac{z^2}{\omega_0^2}}\right)}{dz}\,d\tau\right]\right\} \times s\left(t - \frac{z}{\omega_0}\right) \tag{4.200}$$

$$= \left\{ v_{dd}\omega_0 \left[\frac{1}{\omega_0} e^{-\alpha(z/\omega_0)} - \int_{(z/\omega_0)}^{t} e^{-\alpha\tau} \frac{dI_0\left(\alpha\sqrt{\tau^2 - \frac{z^2}{\omega_0^2}}\right)}{dz} d\tau \right] \right\}$$

$$\times s\left(t - \frac{z}{\omega_0}\right) \tag{4.201}$$

where ω_0 and α follow the notation in expression (4.195). The step function $s(t - (z/\omega_0))$ indicates that voltage wavefront is no faster than the speed of light, that is, $t = (z/\omega_0)$. The first term in the bracket of equation (4.201) is derived by the rising edge of current $i(z, t)$ at time $t = (z/\omega_0)$. The second term is contributed by the expression inside the braces in equation (4.199). Using the operation of the Bessel function, we deduce the following lemma.

Lemma 4.6.2 *The forward voltage solution of step input response in time domain is expressed as:*

$$v(z,t) = \left\{ v_{dd}e^{-\alpha(z/\omega_0)} + v_{dd}\alpha\frac{z}{\omega_0} \int_{(z/\omega_0)}^{t} e^{-\alpha\tau} \frac{I_1\left(\alpha\sqrt{\tau^2 - \frac{z^2}{\omega_0^2}}\right)}{\sqrt{\tau^2 - \frac{z^2}{\omega_0^2}}} d\tau \right\} s\left(t - \frac{z}{\omega_0}\right) \tag{4.202}$$

where $I_1()$ is the modified Bessel function of the first order:

$$I_1(x) = \frac{x}{2} + \frac{x^3}{2^2 \cdot 4} + \frac{x^5}{2^2 \cdot 4^2 \cdot 6} + \cdots \tag{4.203}$$

Or, we can write $I_1(x)$ in a series formula:

$$I_1(x) = \sum_{j=0}^{\infty} u^{(j)}, \text{ where } u^{(0)} = \frac{x}{2}, u^{(j)} = u^{(j-1)}\frac{x^2}{(2j)(2j+2)} \tag{4.204}$$

Proof: We use the derivative operation of the Bessel function,

$$\frac{dI_0(x)}{dx} = I_1(x). \tag{4.205}$$

Hence, the differential term in equation (4.201) can be derived as follows.

$$\frac{dI_0\left(\alpha\sqrt{t^2 - \frac{z^2}{\omega_0^2}}\right)}{dz} = -\alpha\frac{z}{\omega_0^2} \frac{I_1\left(\alpha\sqrt{t^2 - \frac{z^2}{\omega_0^2}}\right)}{\sqrt{t^2 - \frac{z^2}{\omega_0^2}}}. \tag{4.206}$$

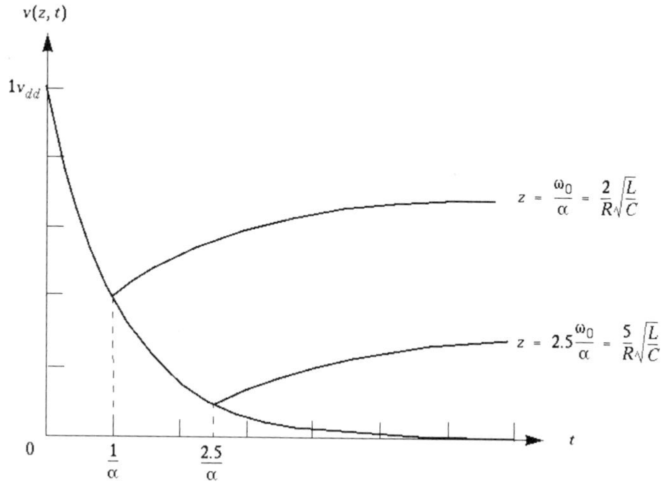

FIGURE 4.9 *Voltage $v(z, t)$ of RLC line.*

Substituting the above expression into equation (4.201), we prove the lemma. □

Figure 4.9 shows the voltage waveform with respect to different z coordinates. The curve that drops monotonically is contributed by the first term in voltage solution (4.202). It is an exponential function of distance z attenuating with rate $\alpha(z/\omega_0) = (R/2)\sqrt{(C/L)}z$. However, for a fixed z, this term is viewed as a step function rise at time $t = (z/\omega_0)$. The magnitude of the step function is $v_{dd}e^{-\alpha(z/\omega_0)}$ which is shown by the dashed lines in the figure. The second term starts to grow from zero at time $t = (z/\omega_0)$ and saturates as time t increases toward infinity. The saturated voltage value of the whole expression is described in the following lemma.

Lemma 4.6.3 *Forward voltage $v(z, t)$ of step input response saturates at source voltage level v_{dd} as time t approaches infinity.*

Proof: We prove the lemma using the operations of Laplace transform. Because $v(z, t)$ is zero at time $t = 0$ for $z > 0$, we have

$$v(z, t)|_{t=\infty} = \int_0^{\infty} \frac{dv(z, t)}{dt} dt = \int_0^{\infty} \frac{dv(z, t)}{dt} e^{-st} dt \bigg|_{s=0} = sv(z, s)|_{s=0} \quad (4.207)$$

Substituting the forward component in equation (4.184) into the above expression, we have

$$v(z, t)|_{t=\infty} = sv(z, s)|_{s=0} = s\frac{v_{dd}}{s} e^{\sqrt{sC(R+sL)}\cdot z}\bigg|_{s=0} = v_{dd} e^{\sqrt{sC(R+sL)}\cdot z}\bigg|_{s=0} = v_{dd} \quad \square \quad (4.208)$$

Exercise Verify that the expressions of current $i(z, t)$, equation (4.191) and voltage $v(z, t)$, equation (4.202), in the above lemmas satisfy the telegrapher equations (4.179) and (4.180).

4.6.2 Attenuation, Phase Shift, and Characteristic Impedance

The attenuation γ of the traveling wave is the real component of the exponential coefficient and the traveling phase shift β is the imaginary component of the exponential coefficient, that is,

$$v(z, s) = v_i e^{-\gamma z - j\beta z} + v_r e^{\gamma z + j\beta z} \tag{4.209}$$

$$\gamma + j\beta = Re\left(\sqrt{sC(R + sL)}\right) + jIm\left(\sqrt{sC(R + sL)}\right) \tag{4.210}$$

For a sinusoidal wave with radian frequency ω, we can replace parameter s with $j\omega$. The attenuation γ and phase shift β are rewritten as:

$$\gamma + j\beta = \sqrt{j\omega C(R + j\omega L)} = j\omega\sqrt{LC}\sqrt{1 - j\frac{R}{\omega L}} \tag{4.211}$$

The decomposition of the real and imaginary components is described in the following lemma.

Lemma 4.6.4

$$\gamma + j\beta = \omega\sqrt{LC}\sqrt{\frac{1}{2}}\sqrt{-1 + \sqrt{1 + \frac{R^2}{\omega^2 L^2}}} + j\omega\sqrt{LC}\sqrt{\frac{1}{2}}\sqrt{1 + \sqrt{1 + \frac{R^2}{\omega^2 L^2}}} \tag{4.212}$$

Proof: Using algebraic process, we square each side of equation (4.211).

$$(\gamma + j\beta)^2 = j\omega C(R + j\omega L) \tag{4.213}$$

Thus, we have

$$\gamma^2 - \beta^2 + j^2\gamma\beta = -\omega^2 CL + j\omega CR \tag{4.214}$$

Separating the equalities of the real and imaginary components, we have

$$\gamma^2 - \beta^2 = -\omega^2 CL \tag{4.215}$$

$$2\gamma\beta = \omega CR \tag{4.216}$$

In the above expressions, we have two variables γ and β and two equations. Using algebraic operations, we thus derive the solution as stated in the lemma. □

The characteristic impedance of the line can be derived as the ratio of transmission voltage and current,

$$Z_0 = \frac{v_i}{i_i} = \frac{v_r}{i_r} = \sqrt{\frac{R + sL}{sC}} \tag{4.217}$$

Similarly, we rewrite the characteristic impedance with radian frequency ω.

$$Z_0 = \sqrt{\frac{R + j\omega L}{j\omega C}} = \sqrt{\frac{L}{C} - j\frac{R}{\omega C}} = \sqrt{\frac{L}{C}}\sqrt{1 - j\frac{R}{\omega L}} \tag{4.218}$$

We decompose the characteristic impedance into real and imaginary components that can be viewed as a resistor R_0 and a capacitor C_0 in series,

$$Z_0 = R_0 + \frac{1}{j\omega C_0} = R_0 - j\frac{1}{\omega C_0} \tag{4.219}$$

The solution of the characteristic resistance R_0 and capacitance C_0 is stated in the following lemma.

Lemma 4.6.5

$$\begin{aligned} R_0 &= \sqrt{\frac{L}{C}}\sqrt{\frac{1}{2}}\sqrt{1 + \sqrt{1 + \frac{R^2}{\omega^2 L^2}}} \\ C_0 &= \frac{1}{\omega\sqrt{\frac{L}{C}}\sqrt{\frac{1}{2}}\sqrt{-1 + \sqrt{1 + \frac{R^2}{\omega^2 L^2}}}} \end{aligned} \tag{4.220}$$

Proof: The proof is similar to the derivation of the attenuation γ and phase shift β. We square each side of equation (4.218). Using algebraic operations, we derive the solution. □

Figure 4.10 shows the effective output load, viewing from the driving gate for the forward wave. The forward voltage on the output pin is derived by

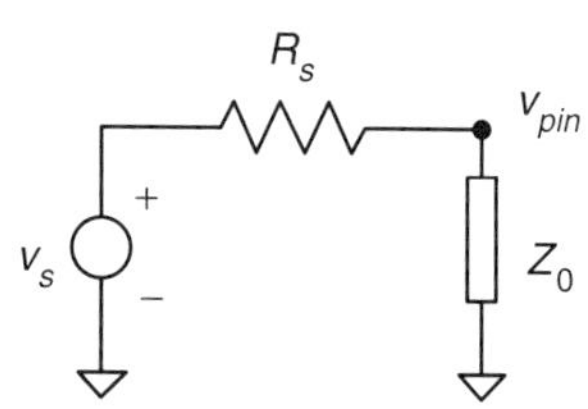

FIGURE 4.10 *Output model at driving pin.*

the impedance divider from the source resistance R_s and the transmission line impedance Z_0.

$$v_{pin} = \frac{Z_0}{R_s + Z_0} v_s. \tag{4.221}$$

Exercise Verify the solution of attenuation and phase shift (4.212) by showing that $(\gamma + j\beta)^2 = j\omega C(R + j\omega L)$.

Exercise Verify the solution of the characteristic resistance and capacitance (4.220) by showing that $Z_0^2 = (R + j\omega L/j\omega C)$.

Summary of Attenuation, Phase Shift, and Characteristic Impedance
Summarizing the above derivations and notations, we can write sinusoidal voltage and current in time domain using complex space representation.

$$v(z,t) = Re\{\tilde{v}(z,t)\} \qquad \tilde{v}(z,t) = v_i e^{-\gamma z + j(\omega t - \beta z)} + v_r e^{\gamma z + j(\omega t + \beta z)} \tag{4.222}$$

$$i(z,t) = Re\{\tilde{i}(z,t)\} \qquad \tilde{i}(z,t) = \frac{v_i}{R_0 + \dfrac{1}{j\omega C_0}} e^{-\gamma z + j(\omega t - \beta z)} - \frac{v_r}{R_0 + \dfrac{1}{j\omega C_0}} e^{\gamma z + j(\omega t + \beta z)} \tag{4.223}$$

The attenuation γ, phase shift β, and impedance Z_0 are functions of radian frequency. For a traveling wave, if time t and traveling distance z keep the relation $\omega t = \beta z$, the phase of the wave remains the same. Thus, $z/t = \omega/\beta$ is defined as the *phase velocity* [18], which is the speed of identical phase at steady state. From equation (4.212), we derive the solution of phase velocity:

$$\frac{\omega}{\beta} = \frac{1}{\sqrt{LC}\sqrt{\dfrac{1}{2}}\sqrt{1 + \sqrt{1 + \dfrac{R^2}{\omega^2 L^2}}}} \tag{4.224}$$

which is slower than the speed of light $(1/\sqrt{LC})$.

Let us analyze the value of attenuation γ as a function of radian frequency ω. Let $x = (\omega L/R)$ and $y = (x/\sqrt{2})\sqrt{-1 + \sqrt{1 + (1/x^2)}}$. We have $\gamma = R\sqrt{(C/L)}y$. The following figure shows function y versus x. The value of $\gamma = R\sqrt{(C/L)}y$ starts to saturate at $x = (\omega L/R) = 1$ or $\omega = (R/L)$. As ω increases beyond (R/L), y saturates toward 0.5. The attenuation value saturates at

$$\gamma = R\sqrt{\frac{C}{L}}y = R\sqrt{\frac{C}{L}}0.5 = \frac{R}{2}\sqrt{\frac{C}{L}} = \frac{\alpha}{\omega_0} \tag{4.225}$$

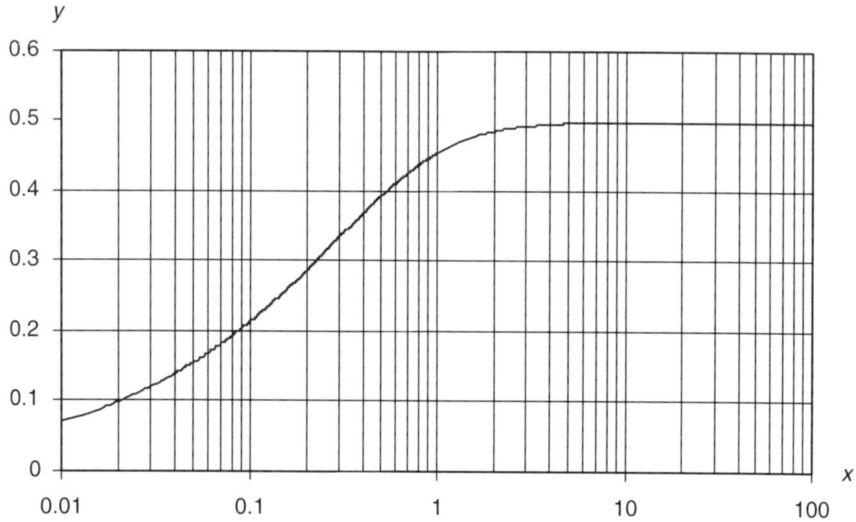

FIGURE 4.11 *The curve shows the function y(x) to analyze the value of attenuation* $r = R\sqrt{(C/L)}\, y$, *where* $x = (\omega L/R)$.

which is equal to the power of the exponent, (α/ω_0), in step input response (4.202). Note that the above derivation assumes a perfect ground return. Otherwise, the attenuation will grow further due to a higher return-line resistance at higher frequency ranges.

Let l be the length of the line. From equations (4.201), (4.222), and (4.225), we can derive that if

$$\gamma l > 1 \tag{4.226}$$

the magnitude of voltage wavefront attenuates more than $e^{-1} = 0.3679$ at the end of the line.

4.6.3 Reflected and Transmitted Waves

At any point of the transmission line, we can decompose the wave into three components: the forward wave, the reflected wave and the transmitted wave (Fig. 4.12). Let us use a cut-line to cross the transmission line. Since the voltage is continuous, the voltage at two sides of the cut-line is equal. In other words, the forward voltage v_i plus the reflected voltage v_r at the input side of the cut-line is equal to the transmitted voltage v_t at the output side of the cut-line.

$$v_i + v_r = v_t \tag{4.227}$$

Similarly, since the current flow is conservative, net current injecting into the cut point is zero. In other words, the forward current i_i minus the reflected current

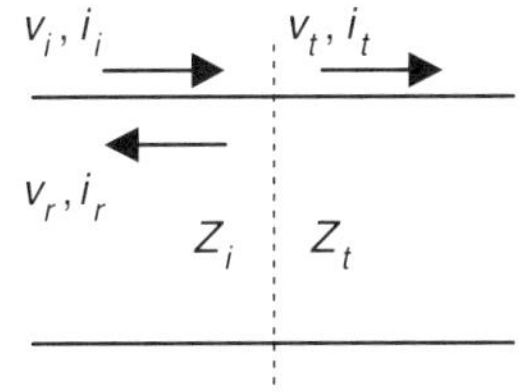

FIGURE 4.12 Illustration of forward wave v_i, i_i, reflected wave v_r, i_i and transmitted wave v_t, i_i. The characteristic impedance at the input side of the cut-line is Z_i and the impedance at the output side of the cut-line is Z_t.

i_r at the input side of the cut-line should be equal to the transmitted current i_t at the output side of the cut-line.

$$i_i - i_r = i_t \tag{4.228}$$

Let us denote the characteristic impedance at the input side of the cut-line to be Z_i and the impedance at the output side of the cut-line to be Z_t. The relation of the voltages and currents can be expressed as

$$v_i = Z_i i_i, \;\; v_r = Z_i i_i, \;\; v_t = Z_t i_t. \tag{4.229}$$

Substituting expressions (4.229) to equation (4.228), we have

$$\frac{v_i}{Z_i} - \frac{v_r}{Z_i} = \frac{v_t}{Z_t}. \tag{4.230}$$

Combining equations (4.227) and (4.230), we deduce the coefficients of reflected and transmitted voltages:

$$v_r = \frac{Z_t - Z_i}{Z_t + Z_i} v_i$$
$$v_t = \frac{2Z_t}{Z_t + Z_i} v_i \tag{4.231}$$

For the case in which the transmission line is continuous, $Z_i = Z_t$, we derive from the above expressions that reflected voltage $v_r = 0$ and transmitted voltage

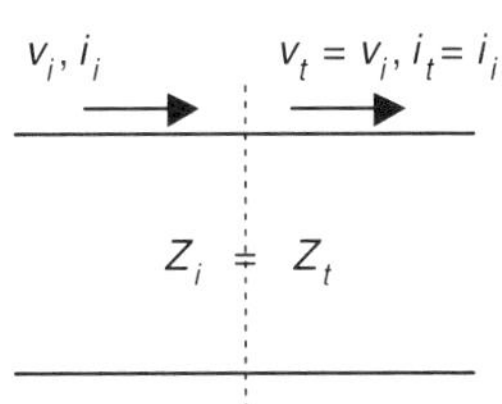

FIGURE 4.13 Illustration of continuous transmission line. The characteristic impedance $Z_i = Z_t$ at the two sides of the cut-line.

$v_t = v_i$. Hence, the reflection caused by the forward wave at that cut-line is zero. The transmitted wave is equal to the forward wave (Fig. 4.13).

Suppose the cut-line is set at the end of the transmission line. The input impedance $Z_i = Z_0$ and the output impedance is defined by the loading capacitance, $Z_t = (1/j\omega C_L)$. Equations (4.231) are rewritten as:

$$v_r = \frac{\dfrac{1}{j\omega C_L} - Z_0}{\dfrac{1}{j\omega C_L} + Z_0} v_i \tag{4.232}$$

$$v_t = \frac{2\dfrac{1}{j\omega C_L}}{\dfrac{1}{j\omega C_L} + Z_0} v_i \tag{4.233}$$

The voltage, expression (4.232), propagates backward toward the driving gate. We set the cut-line at the beginning of the transmission line, that is at the output of the driving gate. The output side is now at the source of the gate. Let us assume that the transition time t_r of the source voltage is smaller than the wave traveling time,

$$t_r < 2\frac{l}{\sqrt{LC}} \tag{4.234}$$

where l is the length of the line.

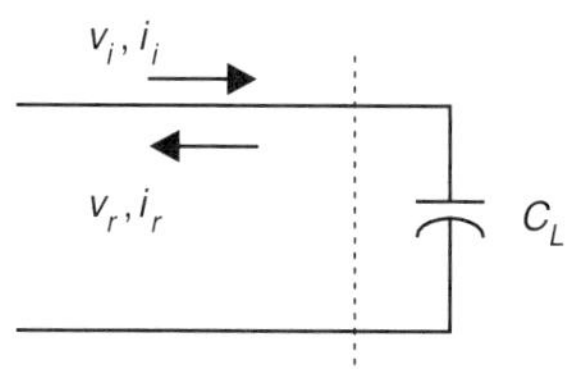

FIGURE 4.14 Interface of transmission line and capacitive load. The cut-line separates the transmission line from the terminal load C_L.

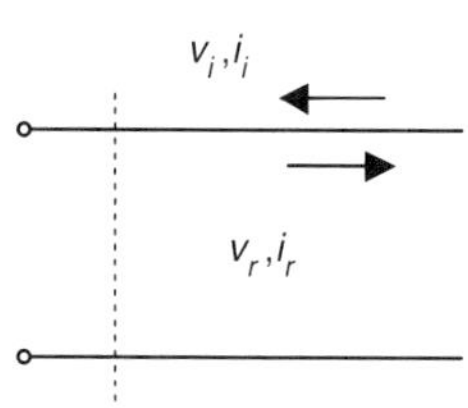

FIGURE 4.15 Interface of transmission line and source node. The cut-line separates the source node from the transmission line.

Thus, we model the source as an open circuit, $Z_t = \infty$:

$$v_r = \frac{Z_t - Z_0}{Z_t + Z_0} v_i = v_i \tag{4.235}$$

$$v_t = \frac{2Z_t}{Z_t + Z_0} v_i = 2v_i \tag{4.236}$$

Therefore, a full amount of voltage reflects and travels from source toward the terminal. At the driving point, we observe twice the amount of voltage because of the superposition of the incoming and reflected waves.

BIBLIOGRAPHY

[1] A. Deutsch, Electrical characteristics of interconnections for high-performance systems, *IEEE Proceedings* 86(2): 315–355, Feb. 1998 .

[2] L. O. Chua, C. A. Desoer, and E. S. Kuh, *Linear and Nonlinear Circuits*, McGraw-Hill, 1987.

[3] L. O. Chua, and P. M. Lin, *Computer-Aided Analysis of Electronic Circuits*, Prentice-Hall, 1975.

[4] G. H. Golub, and C. F. Van Loan, *Matrix Computations*, John Hopkins Press, 2nd ed., 1989.

[5] P. Feldmann, and R. W. Freund, Efficient linear circuit analysis by Pade approximation via the Lanczos process, *IEEE TCAD*, pp. 639–649, May 1995.

[6] L. M. Silveira, M. Kamon, and J. White, Efficient reduced-order modeling of frequency-dependent coupling inductances associated with 3-D interconnect structures, *IEEE/ACM Prof. DAC*, pp. 376–380, June 1995.

[7] K. J. Kerns, and A. T. Yang, Stable and efficient reduction of large, multiport RC networks by pole analysis via congruence transformations, *IEEE Trans. on Computer-Aided Design*, 16(7): 734–44, July 1997.

[8] A. Odabasioglu, M. Celik, and L. T. Pileggi, PRIMA: Passive reduced-order interconnect macromodeling algorithm, *IEEE/ACM Int. Conf. on Computer-Aided Design*, pp. 58–65, Nov. 1997.

[9] E. S. Kuh, and R. A. Rohrer, *Theory of Linear Active Networks*, Holden-Day 1967.

[10] E. Chiprout, and M. S. Nakhla, *Asymptotic Waveform Evaluation*, Kluwer Academic Publishers, 1994.

[11] L. T. Pillage, R. A. Rohrer, and C. Visweswariah, *Electronic Circuit and System Simulation Methods*, McGraw-Hill, 1995.

[12] G. A. Baker, and P. Graves-Morris, *Pade approximants*, 2nd ed., Cambridge University Press, 1996.

[13] R. A. Rohrer, and H. Nosrati, Passivity considerations in stability studies of numerical integration algorithms, *IEEE Trans. on Circuits and Systems*, 28(9); 857–866, Sept. 1981.

[14] R. Kay, and L. Pileggi, Primo: Probability interpretation of moments for delay calculation, *ACM/IEEE Design Automation*, pp. 463–468, 1998.

[15] W. C. Elmore, The transient response of damped linear network with particular regard to wideband amplifiers, *J.Applied Physics*, 19 (1948), pp. 55–63.

[16] F. J. Liu, and C. K. Cheng, Extending moment computation to 2-port circuit representations, *ACM/IEEE Design Automation Conf.*, pp. 473–476, June 1998.

[17] F. J. Liu, J. Lillis, and C. K. Cheng, A new layout-driven timing model for incremental layout optimization, *Asia and South Pacific Design Automation Conf.*, pp. 127–131, 1997.

[18] R. E. Matick, Transmission lines for digital and communication networks, *IEEE Press*, 1995.

[19] F. Bowman, *Introduction to Bessel Functions*, New York, Dover Publications, 1958.

[20] A. Gray, *Treatise on Bessel Functions and Their Applications*, New York, Dover Publications, 1966.

[21] E. Weber, *Lineare Transient Analysis*, Vol. II, subsection 8.2, New York, Wiley, 1956.

[22] W. C. Johnson, *Transmission Lines and Networks*, McGraw-Hill, 1950.

[23] M. S. Ghausi, and J. J. Kelly, *Introduction to Distributed-Parameter Networks*, Holt, Rinehart and Winston, 1968.

5

Inductance and Inductive Coupling for On-Chip Interconnect

5.1 INTRODUCTION

Modern technology has demonstrated a relentless trend toward faster circuits, larger die sizes, shorter rise times, and smaller pulse widths. Simultaneously, the minimal feature sizes have been aggressively decreased. Circuit designers are designing chips operated by Gigahertz clocks on dies larger than 300 m^2 using 0.25 μm technology [1]. Over the years, it has also been shown that interconnect delay and crosstalk have become significant bottlenecks in circuit performance. With these technology trends, on-chip inductance effects, such as delay increase, overshoot, and inductive crosstalk, can no longer be ignored.

Inductance effects have become increasingly significant because:

1. For performance considerations, some global signal and clock wires are routed with large widths and thicknesses at the top levels of the metal to minimize delays. This decreases the resistance of the wires, which makes the wire impedance due to inductance comparable to that due to resistance.

   In [2], Grover gave the following formula for estimating the partial-self-inductance[1] of a straight wire with length, width, and thickness l, w, t respectively.

$$L(nH) = 2l \times \left[\ln\left(\frac{2l}{w+t}\right) + 0.5 - k\right] \tag{5.1}$$

[1] In short, partial self-inductance is the self-inductance of a portion of a current loop with the inductive coupling from the rest return portion ignored. Partial inductance (vs. loop inductance) will be explained in more detail when we discuss the computation of inductance in section 5.1.3.

where k depends on w and t, and $0 < k < 0.0025$, l, w, t in cm. By using this inductance estimation, Figure 5.1 shows the comparisons for lengths ranging from 100 to 10,000 microns between line impedance due to inductance, $2\pi fL$, and resistance, $\rho l/wt$, at $f = 1GHz$ for **Al** wire with the resistivity $\rho = 3.85 \times 10^{-6}\Omega$ cm and with widths of 1.25, 2.5, 5.0, and 10.0 microns[2] with thickness of 1.9 microns corresponding to a typical top-level metallization thickness of a 0.18 μm technology. Also shown in the figure are the inductance values. As can be seen, the inductive line impedance is larger than the line resistance for widths wider than 2.5 μm, and the inductances may be larger than several nH for most cases and are much larger than the typical bonding wire inductances of IC packages.

2. As the clock frequency increases and the rise times decrease, electrical signals comprise more and more high-frequency components, making the inductance effects more significant. For example, for a 600 MHz clock with a 100 ps rise time, at least 15% of its frequency components are above 3.4 GHz. Such frequency is defined as the *significant frequency* of the pulse, for which we will give a detailed definition in section 5.1.2.
3. With the increase of chip size, it is fairly typical that many wires are long (longer than a few cm) and run in parallel, which increases the inductive

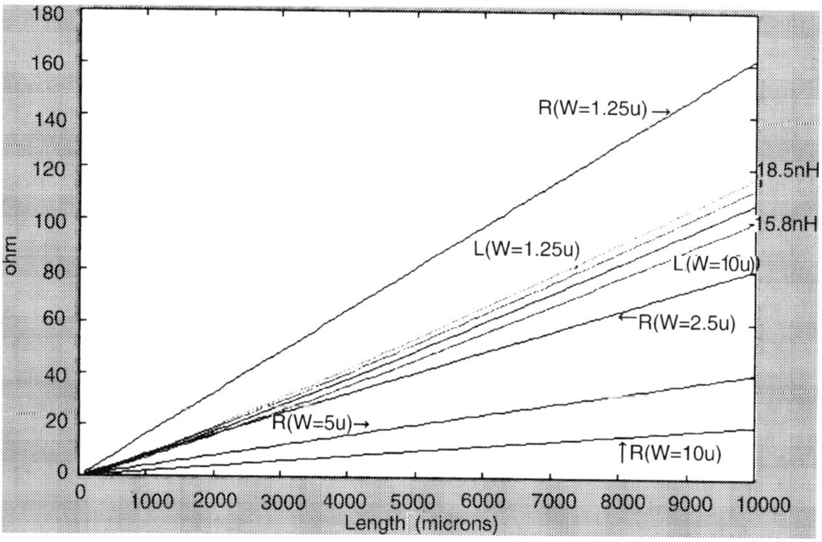

FIGURE 5.1 *Resistance and inductive reactance of wires at 1 GHz for different lengths and widths. (The super-linear behavior of inductive reactances may be seen because as the length gets longer, their curves are further away for $R(w = 2.5u)$ curve).*

[2] Equation (5.1) neglects the effect of skin depth. With higher frequencies, equation (5.1) may overestimate the inductance values if skin depths do not cover the whole cross-section area. At 1 GHz, the skin depth of **Al** wires is about 2u, which is greater than the wire thickness of 1.9u, so no correction is needed.

crosstalk. In section 5.2, we will show that inductive crosstalk increases with the parallel length (wires are not necessarily directly adjacent) and saturates near 1 cm length.

4. With the push of performance, some low-resistivity metals have been explored to replace **Al** in order to minimize wire RC delays. In [3], designers at Motorola demonstrated the success of using copper (**Cu**) wires. This could make the wire indcutive reactance larger than the resistance.[3].

Therefore, circuit designers have to consider on-chip inductance in order to ensure the functional correctness and performance of their designs.

When we compute inductance, the most important thing to know is where the return currents are. The current flowing on a wire has to come back to the wire's driver to close the loop. Impedance of the receiver at the wire's far-end terminal is only one part of the current loop. The electrical behavior of the current loop is determined instead by the whole loop's reactance. Self-inductance of the wire is affected by the inductive coupling from the return currents so the inductive reactance is a function of the size of the loop. The smaller the loop, the less the inductance. In section 5.3, various on-chip techniques will be presented to shorten the loop. And, in the next section, we will show the effectiveness of providing ground planes with nearby high-speed signals to reduce inductance due to multiple low-impedance returns, which is often used by PCB and package designers but is a luxury for on-chip designers. Currents always return through the smallest impedance path. Therefore, minimizing inductance requires the complete consideration of R, L, and C.

In the rest of this section, we discuss the consideration differences between on-chip and off-chip inductance, followed by introduction to the significant frequency of a pulse, the calculation of inductance, and skin-effect resistance. We show figures of merit to characterize the importance of on-chip inductance in section 5.2 and solutions to control inductance effects in section 5.3.

5.1.1 Differences in On-Chip Inductance Consideration

Previously, inductances were considered only for off-chip interconnects, such as those traces on printed circuit boards (PCB) and bonding wires in IC packages. When we shift our attention to the inductance of on-chip wires, some differences in consideration need to be observed.

1. We need to consider the internal inductance for on-chip wires. Because the skin depths at the frequencies that we consider are comparable to the wire thickness or width, most electrical currents flow inside the wires. On the other hand, the skin depths are much smaller than the dimensions of

[3] We ignore the wire capacitance here because the purpose of this discussion is to show the significance of inductance — when the inductive reactance is comparable to resistance, the wire becomes an RLC line and its electrical behavior is significantly determined by the inductance.

off-chip interconnects or planes; hence, we can assume the currents flow on the surfaces of those interconnects or planes. (The skin depth is discussed in detail in section 5.1.4.)

To understand how large the internal inductance can be, let us consider the inductance values determined by *Raphael RC2* and *RI3* tools[4] for the interconnect structure shown in Figure 5.2. RC2 is a two-dimensional field-solver for capacitance, inductance, and resistance; RI3 is a three-dimensional inductance-solver that takes the skin depths at different frequencies into consideration. The per-unit-length inductance matrix $\tilde{L}$ is determined by RC2 through the inversion of the per-unit-length free-space capacitance matrix $\tilde{C}$ as shown in equation (5.2):

$$\tilde{L} = \frac{1}{c^2}\tilde{C}^{-1} \tag{5.2}$$

where c is the speed of light in free space. Therefore, the inductance given by RC2 is only the external inductance. This is because in the capacitance calculation it is assumed that all the fields are outside the conductors. By the inversion of the capacitance matrix to obtain the inductance matrix, it is assumed that signal propagates on the surface of the conductors. Table 5.1 summarizes the inductance values at different

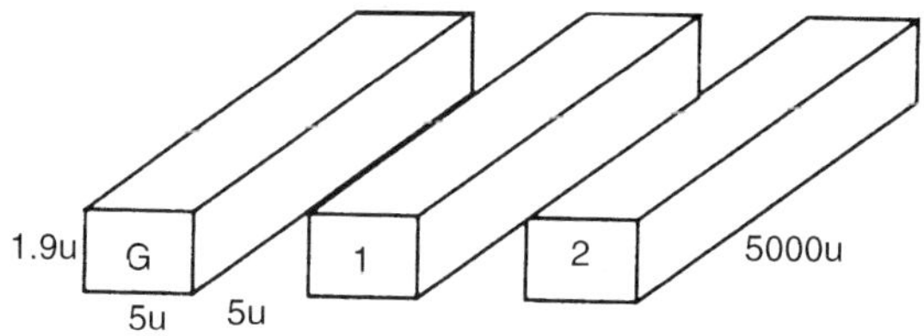

FIGURE 5.2 *Two signal wires and one ground wire of 5000μm length.*

TABLE 5.1 Comparisons between RI3 and RC2 Loop Inductance Results[a]

Frequency	RI3	RC2	% difference
1 GHz	$\begin{bmatrix} 3.768\times10^{-9} & 2.563\times10^{-9} \\ 2.563\times10^{-9} & 5.125\times10^{-9} \end{bmatrix}$	$\begin{bmatrix} 2.983\times10^{-9} & 2.133\times10^{-9} \\ 2.133\times10^{-9} & 4.266\times10^{-9} \end{bmatrix}$	23%
100 GHz	$\begin{bmatrix} 3.538\times10^{-9} & 2.425\times10^{-9} \\ 2.425\times10^{-9} & 4.850\times10^{-9} \end{bmatrix}$	$\begin{bmatrix} 2.983\times10^{-9} & 2.133\times10^{-9} \\ 2.133\times10^{-9} & 4.266\times10^{-9} \end{bmatrix}$	15%
10, 000 GHz	$\begin{bmatrix} 3.203\times10^{-9} & 2.233\times10^{-9} \\ 2.233\times10^{-9} & 4.626\times10^{-9} \end{bmatrix}$	$\begin{bmatrix} 2.983\times10^{-9} & 2.133\times10^{-9} \\ 2.133\times10^{-9} & 4.266\times10^{-9} \end{bmatrix}$	10%

[a] The differences at less than 100 GHz are mainly due to the internal inductances while the difference of 10% at 10,000 GHz might be due to numerical errors.

[4] Raphael is a trademark of Avant! Corporation, whose Web site is http://www.avanticorp.com.

frequencies. As the frequency increases, we see the differences between the RC2 and RI3 results become smaller and smaller, which is due to the fact that the skin depth gets smaller and smaller. At very high frequencies, both results will converge.

2. Due to the lack of highly conductive ground planes on chip[5], the mutual couplings between wires cover very long ranges and decrease very slowly with the increase of spacing.

Figure 5.3 shows two interconnect structures (a) and (b). Structure (a) does not have a ground plane, while (b) has a ground plane at 1.2 μm above the wires. We vary the spacing between wire 1 and wire 2, and we use RI3 to calculate the mutual inductances between them for structure (a) and structure (b). The coupling coefficient results[6] are summarized in Table 5.2. We see that the inductive couplings decrease very slowly without a ground plane for structure (a). Separated by two orders-of-magnitude from 1 μm to 100 μm, the coupling coefficient only decreases by 28%. On the other hand, for structure (b) with a ground plane, the coupling coefficient

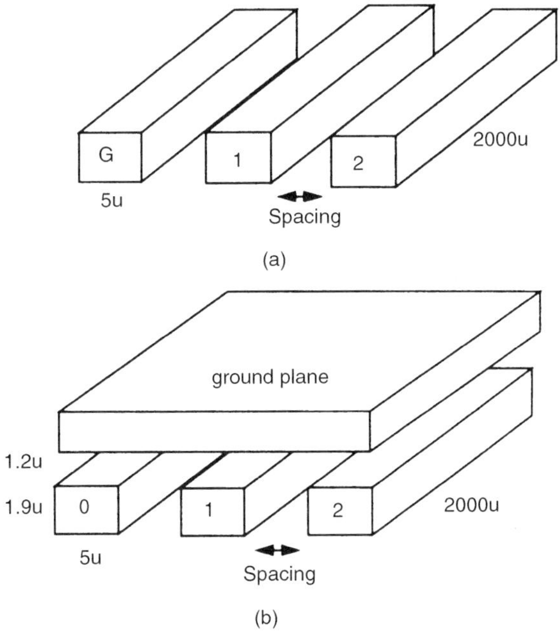

FIGURE 5.3 *Two interconnect structures: (a) has no ground plane and (b) has a ground plane 1.2 μm above the wires. All the wires are 2000 μm long and the spacing is varied. The leftmost wire of (a) is a ground wire.*

[5] Some people mistakenly treat the silicon substrate as a ground plane. Because the resistivity of the lightly doped substrate layer ($\rho = 10\text{–}20\Omega$ cm) is about two million times larger than aluminium, not much high-frequency current may return from that layer.

[6] The coupling coefficient K_{12} is defined as $|L_{12}|/\sqrt{L_{11}L_{22}}$.

TABLE 5.2 Inductive-Coupling-Coefficients versus the Spacing for Figure 5.3*a* and Figure 5.3*b*

Spacing	K_{12} of (a)	Compared with 1 μ*m*-spacing	K_{12} of (b)	Compared with 1 μ*m*-spacing
1 u	0.73	100%	0.29	100%
10 u	0.57	78%	0.034	11%
50 u	0.54	74%	0.0059	2.0%
100 u	0.53	72%	0.0038	1.3%

decreases almost linearly with the increase of spacing. The coefficient at the 100 μ*m* spacing is 1.3% of that at the 1 μ*m* spacing. Furthermore, even at 1 μ*m* apart, the coupling with a ground plane is only 0.29/0.73, or 40% of that without a ground plane to begin with.

Because of this long-range coupling effect for on-chip wires, almost every wire in the same parallel direction couples to every other; hence, considering the inductive couplings becomes difficult. We will end up with a huge and dense $\boldsymbol{L}$ matrix for a thorough circuit analysis or SPICE simulation. However, simply discarding the small mutual inductance terms can make matrix $\boldsymbol{L}$ lose its definite positive property, resulting in an unstable circuit model [5]. Therefore, some clever way of inductance screening on physical layouts is necessary [4].

Moreover, with the long-range coupling effect, the transmission line theory [6], which has been successfully employed both for analyzing traces on PCBs and for determining matched impedance to reduce noise, is no longer applicable to on-chip wires. Strictly speaking, no on-chip wire is a transmission line because of the many nonnegligible couplings along the wire. Some clock wires with designated fat ground wires on both sides may be treated as transmission lines within reasonable approximations, and are the only exceptions.

3. The inductance of on-chip wires is not scalable with length. As demonstrated in equation (5.1), the self-inductance of an on-chip wire increases with length at $n \log n$ rate, which is also shown in Figure 5.1. However, as can be seen in Figure 5.1, beyond a long length (e.g., beyond 8000 μ*m*) the inductance curves look like linear curves owing to the property of the $n \log n$ function. Therefore, it is often mentioned by PCB designers that inductance is scalable. Since most on-chip wires are much shorter than 8000 μ*m*, scalability cannot be applied.

Unfortunately, no good approximation formula exists for mutual inductances of two parallel lines[7] of unequal lengths or unequal dimensions. The mutual coupling for them needs to be determined by a field-solver!

[7] Mutual inductance for two parallel, same-dimension lines of length l and separation D is $\mu_0 l/2\pi$ $[\ln(2l/D) - 1 + (D)/(l)]$ [2].

The non-scalability of inductance is due to the nonuniformity of self- or mutual-coupling near both terminals of the wires. Because the inductive couplings extend to very long ranges, the terminal effect compared to the whole only diminishes with a very long length. This is different from the coupling capacitances of on-chip wires. The couplings between wires or at terminals extend only to very short ranges. Therefore, after a short length, the coupling capacitances become scalable.

5.1.2 Spectrum and Significant Frequency of High-Speed Pulse

Inductive reactance $j\omega L$ is directly proportional to the signal frequency; therefore, in order to estimate the significance of inductance impact, designers need to know the magnitudes of the signal at different frequencies. If most of the signal is in the low-frequency range, the inductance effects may be negligible. On the other hand, inductance effects will be very significant for the opposite situation. In this section, the amplitude spectra of trapezoidal pulses[8] with different pulse widths and rise times are first derived. A spectrum shows the distribution of the components of a pulse over different frequencies. The spectrum indicates to designers whether inductance needs to be considered. Also shown is that for most high-frequency digital pulses, the range of the ratios between pulse width and ramp times is reasonably small. Such pulses can be characterized using a *significant frequency* related only to the ramp time. Designers may use this frequency to estimate the electrical behavior of the pulse and to do design trade-offs. For example, the skin depth may be determined at the significant frequency of the pulse transmitted over the wire. The importance of wire inductive reactance may also be determined by looking at the ratio between $j\omega L$ and R, where ω is 2π times the significant frequency, and L and R are the per-unit-length wire inductance and resistance, respectively. If the ratio is greater than one, inductance effects will be very significant.

Figure 5.4 shows a trapezoidal pulse function $P_{tr}(t)$, where T is the pulse width and tr is the ramp time. This pulse is the convolution result of two step pulses. That is, $P_{tr}(t) = P_S^1(t) \otimes P_S^2(t)$. The two step pulses are depicted in Figure 5.5.

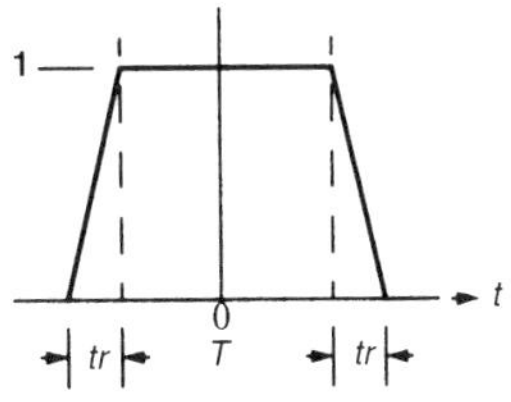

FIGURE 5.4 *A trapezoidal pulse.*

[8] We focus on trapezoidal pulses because they model the signals of digital circuits without much loss of accuracy.

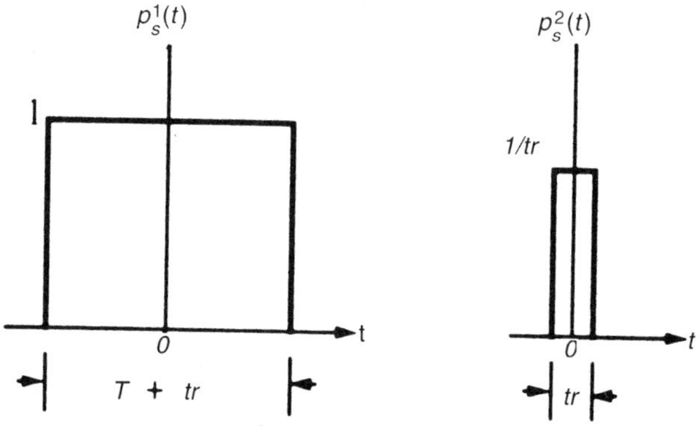

FIGURE 5.5 *Two step pulses.*

The Fourier transform of $P_S^1(t)$ is $(T + tr)\text{sinc} f(T + tr)$, and the Fourier transform of $P_S^2(t)$ is $(tr/tr)\text{sinc} f(tr)$, where $\text{sinc}x \equiv (\sin \pi x)/(\pi x)$. Since the Fourier transform of the convolution of two functions is equal to the product of the Fourier transforms of the two individual functions, the Fourier transform of the trapezoidal pulse will be

$$P_{tr}(f) = (T + tr)\,\text{sinc} f(T + tr)\text{sinc} f(tr) \tag{5.3}$$

If we let $r \equiv T/tr$ and introduce a new variable $z \equiv f \times tr$, then

$$P_{tr}(z) = tr(r + 1)\text{sinc}z(r + 1)\text{sinc}z \tag{5.4}$$

For high-frequency digital pulses, the range of r is quite small mainly because there are physical limitations, such as transistors' turn-on delays and wire loadings, on how small tr can be shrunk, while T is the target that designers try very hard to minimize. Normally, r falls in the range $7 \leq r \leq 13$.[9] The plot of the normalized amplitude spectrum $|P_{tr}(z)|$ in Figure 5.6 shows $r = 7$, $r = 10$, $r = 13$. Most of the frequency components for each curve distribute before $z = 0.34$, and after that the magnitude falls to less than 0.1.

Let us now evaluate the following integral for each curve

$$S(r) = \frac{\int_0^{0.34} |\text{sinc}\, z(r + 1)\,\text{sinc}\, z|dz}{\int_0^{\infty} |\text{sinc}\, z(r + 1)\,\text{sinc}\, z|dz} \tag{5.5}$$

[9] For example, the high-end 600 MHz microprocessor that we saw has about 100 ps rise time on the clock or equivalently $r = 7$, and the existing high-end ASIC chips have about 200 ps rise time on the clock running at 200 MHz to 333 MHz, which gives $r = 12$ to $r = 7$. In general, if the clock cycle time is small, its ramp time cannot afford to be large; on the other hand, if the clock cycle time is large, the ramp time may be accordingly large. Therefore, r falls in a small range.

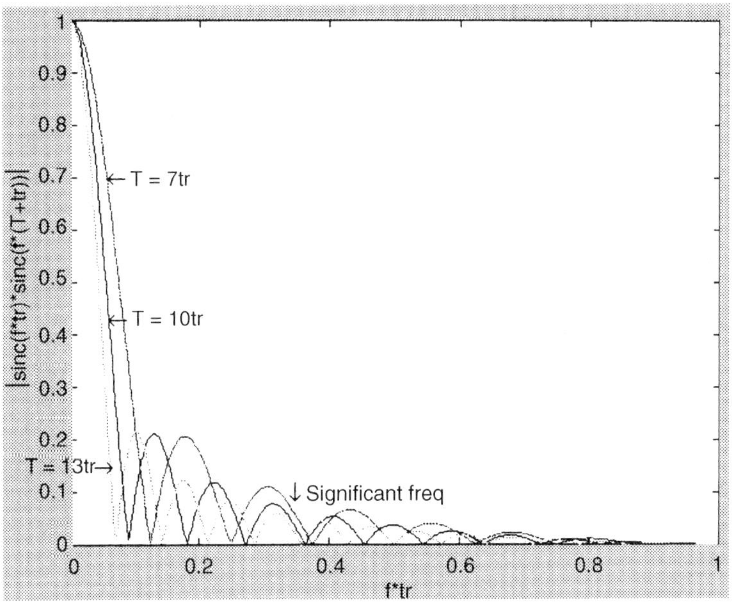

FIGURE 5.6 Pulse spectra for different T over tr.

It can be proven that $S(r)$ is a monotonic function. Through numerical integration, $S(7) = 0.850$, $S(10) = 0.858$, and $S(13) = 0.862$ can be calculated. These values are all very close to 85%, which may be assumed to be true for $7 \leq r \leq 13$ because of $S(r)$ monotonically increasing. With this and the relation $z \equiv f \times tr$, the *significant frequency* of a trapezoidal pulse can be defined.

DEFINITION 5.1.1

Significant Frequency: The *significant frequency* of a trapezoidal pulse is defined as $0.34/tr$, where tr is the ramp time of the pulse.

Properties

1. For each trapezoidal pulse, about *15%* of its frequency components are at frequencies higher than the significant frequency.
2. The magnitude of the pulse's spectrum at a frequency higher than the significant frequency is less than 10% of its maximum value. Beyond the significant frequency, the spectral amplitude rolls off much faster than 20 dB/decade.

These two properties indicate that although 15% of the frequency components of the pulse are at a higher frequency than the significant frequency, the overall magnitude of them is small. Therefore, the significant frequency is a very good representative frequency of a trapezoidal pulse for approximating its high-frequency electrical behavior, especially when deciding whether inductance needs to be considered. Later, in section 5.2.3, we will show the simulation results, proving that the inductance effect is significant if $j\omega L$ is greater than one-fourth of R, where ω is 2π times the significant frequency of the pulse. In section 5.2.1 we also show that the frequency-dependent PEEC inductance extraction at the signal's significant frequency achieves very good accuracy.

5.1.3 Inductance Calculation

In this section, we will introduce a method for magneto-quasistatic analysis of complicated three-dimensional packages and interconnect. This method uses a standard volume-element discretization of an integral formulation from magneto-quasistatic analysis and is a simplified version of the Partial Element Equivalent Circuit (PEEC) method [7,8,9,10]. The volume-element method has the advantage that only the interiors of conductors need to be discretized, as opposed to the finite-element or finite-difference methods, where the entire space needs to be discretized.

5.1.3.1 Loop Inductance Interconnect obeys Maxwell's equations:

$$\nabla \times \boldsymbol{E} = -\frac{\partial \boldsymbol{B}}{\partial t} \tag{5.6}$$

$$\nabla \times \boldsymbol{H} = \frac{\partial \boldsymbol{D}}{\partial t} + \boldsymbol{J} \tag{5.7}$$

$$\nabla \bullet \boldsymbol{B} = 0 \tag{5.8}$$

$$\nabla \bullet \boldsymbol{D} = \rho \tag{5.9}$$

By assuming the medium is homogeneous, isotropic, linear, and time-invariant, we have $\boldsymbol{D} = \varepsilon \boldsymbol{E}$ and $\boldsymbol{B} = \mu \boldsymbol{H}$, where ε and μ are dielectric constant and permeability constants of the medium, respectively. Given this, the Maxwell's equations in the sinusoidal steady state are

$$\nabla \times \boldsymbol{E} = -j\omega\mu \boldsymbol{H} \tag{5.10}$$

$$\nabla \times \boldsymbol{H} = j\omega\varepsilon \boldsymbol{E} + \boldsymbol{J} \tag{5.11}$$

$$\nabla \bullet \boldsymbol{H} = 0 \tag{5.12}$$

$$\nabla \bullet \boldsymbol{E} = \frac{\rho}{\varepsilon} \tag{5.13}$$

Then, applying the magneto-quasistatic assumption that the displacement current, $\varepsilon\omega \boldsymbol{E}$, is negligible everywhere for the frequency of interest, equation (5.11) becomes

$$\nabla \times \boldsymbol{H} = \boldsymbol{J} \tag{5.14}$$

This quasistatic assumption is valid when the dimension of the interconnect is much smaller than the signal wavelengths, usually smaller than one-twelveth of the wavelength for the highest-frequency component of interest. Then, the electric field is somewhat time-invariant over the interconnect.

From equation (5.12), $\nabla \bullet \boldsymbol{H} = 0$, it can be shown that $\boldsymbol{H}$ must be the curl of another vector. Hence we introduce the vector potential $\boldsymbol{A}$, such that

$$\mu \boldsymbol{H} = \nabla \times \boldsymbol{A} \tag{5.15}$$

Applying this to (5.10),

$$\nabla \times (\boldsymbol{E} + j\omega \boldsymbol{A}) = \boldsymbol{0} \tag{5.16}$$

This equality-to-zero implies that there exists a scalar function Φ, such that

$$-\nabla\Phi = \boldsymbol{E} + j\omega \boldsymbol{A} \tag{5.17}$$

By assuming the continuity of $\boldsymbol{E} + j\omega\boldsymbol{A}$, the divergence of $\nabla\Phi$ always equates to zero, that is,

$$-\nabla \bullet \nabla\Phi = \nabla \bullet (\boldsymbol{E} + j\omega \boldsymbol{A}) = \nabla \bullet \boldsymbol{E} + j\omega \nabla \bullet \boldsymbol{A} = 0 \tag{5.18}$$

From equation (5.13) and the fact that no charge exists everywhere in the dielectric, we have $\nabla \bullet \boldsymbol{E} = 0$. Therefore, (5.18) gives

$$\nabla \bullet \boldsymbol{A} = 0 \tag{5.19}$$

From (5.14) and (5.15),

$$\mu \boldsymbol{J} = \mu \nabla \times \boldsymbol{H} = \nabla \times \nabla \times \boldsymbol{A} = \nabla(\nabla \bullet \boldsymbol{A}) - \nabla^2 \boldsymbol{A} \tag{5.20}$$

By using equation (5.19), we have

$$-\nabla^2 \boldsymbol{A} = \mu \boldsymbol{J} \tag{5.21}$$

The equation is composed of Poisson's equations at the x, y, and z directions. By employing the Green's function solution, we have for a point r in the medium,

$$\boldsymbol{A}(r) = \frac{\mu}{4\pi} \int_{V'} \frac{\boldsymbol{J}(r')}{|r - r'|} \, dv' \tag{5.22}$$

where V' is the volume of all conductors.

Let us consider two constant current loops i and j in Figure 5.7. Then, the vector potential at a point of conductor i contributed by the current of loop j,

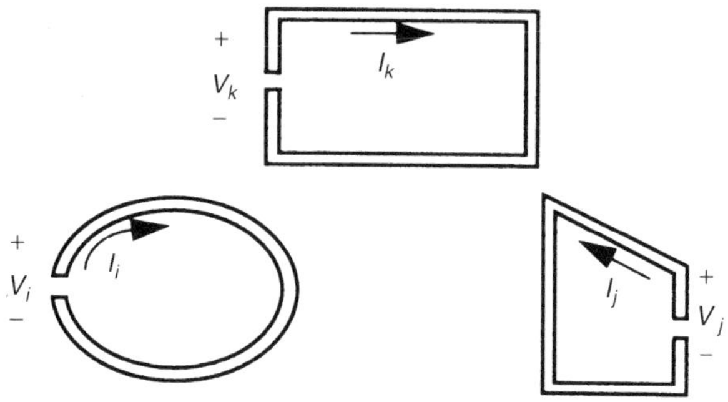

FIGURE 5.7 Systems of coupled conductor loops.

denoted by $\boldsymbol{A}_{ij}$, is

$$\boldsymbol{A}_{ij} = \frac{\mu}{4\pi}\frac{I_j}{a_j}\oint_j \int_{a_j} \frac{d\boldsymbol{l}_\mathbf{j}\, d\mathbf{a}_\mathbf{j}}{r_{ij}} \tag{5.23}$$

Here, the cross-sectional areas of the two loops are assumed to be constant and equal to a_i and a_j, respectively. The current density changes at different points of the cross-sectional areas. r_{ij} is equal to the distance between the point being considered at loop i and the point used for integration at loop j. The average magnetic flux ψ_{ij} in loop i is related to the vector potential $\boldsymbol{A}_{ij}$ as

$$\psi_{ij} = \frac{1}{a_i}\oint_i \int_{a_i} \boldsymbol{A}_{ij} \bullet d\boldsymbol{l}_\mathbf{i}\, da_i \tag{5.24}$$

Then, the coupling inductance $L_{ij} = \Psi_{ij}/I_j$ is

$$L_{ij} = \frac{1}{a_i a_j}\frac{\mu}{4\pi}\oint_i \int_{a_i} \oint_j \int_{a_j} \frac{d\boldsymbol{l}_\mathbf{i} \bullet d\boldsymbol{l}_\mathbf{j}}{r_{ij}}\, da_i\, da_j \tag{5.25}$$

$\boldsymbol{dl}_i \bullet \boldsymbol{dl}_j$ is the vector-inner-product of two vectors, $\boldsymbol{dl}_i$ and $\boldsymbol{dl}_j$. Equation (5.25) also shows that the coupling inductance is reciprocal, that is, $L_{ij} = L_{ji}$. Self-loop-inductance L_{ii} can also be calculated by equation (5.25) with index j being replaced by i.

5.1.3.2 Partial Inductance Although inductance is defined for current loops, it is difficult to determine each loop (or return path) for on-chip interconnects, mainly due to the lack of a ground plane. Almost every wire couples to many other wires. To avoid this difficulty, the concept of partial inductances has been introduced by Albert Ruehli in his Partial Element Equivalent Circuit (PEEC) method [7,8]. Partial inductance is well-suited for computer circuit simulation.

Let us consider the two current loops, a-b-c-d and e-f-g-h in Figure 5.8. From (5.25), the mutual inductance L_{ae} between these two loops may be calculated as follows:

$$
\begin{aligned}
L_{ae} = \frac{1}{a_a a_e}\frac{\mu}{4\pi}\Bigg(& \int_a^b \int_{a_a} \int_e^f \int_{a_e} \frac{dl_a \bullet dl_e}{r_{ae}} da_a da_e \\
& + \int_a^b \int_{a_a} \int_f^g \int_{a_e} \frac{dl_a \bullet dl_e}{r_{ae}} da_a da_e \\
& + \int_a^b \int_{a_a} \int_g^h \int_{a_e} \frac{dl_a \bullet dl_e}{r_{ae}} da_a da_e + \int_a^b \int_{a_a} \int_h^e \int_{a_e} \frac{dl_a \bullet dl_e}{r_{ae}} da_a da_e \\
& + \int_b^c \int_{a_a} \int_e^f \int_{a_e} \frac{dl_a \bullet dl_e}{r_{ae}} da_a da_e + \int_b^c \int_{a_a} \int_f^g \int_{a_e} \frac{dl_a \bullet dl_e}{r_{ae}} da_a da_e \\
& + \int_b^c \int_{a_a} \int_g^h \int_{a_e} \frac{dl_a \bullet dl_e}{r_{ae}} da_a da_e + \int_b^c \int_{a_a} \int_h^e \int_{a_e} \frac{dl_a \bullet dl_e}{r_{ae}} da_a da_e \\
& + \int_c^d \int_{a_a} \int_e^f \int_{a_e} \frac{dl_a \bullet dl_e}{r_{ae}} da_a da_e + \int_c^d \int_{a_a} \int_f^g \int_{a_e} \frac{dl_a \bullet dl_e}{r_{ae}} da_a da_e \\
& + \int_c^d \int_{a_a} \int_g^h \int_{a_e} \frac{dl_a \bullet dl_e}{r_{ae}} da_a da_e + \int_c^d \int_{a_a} \int_h^e \int_{a_e} \frac{dl_a \bullet dl_e}{r_{ae}} da_a da_e \\
& + \int_d^a \int_{a_a} \int_e^f \int_{a_e} \frac{dl_a \bullet dl_e}{r_{ae}} da_a da_e + \int_d^a \int_{a_a} \int_f^g \int_{a_e} \frac{dl_a \bullet dl_e}{r_{ae}} da_a da_e \\
& + \int_d^a \int_{a_a} \int_g^h \int_{a_e} \frac{dl_a \bullet dl_e}{r_{ae}} da_a da_e + \int_d^a \int_{a_a} \int_h^e \int_{a_e} \frac{dl_a \bullet dl_e}{r_{ae}} da_a da_e \Bigg)
\end{aligned}
\tag{5.26}
$$

In total, there are 16 terms because each of the four sides of one loop needs to be integrated with each of the four sides of another loop. We use the subscript indices ***a*** and ***e*** to denote the a-b-c-d loop and the e-f-g-h loop, respectively.

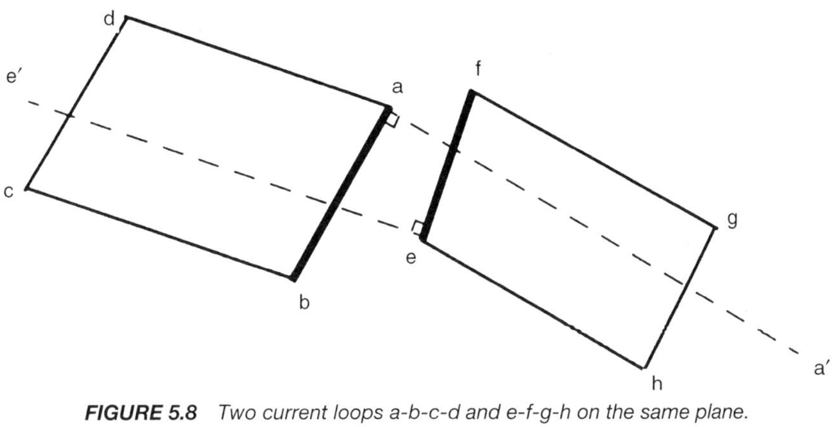

FIGURE 5.8 *Two current loops a-b-c-d and e-f-g-h on the same plane.*

Let us assume that **ad** is perpendicular to **ef** and so is **bc** (the dash line **ee'** is drawn to pictorially show that, since both **ad** and **bc** are parallel to **ee'**, they are perpendicular to **ef**). Similarly, we assume **fg** and **eh** are perpendicular to **ab**. Furthermore, we assume that the two loops extend to infinity, meaning that the distance between any point on **cd** and any point on **ef** is infinite, as is the distance between any point on **ab** and any point on **gh**. Based on these assumptions, most of the terms in equation (5.26) evaluate to zero because they either are vector-inner-products of two orthogonal vectors or have the distances at denominator equal to infinity. Therefore, we have

$$L_{ae} = \frac{1}{a_a a_e}\frac{\mu}{4\pi}\int_a^b \int_{a_a} \int_e^f \int_{a_e} \frac{dl_a \bullet dl_e}{r_{ae}}\, da_a\, da_e \tag{5.27}$$

That is, L_{ae} depends only on the geometry of the wire segments **ab** and **ef** under these assumptions; no loop information is required. Equation (5.27), in fact, is the definition of partial inductance for conductor segments. The loop inductance of (5.25) may be rewritten as summations over the straight loop segments (which may be infinite in number for curved conductors) with all segments allowed to have a different cross section:

$$L_{ij} = \sum_{k=1}^{K}\sum_{m=1}^{M} \frac{\mu}{4\pi}\frac{1}{a_k a_m}\int_{b_k}^{c_k} \int_{a_k} \int_{b_m}^{c_m} \int_{a_m} \frac{dl_k \bullet dl_m r_{km}}{r_{km}}\, da_k\, da_m \tag{5.28}$$

Here, the ith loop is assumed to consist of K segments while the jth loop is divided into M segments. The limits in the integrals are the starting points b_k, b_m and the end points c_k and c_m of the segments. Equation (5.28) is the definition of partial inductance for conductors of general geometry when either the K segments or the M segments do not close a loop.

With this definition, it is easy to calculate the inductances of interconnects for circuit simulation because no loop information is required and the partial inductances depend only on geometry. We basically let the circuit simulator determine the current loops (or the least-impedance return paths). The correctness of doing so can be proven from the example shown in Figure 5.9.

Figure 5.9(a) shows a current loop, where segment 1 is assumed to be tilted for generality. For circuit simulation without knowing this loop a priori, let us assume that we calculate only the partial inductances. Equation (5.28) gives the total loop inductance in terms of these partial inductances. Each of the partial inductances has a flux area associated with it in accordance with Figure 5.8. Specifically, the areas associated with conductors 2 and 4 are considered. The tilted conductor segment 1 must be approximated by an infinite number of minute steps. In this example, only a single step is shown for clarity. The flux area associated with the partial self-inductance of segment 2 extends from somewhere near the conductor to infinity, as shown by the backward slash area of Figure 5.9(b). The flux area associated with the partial mutual-inductance of segment 4 is shown by the forward slash area of Figure 5.9(b). Therefore, the flux areas cancel outside

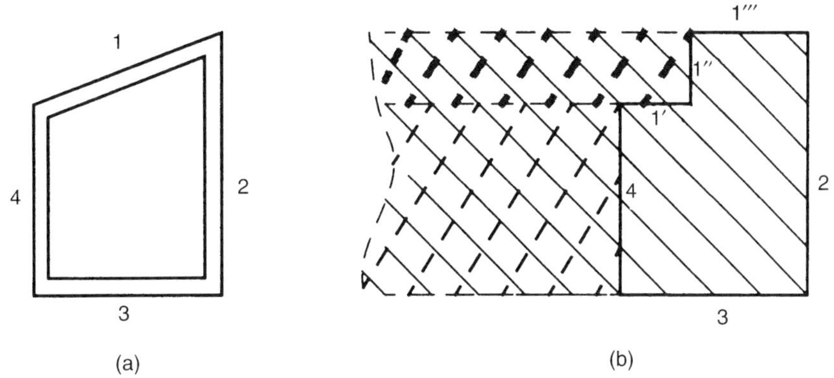

FIGURE 5.9 *(a) A closed loop with a tilted segment. (b) Partial inductance flux areas associated wigh segment 1", segment 2, and segment 4.*

of conductor 4, and the only remaining flux area is restricted to the inside of the loop, as expected. The same principle can be applied to conductor 1, where $1''$ cancels the flux area outside the loop. If this concept is applied to all partial inductances in (5.28), it is found that the only remaining flux area is restricted to the inside of the loop.

If nonplanar loops are considered, canceling pairs of currents can be introduced by reducing the general problem into a new set of locally planar loops. Then, the above analysis is applied to each loop. Therefore, the loop inductance derived from equation (5.28) by using partial inductances of conductor segments will be the same as that given by (5.25); hence, we will have the correct simulation results.

5.1.3.3 Evaluation of Partial Inductances Given the magneto-quasistatic assumption, the current within a long thin conductor can be assumed to flow parallel to its surface, as there is no charge accumulation on the surface. Thus, a long thin interconnect can be divided into rectangular filaments, and the current is assumed to flow along the length of the filament with constant density [11]. In order to properly capture skin and proximity effects in these long, thin conductors, the cross section of the conductors can be further divided into a bundle of parallel filaments, as shown in Figure 5.10. The widths or heights of filaments increase exponentially from boundary to center because the current density decreases exponentially from boundary to center due to skin effect and because we assume the current density within each filament is constant. To achieve good accuracy, it is desired that the smallest width or height be smaller than one skin depth, or

$$\frac{W}{2\left(1+e+e^2+\ldots+e^{\lfloor N/2\rfloor-1}\right)+e^{\lfloor N/2\rfloor}} \leq \delta \tag{5.29}$$

$$\frac{H}{2\left(1+e+e^2+\ldots+e^{\lfloor M/2\rfloor-1}\right)+e^{\lfloor M/2\rfloor}} \leq \delta$$

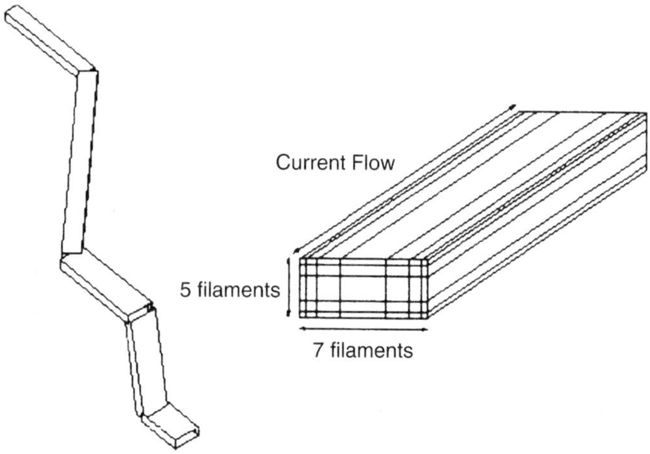

FIGURE 5.10 *An interconnect divided into 5 sections, each of which is a bundle of 35 filaments.*

where W and H are the width and the height of a straight conductor, and N and M are the number of horizontal and vertical divisions, respectively. δ is the skin depth at *significant frequency*. $\lfloor x \rfloor$ is the largest integer less than or equal to x. The multiplier of two at the denominator is there due to the fact that the widths (or the heights) of filaments are symmetric at both sides. The length of a filament is determined according to the shape of the conductor (piecewise-straight approximation is used) and by the changes of magnetic field distribution. For example, if there exist orthogonal wires, more sections are required around the crossings.

It is also possible to use the filament approach for planar structures, such as ground planes, where the current distribution is two-dimensional. In such cases, a mesh of filaments must be used, as in Figure 5.11. Once an interconnect is discretized into filaments as in Figure 5.12, it can be represented with a planar graph, in which the n nodes in the graph are associated with the points connecting two bundles of filaments, and the b branches in the graph represent the filaments into which each conductor's segments are discretized. An example graph or circuit for a single interconnect is shown in Figure 5.12.

In Figure 5.12, V_i and I_i are the branch voltage and the current for each filament. V_s and I_s are the source voltage and the total current applied at the terminals of the interconnect. Therefore, we have

$$V_i = \left(\frac{l_i}{\sigma a_i}\right) I_i + j\omega \sum_{j=1}^{b} \left(\frac{\mu}{4\pi} \int_i \int_{a_i} \int_j \int_{a_j} \frac{dl_i \bullet dl_j}{r_{ij}} \, da_i \, da_j\right) I_j \tag{5.30}$$

where $l_i/\sigma a_i$ is the resistive reactance of the filament. The inductive reactance is determined by the partial inductance equation given in (5.27). In matrix form,

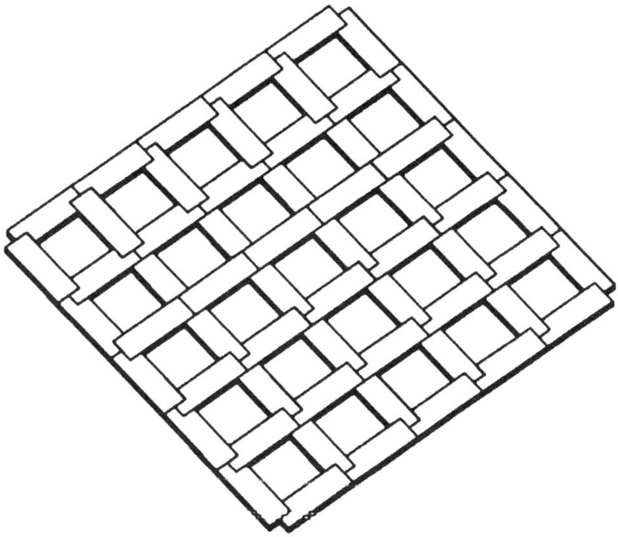

FIGURE 5.11 *Discretization of a ground plane. Segments are drawn at one-third of the actual width.*

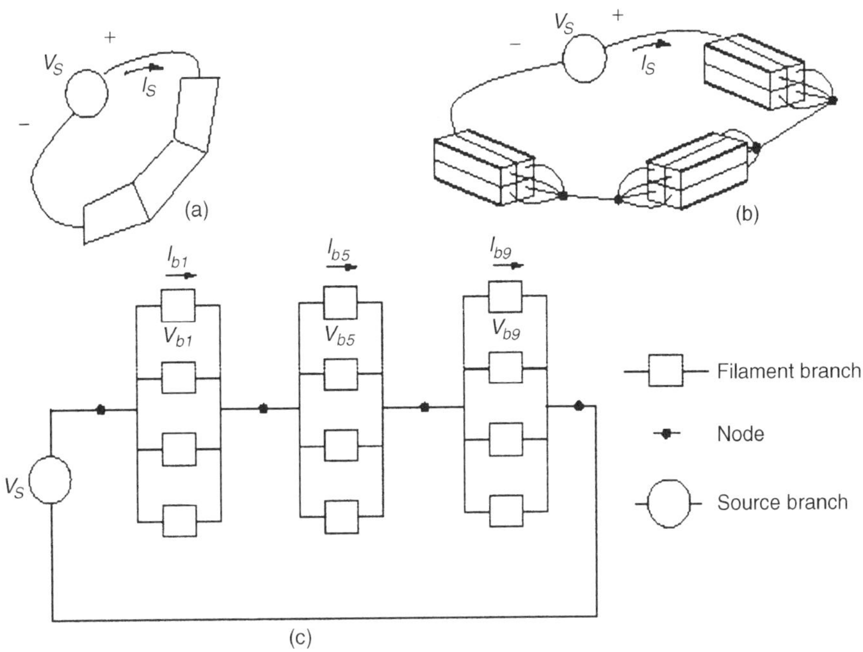

FIGURE 5.12 *One interconnect, (a) as piecewise-straight sections, (b) discretized into filaments, (c) modeled as a circuit.*

(5.30) becomes

$$(\boldsymbol{R} + j\omega\boldsymbol{L})\boldsymbol{I}_b = \boldsymbol{V}_b \tag{5.31}$$

where $\mathbf{I}_b$ and $\mathbf{V}_b$ are the vectors of all branch currents and voltages. Also, from basic circuit theory, we have

$$\boldsymbol{A}\boldsymbol{I}_b = I_s \quad \text{and} \quad \boldsymbol{A}^t\boldsymbol{E}_n = \boldsymbol{V}_b \tag{5.32}$$

where the matrix $\boldsymbol{A}$ is the branch incidence matrix, with $\mathbf{A}^t$ being its transpose, and $\mathbf{E}_n$ and $\mathbf{I}_s$ are the vector of node voltages and the vector of source currents, which happen to be all zeros except I_s and $-I_s$ at the two terminal nodes of the interconnect. Combining (5.31) and (5.32), we have

$$\boldsymbol{A}(\boldsymbol{R} + j\omega\boldsymbol{L})^{-1}\boldsymbol{A}^t\boldsymbol{E}_n = \boldsymbol{I}_s \tag{5.33}$$

Based on equation (5.33), we can solve for the total current I_s flowing into the interconnect, given any voltage difference between its terminals. The ratio of the voltage to the current gives the total effective reactance of the interconnect, whose real part is the resistance and whose imaginary part divided by ω gives the inductance. Clearly, due to the matrix inversion in (5.33), both resistance and inductance are frequency-dependent. That demonstrates the skin effect. For a system with multiple interconnects, the reactance derivation is similar. The only difference is that the indices i and j are counted across all the filaments of all the interconnects, b and n are the total numbers of filaments and nodes in the system, respectively.

5.1.4 Skin-Effect and Proximity-Effect Resistance Calculation

We often heard that skin-effect and proximity-effect induced resistance is not a concern for the current high-speed digital designs on-chip. However, due to modern CPU designs extending the clock frequencies into GHz range, the skin-effect and proximity-effect impact on resistance may not be ignored. In this section, we clarify and illustrate when these effects will become significant for high speed VDSM (very-deep-submicron) digital designs.

First, skin effect and proximity effect need to be clearly defined. Skin effect is the exponential increase of the current concentration near the conductor surface as the frequency, the conductivity, and the permeability of the conductor are increased. This causes an increase of resistance and a decrease of internal inductance compared to the DC resistance and inductance at the frequencies when this effect is significant. The skin depth is defined as the distance to the point inside the conductor at which the current density is reduced to 37% (i.e., 1/e) of the current density on the surface. The skin depth, δ, into the conductor can be estimated using the expression in equation (2.1) repeated here in (5.34).

$$\delta = \frac{1}{\sqrt{\pi f \mu \sigma}} \tag{5.34}$$

where μ and σ are the permeability and conductivity of the conductor, and f is the frequency of the current in the conductor. The skin depth versus frequency for Aluminum (Al) and Copper (Cu) interconnects are shown in Table 5.3 and plotted in Figure 5.13.[10] Figure 5.14 shows the current distribution versus frequency in an isolated conductor. However, in a microstrip configuration where a conductor is on top of a ground plane as shown in Figure 5.15, the current distribution tends to concentrate near the bottom side of the conductor because of the proximity of the ground. This effect is referred to as proximity effect.

TABLE 5.3 Skin Depth (δ) versus Frequency for Aluminum and Copper Interconnects

Frequency (GHz)	δ for Al (micrometer)	δ for Cu (micrometer)
0.5	3.651	2.955
1	2.581	2.089
2	1.825	1.478
3	1.490	1.206
4	1.291	1.045
5	1.154	0.934
6	1.054	0.853
7	0.976	0.790
8	0.913	0.739
9	0.860	0.697
10	0.816	0.661

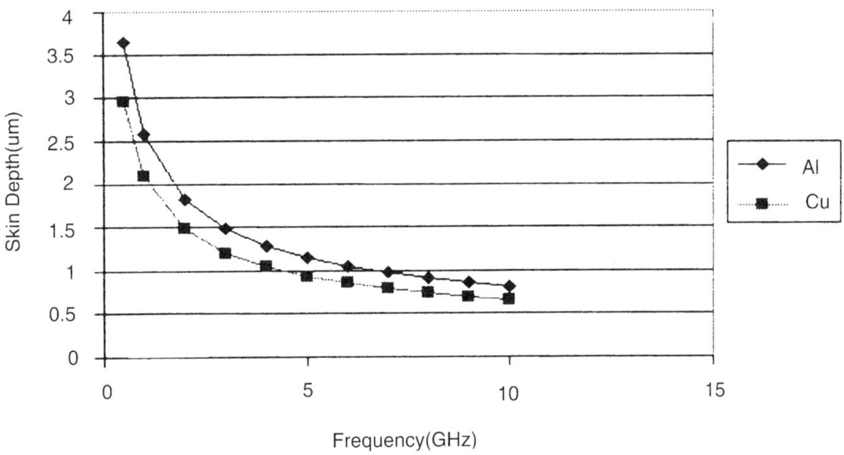

FIGURE 5.13 *Skin depth versus frequency of Al and Cu interconnects on-chip.*

[10] $\sigma_{Cu} = 5.8 \times 10^7$ mho/m, $\sigma_{Al} = 3.8 \times 10^7$ mho/m, $\mu_{Cu} = 1.257 \times 10^{-6}$ henry/m, and $\mu_{Al} = 1.257 \times 10^{-6}$ henry/m.

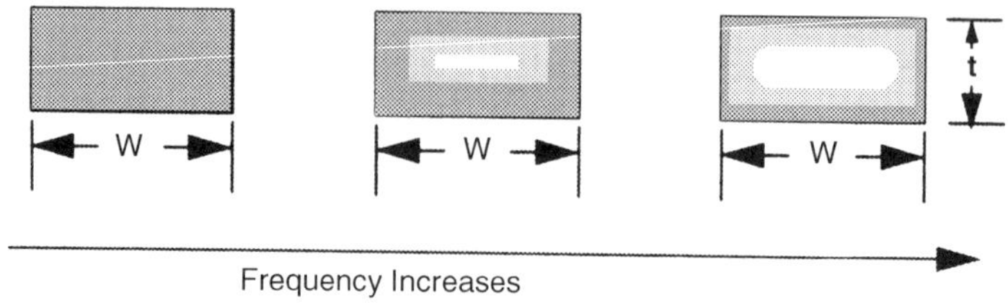

FIGURE 5.14 *Current distribution versus frequency in an isolated conductor. The darker the shading, the higher the current density is. w is the width and t is the thickness of the conductor.*

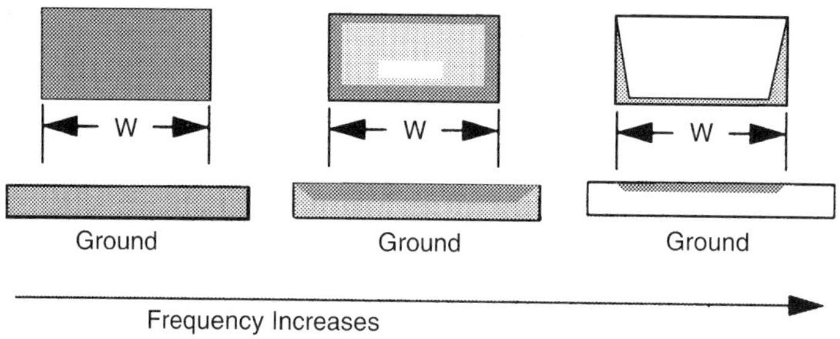

FIGURE 5.15 *Current distribution versus frequency in a conductor over a ground plane.*

There have been various techniques developed to obtain $R(f)$, that is, frequency-dependent resistance of the conductors. One example is the popular incremental inductance rule by [28], which states that the series skin impedance, Z per unit length,

$$Z = R + jX = R + jwL_i \tag{5.35}$$

has equal real and imaginary parts at high frequencies. L_i is calculated by the value of the change in external inductance caused by an incremental recession of all metallic walls carrying a skin current. However, this method requires that Wheeler's incremental rule be applied when the cross-sectional dimensions of the conductor are large compared to the skin depth, for example, when the dimensions are four times the skin depth. For on-chip applications, skin depth is typically comparable to conductor thickness, and therefore this rule does not hold. Other methods include: separation of variable and method of moments in [18], integral equation and method of moments in [17], variational method in [26], finite-element method in [19], full-wave analysis using finite difference method in [24], solving vector potential using boundary integral equation in [22],[29], current density method in [25], and network approach using partial element equivalent circuits (PEEC) in [27]. A summary of some of the techniques listed above is given in [39]. In this section, we use the PEEC method to obtain $R(f)$. However, applying this method to obtain accurate resistances at very high frequencies (larger than 20 GHz) is not efficient due to the need in PEEC for an extremely large number of segments (or meshes) or calling for a more efficient meshing scheme [38]. Luckily, the onchip interconnect resistance $R(f)$ of concern will probably remain below 20 GHz for the next few years.

Following Wheeler's incremental inductance rule, we can derive an approximate formula for computing the AC resistance for conductors that have high-frequency skin-depth much smaller than the dimensions, given the low-frequency R_{dc} and the high-frequency R_{hf} of an isolated rectangular conductor with an error of less than 5% [46]. At low frequencies, where the skin depth is of the order of the conductor thickness (say $t < 2\delta$) a near-uniform current distribution is observed across the conductor cross section. At high frequencies (say $t > 4\delta$), the current distribution is nearly at exponential decay from the surface of the conductor.

$$\lim_{f\to 0} R_{ac} = R_{dc} = \frac{1}{\sigma wt} \tag{5.36}$$

$$\lim_{f\to\infty} R_{ac} = \kappa R_{hf} = \kappa \frac{1}{2\sigma\delta(w+t)} \tag{5.37}$$

where κ varies between $1.1 \sim 1.2$ for different values of w/t. However, the best overall approximation for all values of w/t can be realized at $\kappa = 1.2$, which accounts for the effective path of current at high frequency in Figure 5.16. Therefore, we can approximate R_{ac} per unit length at different frequencies of an isolated rectangular conductor by the following formula.

$$R_{ac} = \sqrt{(R_{dc})^2 + (\kappa R_{hf})^2} \tag{5.38}$$

However, as shown in Figure 5.15, if there is a ground plane nearby, the value of R_{ac} can increase from one to up to two times the R_{ac} of an isolated rectangular conductor, if the width of the conductor is much larger than the distance from the conductor to the ground plane. The ground plane AC resistance, R_g, can be approximated as in (5.39).

$$R_g = \frac{1}{2\pi h\sigma\delta} \tag{5.39}$$

where h is the distance from the conductor to the ground plane. So the total AC resistance of a conductor on top of a ground plane can be four times higher (i.e., $R_{Tac} = 4R_{dc}$) than the DC resistance at high frequencies with large w/h.

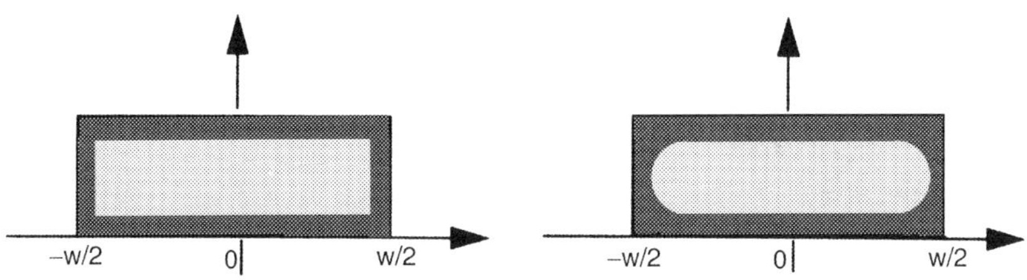

FIGURE 5.16 *High-frequency model of current used in R_{hf} computation on the left; the effective path of current at high frequency on the right.*

However, an isolated conductor or an isolated conductor on top of a ground plane are idealized situations on-chip. The more realistic cases are multiconductors (i.e., arrays of conductors) on the same layer while the neighboring vertical layers are orthogonal arrays of conductors. Since there is no inductive coupling between two orthogonal arrays of conductors, we examine only the array of conductors on the same layer. To consider a special case, what is the AC resistance for two identically coupled conductors as shown in Figure 5.17?

Conceptually, for even-mode coupling of the two identical conductors, the current distribution is concentrated at the far edge of the conductors when the two conductors are brought closer to each other. The AC resistance of each conductor therefore increases for large values of w/s ratio or small value of s with a fixed w, where s is the spacing between the two conductors. On the contrary, for odd-mode coupling of two identical conductors, the current distribution is concentrated at the near edge of the conductors when the two conductors are brought closer to each other. The AC resistance of each conductor in the odd-mode coupling case increases much faster than in the even-mode coupling case due to the closer interaction of the generated magnetic fields and therefore the higher current density at the near edge. The rate of AC resistance increase in the odd mode can be two times larger than in the even mode.

When there are more than two conductors on the same layer with different current-flowing directions/phases at high frequencies, the AC resistance is much more complicated. A numerical method has to be applied in such cases.

Using the PEEC method and SPICE simulation, we can compute and generate resistance and current density plots for the interconnect structures in VDSM technology. Figure 5.18 shows the resistance versus frequency for an isolated rectangular conductor. We can see that the resistance does not change much (maximum of 6%) with frequency varying from 500 MHz to 10 GHz. Also the resistance versus frequency of an isolated rectangular conductor on top of a ground plane is pretty much the same as an isolated rectangular conductor. This last observation shows that the change in resistance induced by the proximity-effect is not significant for frequencies up to 10 GHz. Therefore, for example, in high-frequency clock systems, the resistance changes only slightly (less than 6% increase up to 10 GHz). However, for multiple conductors, proximity-effect induced resistance change may not be ignored as shown below. Equation (5.38)

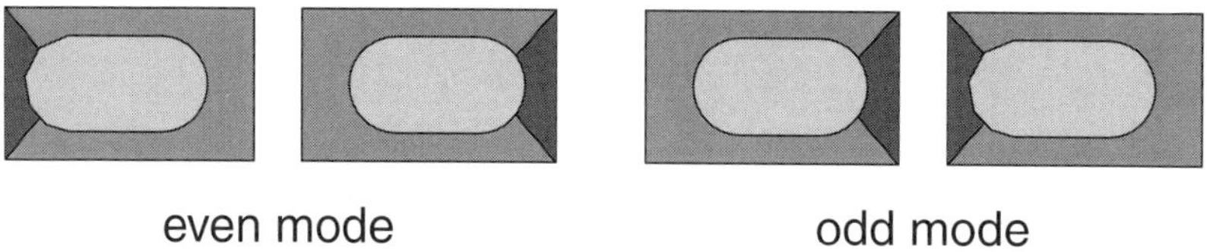

FIGURE 5.17 *On the left shows the current distribution at high frequency for even-mode coupling (current flows in the same direction); on the right the current distribution is of odd-mode coupling (current flows in the opposite direction). The darker color means higher current density. With the same dimensions, the maximum current density of odd mode is higher than the one of even mode.*

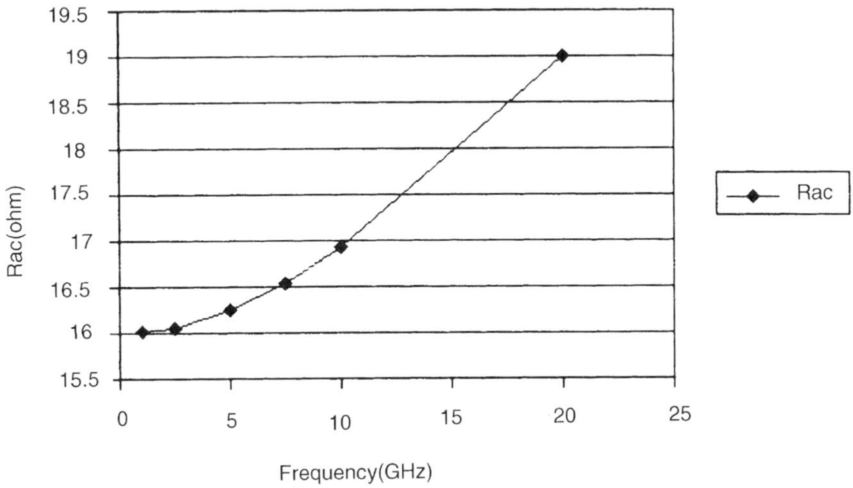

FIGURE 5.18 *AC resistance versus frequency for an isolated conductor with width 2.5μm, thickness 1.9μm, and 10,000μm in length.*

is not applicable here because the assumption ($t > 4\delta$) is not easily satisfied for on-chip interconnects, but it can be a good approximation for PCB and MCM interconnects.

It is interesting to observe that the current density plot for two coupled rectangular conductors on top of an orthogonal array at a relatively high frequency (say, 4 GHz). One method to model and simulate the skin-effect and proximity-effect induced current density in a multiconductor system is to use the direct-filament-extraction method (explained in detail in section 5.2.1). We model in Raphael RI3 each rectangular conductor with multiple single bars. An example of the model for the six-conductor case is shown in Figure 5.19. Current value is monitored

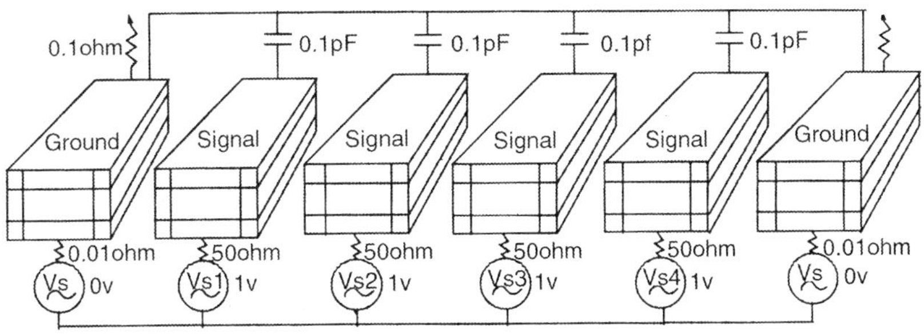

FIGURE 5.19 *Multiple single bar composition of each rectangular conductor in Raphael RI3 is used to model the skin-effect and proximity-effect induced current density on the six-conductor case, for example. Current value is monitored during SPICE simulation of the formulated RLC desk of this six-conductor case. All the other examples such as two conductors or three conductors are modeled and simulated in a similar fashion. In this case, the current density is monitored at the far end of the conductors.*

during SPICE simulation of the formulated RLC partial element circuit model of this six-conductor case. All the other examples presented below, such as two conductors or three conductors, are modeled and simulated in a similar fashion.

Comparing Figure 5.20 and Figure 5.21, one can see that the current density is higher at the near proximity when the currents are in opposite directions. The

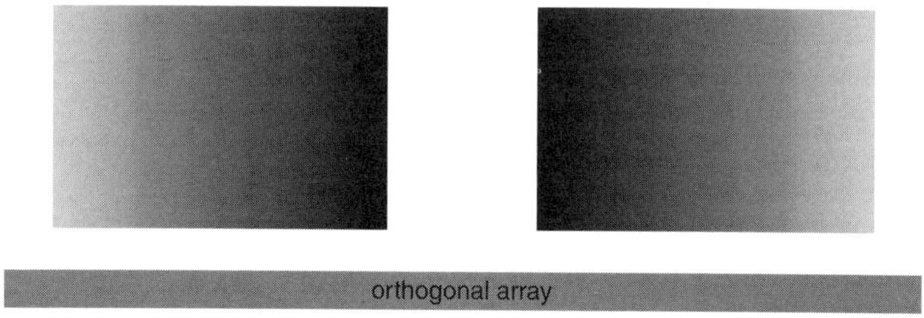

FIGURE 5.20 Current density plot for odd-mode two-coupled conductors on top of an orthogonal array with width 2.5μm, thickness 1.9μm, spacing 1.2μm, distance to orthogonal array 1.2μm, and 2000μm in length. The current density is higher (5%) at the near end of two-coupled conductors at 4 GHz due to the current flowing in the opposite direction.

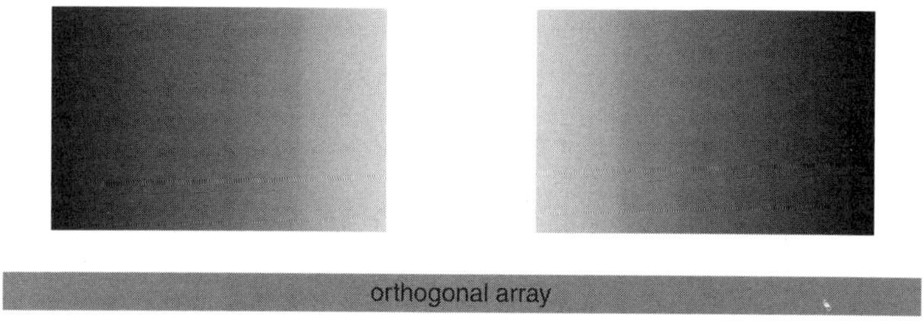

FIGURE 5.21 Current density plot for even-mode two-coupled conductors on top of an orthogonal array with width 2.5 μm, thickness 1.9 μm, spacing 1.2 μm, distance to orthogonal array 1.2 μm, and 2000 μm in length. The current density is higher (4%) at the far end of two-coupled conductors at 4 GHz due to the current flowing in-phase.

FIGURE 5.22 Current density plot for even-mode two-coupled conductors on top of an continuous ground plane with width 2.5 μm, thickness 1.9 μm, spacing 1.2 μm, distance to the ground plane 1.2 μm, and 2000 μm in length. The current density is higher at the bottom end of the conductor due to proximity effect. Since the current is flowing in the same direction in the two conductors, the current density is also higher at the outer edge of the conductors (about 5% maximum difference at 4 GHz).

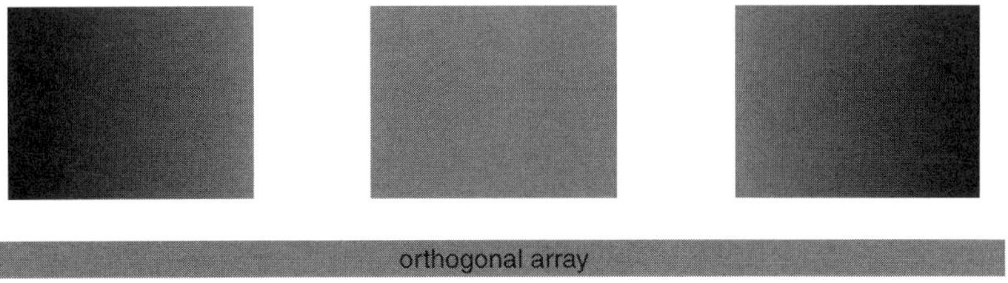

FIGURE 5.23 *Current density plot for three-coupled conductors on top of an orthogonal array with 2.5 μm, thickness 1.9 μm, spacing 1.2 μm, distance to the ground plane 1.2 μm, and 2000 μm in length. The currents flow in the same phase in the three conductors. The current density of the center conductor is quite uniform due to the conflicing forces of the skin-effect pushing the current outward from the center and the proximity-effect given by the two outer conductors pushing the current inside toward the center. The maximum difference in current density is about 4% at 4 GHz in this case.*

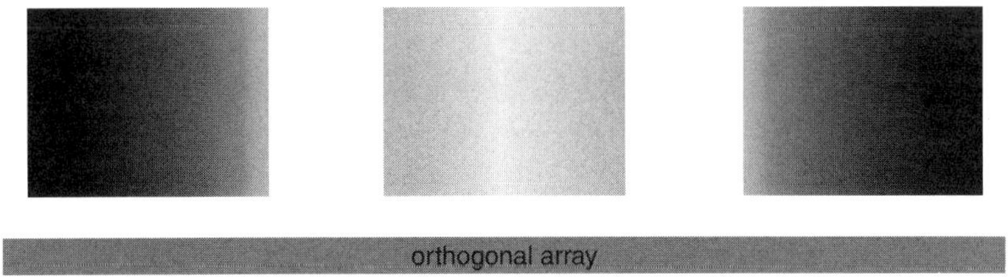

FIGURE 5.24 *Current density plot for three-coupled conductors on top of an orthogonal array with width 2.5 μm, thickness 1.9 μm, spacing 1.2 μm, distance to the orthogonal array 1.2 μm, and 2000 μm in length. The currents flow in the same phase in the outer two conductors but in the opposite phase in the center conductor. Interestingly, the current density can be differed by more than 7% in this case at 4 GHz.*

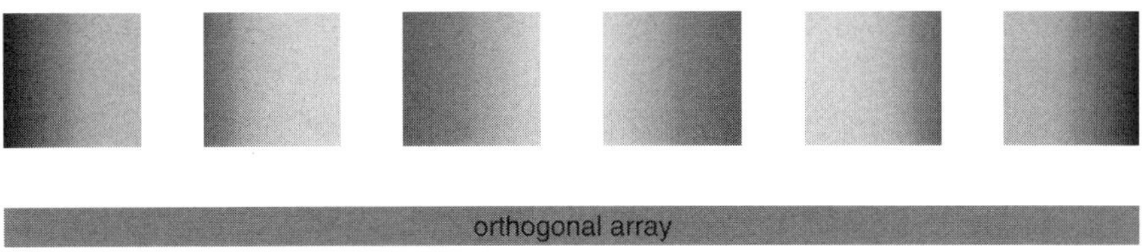

FIGURE 5.25 *Current density plot for six-coupled conductors on top of an orthogonal array with width 2.5 μm, thickness 1.9 μm, spacing 1.2 μm, distance to the ground plane 1.2 μm, and 2000 μm in length. The same-phase ac voltage is applied in all the conductors. The maximum difference in the current density is 8% in this case at 4 GHz.*

opposite result is observed when the currents flow in the same direction. This is in good agreement with the hypothetical condition in Figure 5.17. Therefore, for two coupled rectangular conductors on top of an orthogonal array Rac is higher in the odd mode than in the even mode. When a continuous ground plane is present near the two conductors, the current density is higher near the plane as shown in Figure 5.22. Figures (5.23) and (5.24) show three conductors on

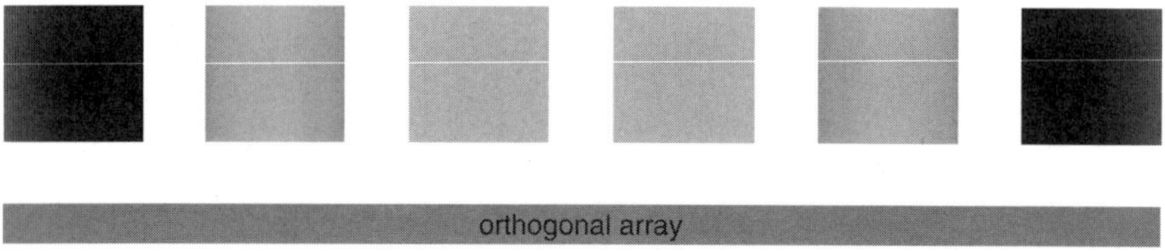

FIGURE 5.26 *Current density plot for six-oupled conductors on top of an orthogonal array with width 2.5 μm, thickness 1.9 μm, spacing 1.2 μm, distance to the ground plane 1.2 μm, and 2000 μm in length. The same ac voltage is applied in the center four conductors while the outer two conductors grounded. This plot shows the current density of the conductors. The maximum difference in the current density is about 34% in these six conductors.*

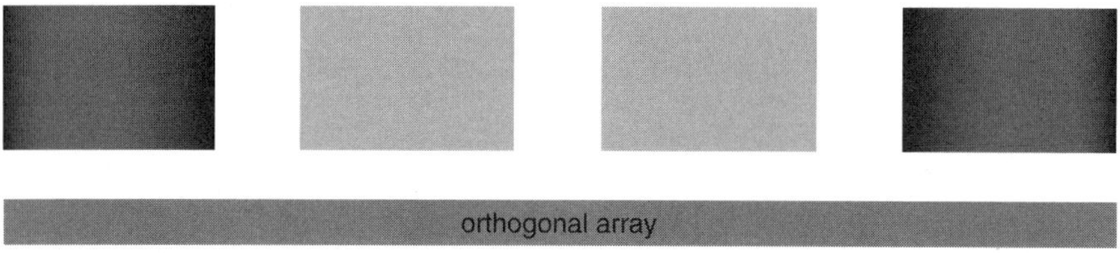

FIGURE 5.27 *Current density plot for the center four conductors of the six-coupled conductors on top of an orthogonal array as shown in Figure 5.26 with width 2.5 μm, thickness 1.9 μm, spacing 1.2 μm, distance to the ground plane 1.2 μm, 2000 μm in length. The same ac voltage is applied in the center four conductors while the outer two conductors are grounded. The maximum difference in current density among the center four conductors is about 12% at 4 GHz.*

top of an orthogonal array of conductors with the incident voltage in the same direction (in Figure 5.23) and not in the same direction (in Figure 5.24) to the three conductors. Figures (5.25) and (5.26) illustrate the six-coupled conductors on top of an orthogonal array. Two outer conductors are grounded and provide the ground return paths for the center four conductors. Since all the currents are returned on the two outer conductors, the current density is much higher on the two outer conductors than on the center four conductors as shown in Figure 5.25. One interesting phenomenon to notice is that the current density is about 12% higher on the conductors near the ground conductors (the outer conductors) than on the center two conductors as shown in Figures (5.26) and (5.27). Therefore, the current density is slightly higher (12%) on the outer two conductors among the center four conductors in high-frequency designs (4 GHz as *significant frequency* in this case).

Our simulation results show that the change in the AC resistance due to skin effect and proximity effect does not seem to be significant (only about 10%). Such results refer to the multiconductor system with signal at 1-GHz clock frequency and about 4-GHz significant frequency. But in the upcoming two or above GHz designs, the resistance may change by more than 20% due to skin effect and proximity effect. In that case, this phenomenon may deserve a detailed simulation with frequency-dependent RLC (such as W model in HSPICE [48]), or S-parameter [43,44] to understand its potential impact.

5.2 ON-CHIP INDUCTANCE CONSIDERATION

As we have explained in previous sections, the interconnect reactance is frequency dependent, resulting in problems for conventional SPICE simulations, which only consider frequency independent RLC elements. In this section, we will first explore the solution to this problem. In section 5.2.1 we empirically show that using the RLC extracted at the *significant frequency* gives correct results. Then, we will examine the inductance impact on delay and crosstalk noise for those interconnect structures that may exist in future advanced circuit designs. We use Raphael RI3 and RC3 to extract the resistances, inductances, and capacitances of those structures and simulate, by using SPICE, their corresponding equivalent RLC circuits, including devices of the industrial CMOS 0.18 μm technology. The results will be shown in section 5.2.2. In section 5.2.3, we try to answer the question "When do we need to consider on-chip inductance?" based on a series of simulations of different interconnect structures at different frequencies.

5.2.1 Handling Frequency-Dependent Resistance and Inductance

In section 5.1.3.3, we showed that both the resistance and the inductance of an interconnect are frequency dependent due to the matrix inversion taken in equation (5.33). When extracting inductance and resistance with PEEC tools, such as Raphael RI3 [47], or FASTHENRY, developed at MIT [12], we need to specify the frequency of interest. The inductances and resistances at that frequency will be determined. For the fast-transient portion of circuit simulations, the high-frequency values must be used, and for the slow-transient portion, the low-frequency ones must be used. However, a conventional circuit simulator, such as SPICE, is not capable of considering frequency-dependent RLC elements. Only the resistance and the inductance at one frequency can be considered. To cope with this difficulty, we are trying to find the best representative frequency, giving minimal errors. We empirically show that the *significant frequency* introduced in section 5.1.2 serves this purpose. We create the equivalent RLC circuit for a typical interconnect structure, extracted at the *significant frequency*. The SPICE simulations of this circuit are compared with the simulations of the circuit with its equivalent RLC elements extracted directly from the filaments forming all the interconnects, hence avoiding the matrix inversion in equation (5.33). Although the direct extraction from filaments gives a circuit that is correct across the entire frequency spectrum, it produces too many elements and is impractical for many circuit designs. Some approaches [43,44] circumvent this frequency-dependence problem by extending SPICE to simulate frequency-domain data (i.e., S-parameter) in time-domain. Before discussing the results of our simulation, we first describe the creation of an equivalent RLC circuit for an interconnect structure and identify a few mistakes that people commonly make.

The interconnect structure that we simulated is shown in Figure 5.28(a). There are four signal wires, each of which is 5 μm wide, 1.5 μm thick, and 2000 μm long and all of which lie on top of a layer of orthogonal wires with equal width

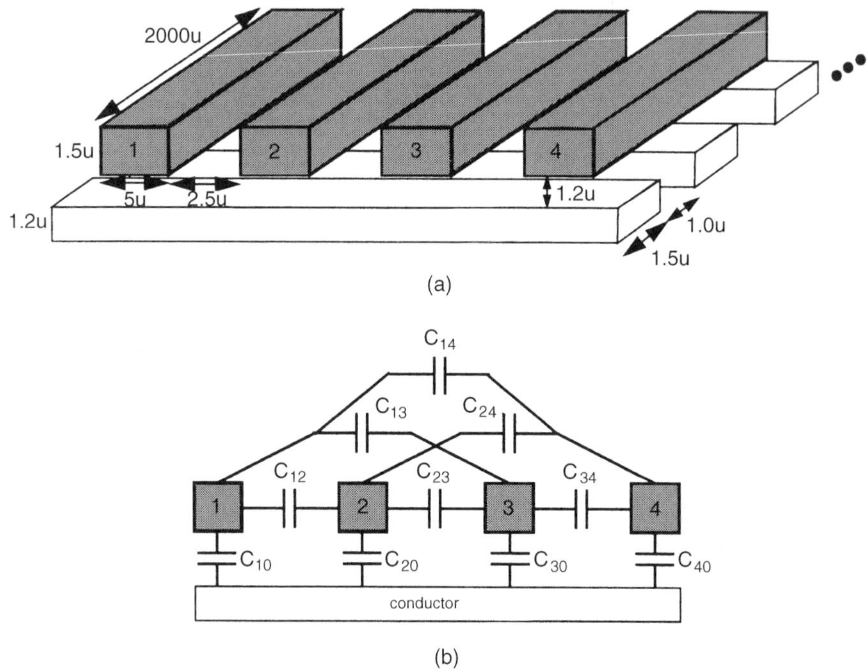

FIGURE 5.28 *(a) Four signal wires on top of orthogonal wires; (b) Cross-section view with parasitic capacitances.*

and spacing, representing a layer of almost 100% routing density. Since the width of each orthogonal wire is only 1.5u, no eddy current will be formed[11], so any inductance coupling between the signal wires and those orthogonal wires may be completely neglected. Due to this high-wire density on the orthogonal layer, it can be assumed that the layer completely blocks the capacitive coupling from the signal wires to anything below it. Figure 5.28(b) shows the cross-sectional view of the structure including its parasitic capacitances. RI3 is run to get the partial resistances and inductances of this four-wire system (neglecting the orthogonal layer) at 4 GHz[12]. A common mistake that people make is to assume the inductance is scalable; for example, using RI3 to get the inductances at 1000 μm and then multiplying the values by two to get the inductances at 2000 μm. As demonstrated in section 5.1, RI3 must be run for the exact length to get the accurate inductances. Capacitance, on the other hand, is scalable. Therefore, we may run RC3, a three-dimensional field-solver, for 100 μm along the signal wires, including the orthogonal layer, to get the parasitic capacitances per unit length, and assume the values apply for the whole length, 2000u, of the signal

[11] From our RI3 simulations, we found that the coupling between orthogonal wires is almost zero up to the wire width of 25 μm.

[12] With the significant frequency defined as $0.34/tr$, 4 GHz corresponds to 85 ps ramp time.

wires. Combining the resistances, inductances, and capacitances, we may create the equivalent RLC circuit as shown in Figure 5.29, which does not show the parasitic capacitances from signal wires to the orthogonal layer. These will be discussed later.

To combine the resistances, inductances, and capacitances, the wires have to be partitioned into sections. The capacitors are inserted at section points. Figure 5.29 shows N equal-length sections. To achieve enough accuracy, the length should not be too long. We use 250 μm or eight sections for our example.[13] As shown in Figure 5.29, the R and L for each section are equal to the wires' partial reactances divided by N. Notice that L is obtained from inductance-solver given the total length of the interconnect structure since L is not scalable, as mentioned in section 5.1. However, once L for a given total length is determined, we can section L by N for RLC netlist formulation. The C for each section is equal to the per-unit-length parasitic capacitances multiplied by the section length. Half of the values are inserted at the two terminals of each section.

One question remains. What about the parasitic capacitances to the orthogonal layer, C_{10}, C_{20}, C_{30}, and C_{40} in Figure 5.28(b)? Some people turn these capacitances into grounded capacitors, treating the whole orthogonal layer as a ground node connecting to the drivers or the receivers of the signal wires. In reality, most current flowing into the orthogonal wires has to travel a long way to their drivers or receivers and, then through these devices to the ground node, which makes the ground node assumption too optimistic. We use the

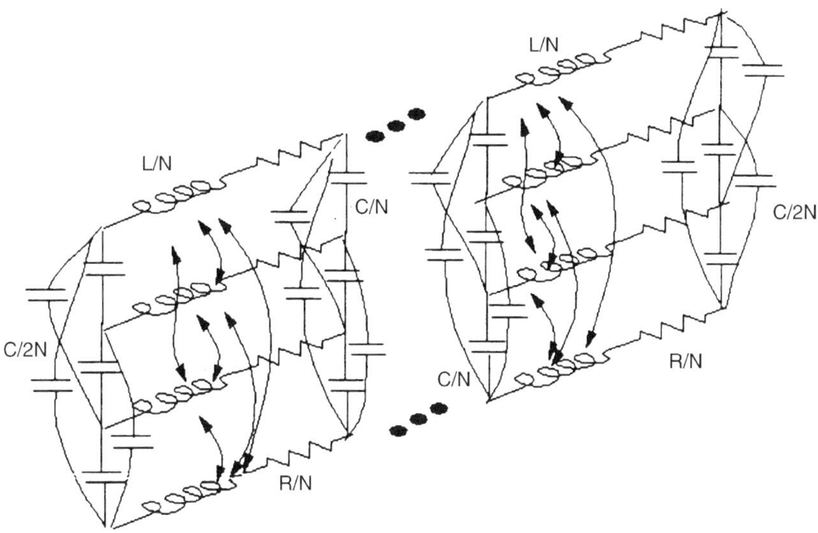

FIGURE 5.29 *The equivalent RLC circuit for the four-wire structure.*

[13] The suitable section length also depends on signal frequency. Another consideration is the complexity. We found the length may be between 250 μm and 500 μm for around 4-GHz signals.

approach demonstrated by Figure 5.30 to eliminate the orthogonal layer and to distribute the capacitances into extraparasitic capacitances between signal wires. Figure 5.30(b) shows that in reality an orthogonal wire is a series of resistors. Since the wire spacing is small, the resistances are small and can be neglected. Therefore, we can treat the orthogonal layer as a supernode, as in Figure 5.30(c). This supernode in (c) may be eliminated through network reduction, resulting in extraparasitic capacitors C'_{ij} between wires, as in Figure 5.30(d) or

$$C'_{ij} = \frac{C_{i0} \times C_{j0}}{C_{10} + C_{20} + C_{30} + C_{40}}, i, j = 1 \ldots 4. \tag{5.40}$$

Signal wire 1 is treated as a ground wire for signal return with both of its terminals connected to the ground node. The rest of the signal wires are driven by CMOS buffers and terminated at CMOS buffers, all in 0.18 μm technology. The signal ramp times at the drivers' outputs are about 85 ps, corresponding to the 4-GHz *significant frequency*. We also create another equivalent circuit with about 170 ps ramp time by performing RI3 extraction at a 2-GHz *significant frequency*. The simulations of these two circuits are compared with the simulations of the

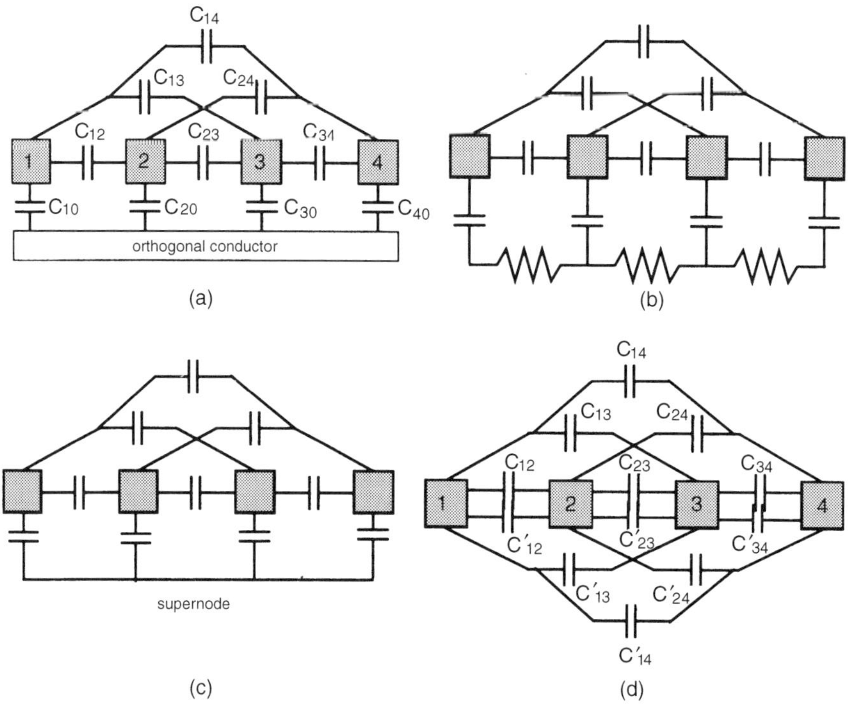

FIGURE 5.30 *(a) Original structure with orthogonal layer; (b) the conductor is replaced by resistors; (c) the resistors are neglected; (d) extra caps are introduced between wires after network reduction.*

corresponding circuits extracted directly from the wires composed of filaments, denoted by direct-filament-extraction circuits.

To extract the resistances and inductances directly from filaments, each wire is replaced by a bundle of 3-by-10, 0.5 μm-wide and 0.5 μm-thick smaller wires. As long as the width and the thickness are smaller than the skin depth for the frequency of interest, the extracted resistance and inductance for each filament is frequency independent under the constant current density in each filament assumption; 0.5 μm corresponds to the skin depth at about 16 GHz, which provides the desired accuracy for frequency components up to that range. RI3 extraction is then performed on this bundled structure with its filament division turned off, hence by passing the matrix inversion. The parasitic capacitors at each section point are the same as those determined earlier for Figure 5.28. We end up with about 960 inductors and 960 resistors to model these four signal wires in the equivalent circuit. The inductors all couple to each other.

Figure 5.31 shows the comparison between the waveforms at the far ends of signal wires 2 and 3 using simulations of both the extraction-at-4-GHz circuit and the direct-filament-extraction circuit. Figure 5.32 compares the waveforms at the far ends of wires 2 and 3 using simulations of both the extraction-at-2-GHz circuit and the direct-filament-extraction circuit. We observe very good agreements between corresponding waveforms and conclude that the resistances and inductances extracted at the significant frequency achieve an accuracy of frequency-dependent resistance and inductance.

Please note that the circuit modeling employed the PEEC approach and we did not guess the return paths (or current loops). The SPICE simulations determined the loops. The authors of paper [13] observed significant proximity effects on their structures and claimed that inductances were very frequency dependent and no single frequency approximation was feasible. Their conclusions came from their loop inductance approach. For complex structures, the loops change a lot at different frequencies. Electrical currents always try to return through the smallest impedance path. If [13] had used partial inductances and had allowed SPICE to determine the return paths, the inaccuracy problem introduced by single-frequency approximation would not have been that severe. For on-chip analysis, partial inductance should always be preferred.

5.2.2 Worst Case and Inductance Impact on Delay and Crosstalk

Now we consider the interconnect structure shown in Figure 5.33. The structure has N signal wires sandwiched between ground and vdd wires. The wires are 1.9 μm thick and 1000 μm to 6000 μm long, representing typical top-level global wires between repeaters or gates in advanced high-speed designs, where repeaters should be inserted for wires longer than 6000 μm to minimize delays and control crosstalk. Different N, width W, and spacing S are used in our analyses. All the signal wires are driven by and terminated at CMOS drivers of 0.18 μm technology with 1.3 V Vdd. Drivers with resistances of about 46 Ohm and 72 Ohm are used. Each driver's Gnd and Vdd terminals are connected to the ground and vdd wires,

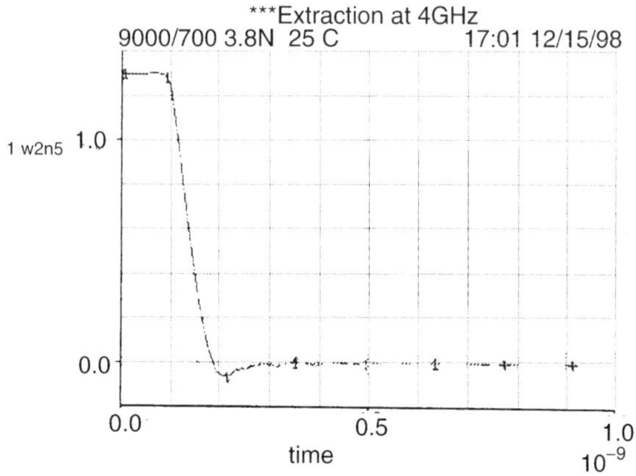

(a)

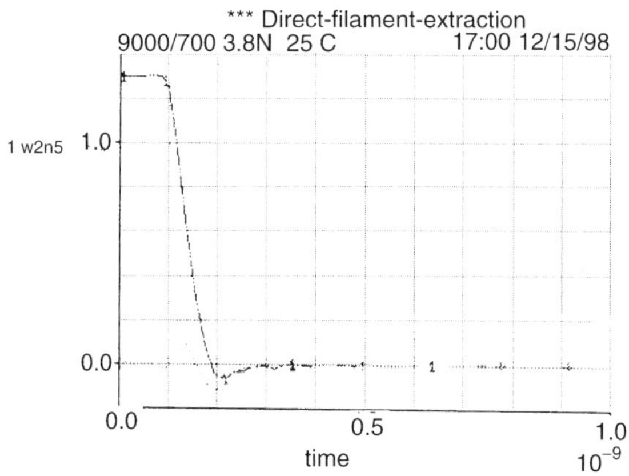

(b)

FIGURE 5.31 *Comparisons between waveforms at the far ends of wire2 and wire3 from the (a) extraction-at-4 GHz circuit; (b) direct-filament-extraction circuit.*

whose terminals are connected to the power source through small reactances. In this section, we first investigate the worst delay and crosstalk scenarios in terms of aggressors' switching directions, and then study the inductance impacts on delay and crosstalk by comparing the simulations with and without considering inductance. Stochastic worst cases due to signal-switching conditions and arrival times will also be studied. Other stochastic worst cases due to manufacturing variations are not included here but are discussed in [45].

Understanding the aggressors' switching directions in the worst case is important to designing a verification tool to identify potential signal integrity problems. Two different sets of signal wires with $N = 8$ and 16 are considered, as are three different width-and-spacings with ($W = 0.8u$, $S = 0.6u$), ($W = 2.5u$, $S = 1.25u$), and ($W = 5u$, $S = 2.5u$). The large inductive noise and overshoots prohibit the employment of wires wider than 5 μm in high-speed designs, because the wire

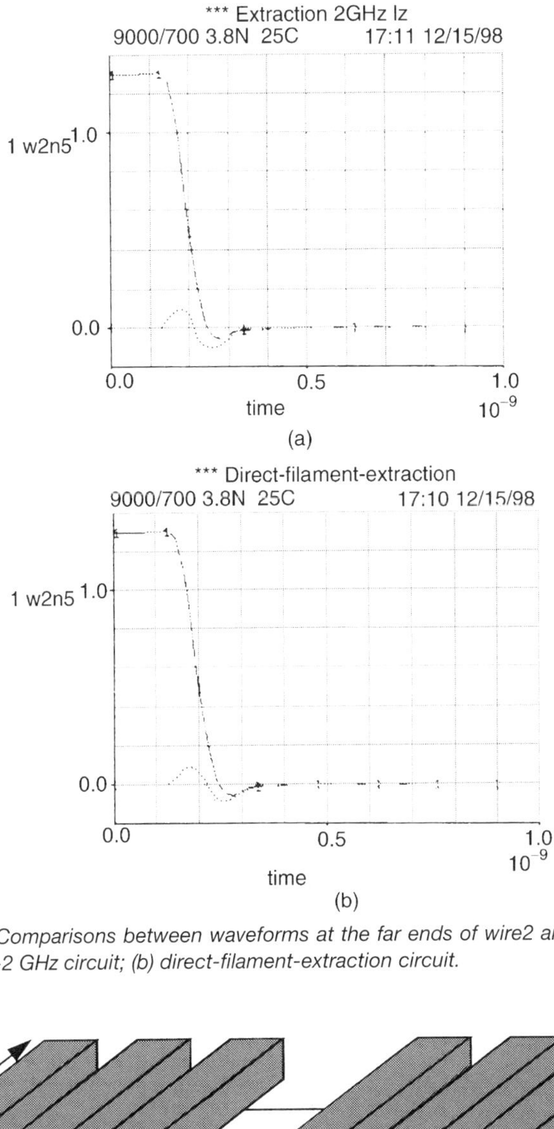

FIGURE 5.32 *Comparisons between waveforms at the far ends of wire2 and wire3 from the (a) extraction-at-2 GHz circuit; (b) direct-filament-extraction circuit.*

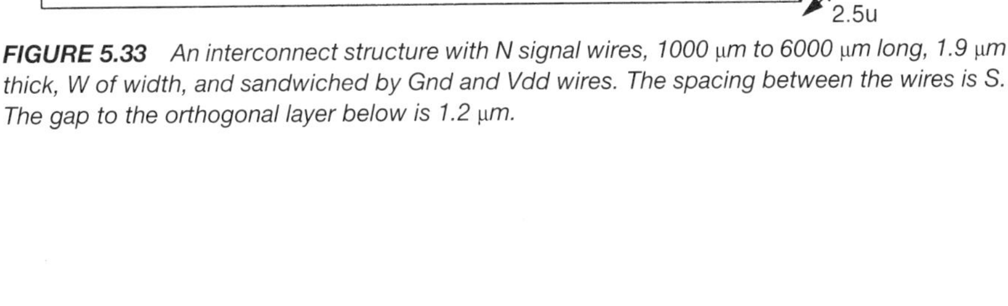

FIGURE 5.33 *An interconnect structure with N signal wires, 1000 μm to 6000 μm long, 1.9 μm thick, W of width, and sandwiched by Gnd and Vdd wires. The spacing between the wires is S. The gap to the orthogonal layer below is 1.2 μm.*

reactances are too inductive. Wires wider than 5 μm should be split into multiple narrow wires to reduce inductance effects[14], which will be discussed in more detail in section 5.3. Similarly, N greater than 16 (or more than 16 wires in between dedicated Gnd or Vdd wires for return) should not be considered for high-speed design unless with extreme caution. Their insufficient ground return results in large inductive coupling noise and overshoots. The three widths are chosen to cover different situations for inductance consideration. A 2.5-μm structure should show some inductive effects. A 5-μm structure should be strongly inductive, while the 0.8-μm structure would have minimal inductance effects. In fact, interconnects of about 0.8 μm width are frequently used at the top-level metal layers in today's advanced chip designs. Their inductance is often ignored. The per-unit-length resistance for the 0.8-μm-wide wires is 36 Ohm/mm, while it is 11 Ohm/mm for 2.5-μm-wide wires, and 5.4 Ohm/mm for 5-μm-wide wires.

From our simulations, the worst position for the victim wire is always in the middle (i.e., wire $\lfloor N/2 \rfloor$ or wire $\lfloor N/2 \rfloor + 1$). This is easy to explain because it is farthest from the ground or vdd wires. We also observed that the worst case of the more inductive 5 μm- and 2.5 μm-structures corresponds to the situation in which all aggressors switch in the same direction, exactly the same as in the capacitive worst case, where the polarity of the noise is in the same direction of the aggressors. This point is against intuition given the opposite polarities of inductive and capacitive couplings. In fact, when a wire switches in one direction, its inductive coupling noise on other wires is in the other direction; the opposite happens with the capacitive coupling noise. In addition, because the capacitive coupling effect has a shorter range, we might guess that the worst case is for the two closest neighbors of the victim line to switch in one direction while all other aggressors switch in the other direction. In this way, their effects on the victim would be all in the same direction. Figures 5.34 and 5.35 show that this guess is wrong. Figure 5.34(a) shows the crosstalk noise waveform at the far end of the fourth signal wire (the victim) of the eight signal wires of 5 μm width and 4000 μm length with all the aggressors switching low, while Figure 5.34(b) shows the noise waveform at the far end of the fourth signal wire with all the aggressors switching low, except the third and the fifth wires, which are switching high. Clearly, the noise of Figure 5.34(a) is larger. The maximum noise is 0.6 V, almost half the value of Vdd, and it is able to faultily trigger the receiver. Figures 5.35(a) and (b) show the waveforms at the victim line while it is switching high. The aggressors of Figure 5.35(a) are all switching low while the aggressors of Figure 5.35(b) are switching high, except the two closest neighbors of the victim line, which are switching low, presumably providing a consistent dragging on the victim. However, the delay increased by the couplings for Figure 5.35(a) is 87 ps while the increased delay for Figure 5.35(b) is −4 ps (actually faster). Because the original victim interconnect delay, measured as the

[14] The clock lines of some high-speed designs use wires wider than 5 μm, but neighboring wide dedicated ground wires running in parallel are needed for return.

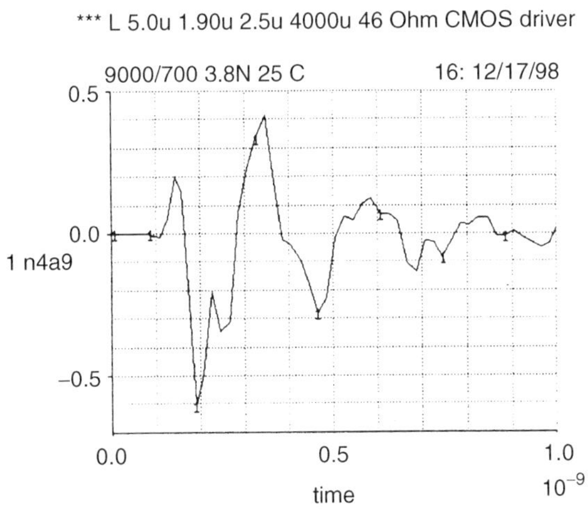

(a)

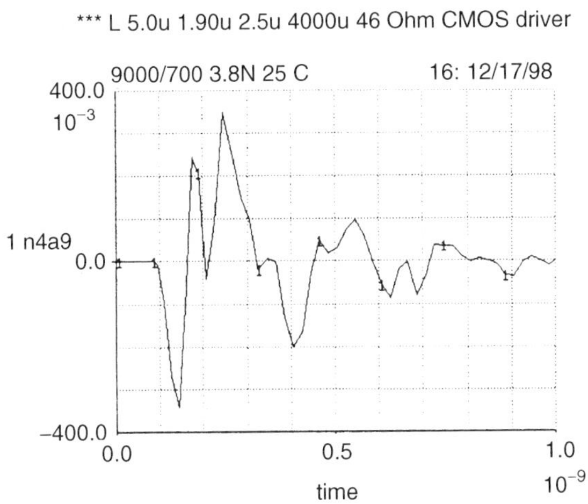

(b)

FIGURE 5.34 *(a) Noise at the victim wire when all aggressors switch low. (b) Noise at the victim wire when all aggressors switch low except the two neighbors of the victim switch high.*

delay from the time the near end first reaches 50% to the time the far end finally is stable above 50%, with all the other wires not switching, is only 20.5 ps, the delay increase due to inductive coupling in Figure 5.35(a) is more than 250%.

From our simulations, the worst delay case happens in exactly the same situation as Figure 5.35(a) with the victim, which can be either of the middle wires, switching in the direction opposite to the remaining wires. This is the same as in the worst delay case for which only the capacitive coupling is considered. This is again counterintuitive, because the inductive coupling should have increased the switching rate of the victim for this scenario. Do these two surprises indicate to us

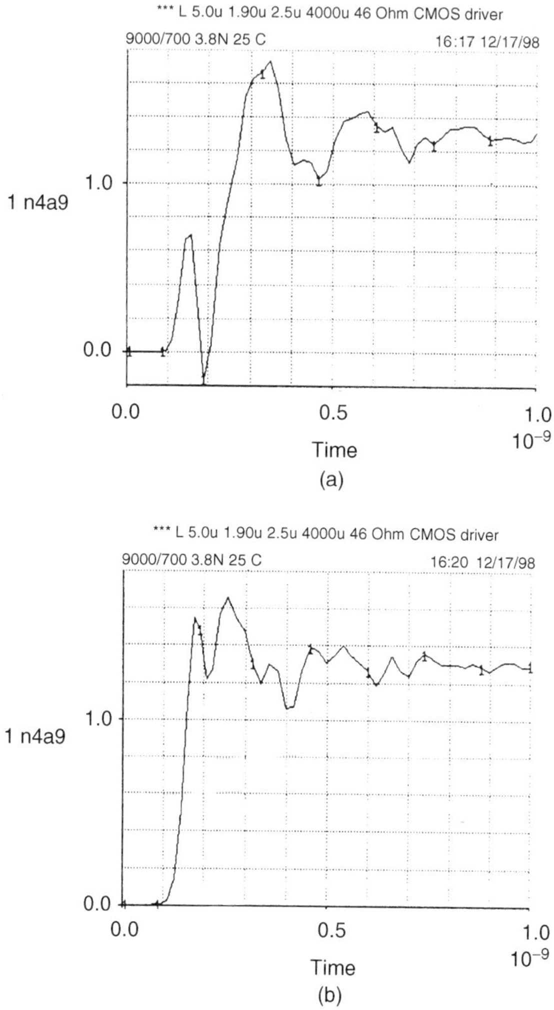

FIGURE 5.35 *(a) Raise-switching waveform at the victim wire when all aggressors switch low. (b) Waveform at the victim wire when all aggressors switch high, except the two neighbors of the victim switch low.*

that the capacitive coupling is dominant, since the results coincide with capacitive worst cases? However, this could not be the reason because we observe significant inductive effects, such as large overshoots and oscillations, in Figure 5.34(a) and Figure 5.35(a). A qualitative explanation is that, although the inductance is dominant, the inductive effects are caused by the change of currents, which is due to capacitive coupling. Therefore, if we try to create a "worst" case primarily based on inductive coupling, the case will not be truly the worst.

For the less-inductive 0.8-μm-wide structure, the worst crosstalk case happens when all the aggressors, except the two neighbors of the victim line, switch in one direction and the neighboring lines switch in the other direction. This agrees with one's intuition, although the magnitude is only slightly larger. The polarity of the noise is the same as the switching direction of the victim's neighbors. However, the worst delay case is still the same as the 5 μm or 2.5 μm case, as in the situation of Figure 5.35(a). Therefore, it is advised that for less-inductive structures, we need to check both cases. In section 5.2.3, we will discuss the figures of merit to determine how inductive an interconnect structure is.

So far, for the delay impact analysis, we have assumed that all the aggressors are switching at the same time as the victim line. Actually, we will experience much larger delay increase if the aggressors switch synchronously some time later than the victim line. Figure 5.36 shows the waveforms at the far ends of the fourth line as the victim, and the third line as one of the aggressors, of an 8-signal-wire, 2000-μm-long, and 2.5-μm-wide structure. The victim line switches up at time 0 while all the aggressors switch down at 100 ps. The time when the victim is finally stable above 50% of Vdd is 229 ps, or 84 ps increase over the delay where all the aggressors switch at the same time with the victim, or 141 ps delay increase over the scenario where no other lines switch. From the simulation results, we discover that the reason for this large increase is because the aggressors switch down almost at the same time as the downward oscillation of the victim line and, hence, aggressors' capacitive couplings pull the victim down even further, causing a large undershoot. Table 5.4 lists some victim delay increases versus delayed switching time of the aggressors for this 8-signal-wire, 2.5-μm-wide, 2000-μm-long structure. When the delayed switching time of aggressors is 100 ps, it results in the lowest point of the undershoot higher than or roughly equal to half Vdd, so the victim reaches 50% at the first upward transition, earlier than 100 ps, which is faster than the zero delayed switching time case. It is difficult to determine

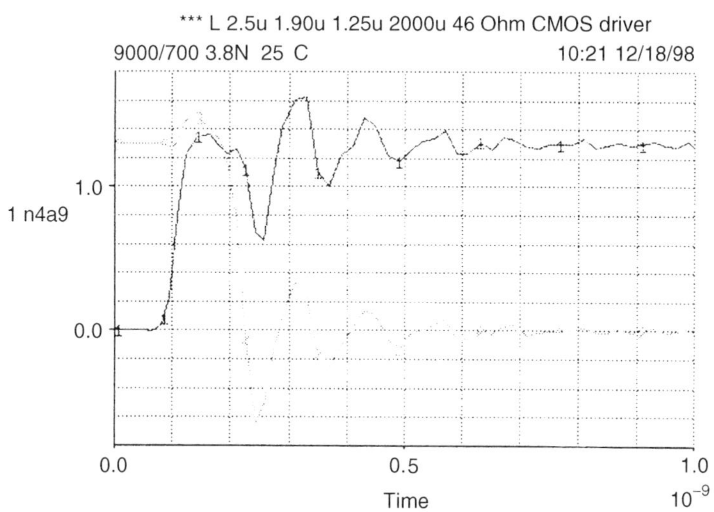

FIGURE 5.36 *The waveforms at the far ends of line 4, the victim, and line 3, one of the aggressors. The aggressors switch 100 ps later than the victim.*

TABLE 5.4 Victim Delay Increase versus Delays of the Aggressors

Delayed switching time of aggressors	Near end of the victim reaching 50%	Far end of the victim reaching 50%	Delay increase over aggressors switching same time with victim
0 ps	111.1 ps	145.7 ps	0
6 ps	112.6 ps	149.4 ps	3.7 ps
10 ps	113.8 ps	149.6 ps	3.9 ps
15 ps	115.1 ps	162.0 ps	16.3 ps
25 ps	115.9 ps	165.1 ps	19.4 ps
35 ps	70.1 ps	178.9 ps	33.2 ps
45 ps	70.9 ps	194.2 ps	38.5 ps
55 ps	70.6 ps	198.6 ps	42.9 ps
100 ps	71.7 ps	229.3 ps	83.7 ps

the worst case in terms of the delayed switching time of the aggressors because it depends on the driver resistance, line reactances, coupling inductances and capacitances, number of aggressor lines, and the ramp rates of all lines. We have to rely on simulations to determine the worst case. Because of this stochastic nature, we call this the *stochastic worst case*. The 2.5 μm-wide, 2000 μm-long, 16-wire structure has its stochastic worst case when the aggressor has delayed by 175 ps. Since the coupling noise of 5-μm structures with simultaneously switching aggressors is larger than half Vdd, their stochastic worst- case analyses do not make sense. On the other hand, there are no significant inductance effects for 0.8-μm structures; their stochastic worst cases are primarily determined by capacitive couplings, which happen when aggressors switch at the time and the victim driver is about half turned-on. The victim wire's charge-accumulating momentum would be knocked down and it takes a while for the victim wire to regain that momentum. We still need to rely on simulations to determine the worst case in this situation but its delay increase is much smaller than the inductive stochastic worst case.

Besides modeling the interconnect structures using RLC equivalent circuits, the corresponding RC equivalent circuits were also created by totally neglecting the inductances. By comparing the simulations of the RLC circuits and their corresponding RC circuits, we would be able to understand the impact of inductance and the possible errors that might have occurred if the inductances in our circuit designs were neglected. Table 5.5 summarizes the crosstalk comparisons for different structures. All the aggressors switch low and are driven by inputs of 80 ps fall time. Table 5.6 summarizes the interconnect delay comparisons for different structures. The aggressors switch low while the victim switches high. All drivers' inputs of 80 ps ramp time are applied at the same time.

The capacitive crosstalk noise is approximately proportional to the length of the coupled wires, and the RC interconnect delay is roughly proportional to the square of length. These two are the most important properties for RC lines.

TABLE 5.5 Crosstalk Comparisons Between the Cases With and Without the Consideration of Inductance

Structures	With inductance	Without inductance	% Error
1000 μm-long	$Z_{DRV} = 72$ Ohm		
0.8 μm-wide 8-signal-wire	377 mV	340 mV	10%
2.5 μm-wide 8-signal-wire	393 mV	302 mV	23%
5.0 μm-wide 8-signal-wire	452 mV	259 mV	43%
2000 μm -long	$Z_{DRV} = 46$ Ohm		
0.8 μm-wide 8-signal-wire	457 mV	570 mV	25%
0.8 μm-wide 16-signal-wire	479 mV	604 mV	26%
2.5 μm-wide 8-signal-wire	531 mV	312 mV	41%
2.5 μm-wide 16-signal-wire	693 mV	326 mV	53%
5 μm-wide 8-signal-wire	619 mV	246 mV	60%
5 μm-wide 16-signal-wire	741 mV	269 mV	64%
4000 μm-long	$Z_{DRV} = 46$ Ohm		
0.8 μm-wide 8-signal-wire	664 mV	798 mV	20%
0.8 μm-wide 16-signal-wire	729 mV	891 mV	22%
2.5 μm-wide 8-signal-wire	482 mV	615 mV	28%
2.5 μm-wide 16-signal-wire	501 mV	677 mV	35%
5 μm-wide 8-signal-wire	607 mV	464 mV	24%
5 μm-wide 16-signal-wire	677 mV	533 mV	21%
6000 μm -long	$Z_{DRV} = 46$ Ohm		
0.8 μm-wide 8-signal-wire	764 mV	871 mV	14%
2.5 μm-wide 8-signal-wire	559 mV	701 mV	25%
5 μm-wide 8-signal-wire	485 mV	567 mV	17%

However, with the consideration of inductance, the crosstalk noise tops off (or roughly saturates) at lengths of around 4000 μm because the inductive crosstalk noise is primarily caused by the aggressors' overshooting spikes, which then pull the victim with them. And, the spikes decrease as the inductance effects decrease with the increase of length. Although the interconnect delays increase with length, no relation can be derived. The short-range effect of capacitive coupling explains why the delays and crosstalk noise are almost the same for the 8-wire structure and the 16-wire structure. In general, the capacitive couplings are completely shielded by a couple of wires in between. However, it is not true for inductive coupling, where the delays and crosstalk increase from an 8-wire structure to a 16-wire structure. The delay errors introduced by overlooking inductance for the 5 μm- and 2.5 μm-wide structures are more than 40%. The crosstalk errors are more than 40% for the 2000-μm-long structures and more than 20% for the 4000-μm-long structures. The decrease of errors is due to the capacitive crosstalk double, making their differences smaller. The 0.8 μm-wide structures behave differently because the inductance effects in 0.8-μm-wide structures are not significant. Figure 5.37 shows the waveforms at the far ends of the third and the fourth lines of the 8-wire, 2000-μm-long structure. There is no significant overshoot or oscillation. Their minimal inductance effects, such as the minimal overshoot, actually shorten signal transition and

TABLE 5.6 Interconnect Delay Comparisons Between the Cases With and Without the Consideration of Inductance

Structures	With inductance	Without inductance	% Error
1000 μm-long	$Z_{DRV} = 72$ Ohm		
0.8 μm-wide 8-signal-wire	6.0 ps	6.8 ps	13%
2.5 μm-wide 8-signal-wire	9.0 ps	1.8 ps	80%
5 μm-wide 8-signal-wire	11 ps	1.0 ps	91%
2000 μm-long	$Z_{DRV} = 46$ Ohm		
0.8 μm-wide 8-signal-wire	40.6 ps	27.2 ps	33%
0.8 μm-wide 16-signal-wire	43.1 ps	27.9 ps	35%
2.5 μm-wide 8-signal-wire	38.1 ps	6.6 ps	83%
2.5 μm-wide 16-signal-wire	44.2 ps	6.7 ps	85%
5 μm-wide 8-signal-wire	36.8 ps	3.3 ps	91%
5 μm-wide 16-signal-wire	41.3 ps	3.4 ps	92%
4000 μm-long	$Z_{DRV} = 46$ Ohm		
0.8 μm-wide 8-signal-wire	125 ps	105 ps	16%
0.8 μm-wide 16-signal-wire	127 ps	103 ps	19%
2.5 μm-wide 8-signal-wire	50.0 ps	26.7 ps	47%
2.5 μm-wide 16-signal-wire	67.0 ps	26.9 ps	60%
5 μm-wide 8-signal-wire	48.5 ps	13.6 ps	72%
5 μm-wide 16-signal-wire	61.0 ps	13.0 ps	79%
6000 μm-long	$Z_{DRV} = 46$ Ohm		
0.8 μm-wide 8-signal-wire	381 ps	329 ps	14%
2.5 μm-wide 8-signal-wire	96 ps	73 ps	24%
5 μm-wide 8-signal-wire	70 ps	36 ps	49%

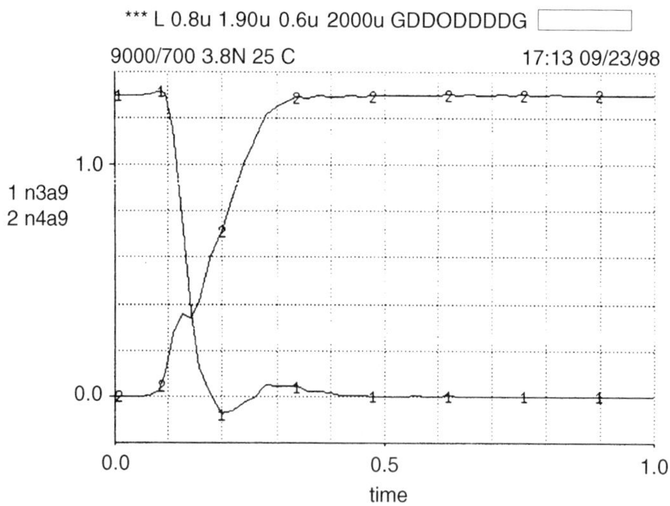

FIGURE 5.37 The waveforms at the far ends of line 4 and line 3 of the 0.8 μm- wide, 2000 μm-long 8-wire structure.

decrease noise, which is due to the opposite polarity of the inductive couplings and capacitive couplings. Neglecting inductance (by only considering R and C) tends to overestimate crosstalk noise and delays for long or minimal inductive lines.

5.2.3 When Do We Need to Consider On-Chip Inductance?

A couple of methods or figures of merit to characterize the importance of on-chip inductance have been published in [14,15,16]. They can be used to answer the question as to when we need to consider on-chip inductance. Circuit designers rely on these rules to know whether the consideration of inductance is necessary in their design verifications. From the significant inductance impacts shown in the previous section, we definitely do not want to overlook the potential inductance problems in our designs. Because accurate and stable RLC delay and crosstalk estimations still rely on computation-intensive simulations, in order to efficiently verify a chip's signal integrity, determining which nets require the special consideration of inductance is important. Based on an effective inductance screening, designers may apply those RC delay and crosstalk prediction methods, proven to be fast and reliable [32, 33, 40, 41, 42], to as many nets as possible.

The interconnect is treated as a uniform RLC transmission line. Based on the transmission line analysis, the authors of [14] introduced the following figures of merit for an interconnect of length l. Inductance needs to be considered if

$$\frac{t_r}{2\sqrt{LC}} < l < \frac{2}{R}\sqrt{\frac{L}{C}} \tag{5.41}$$

where R, L, C are per-unit-length resistance, self-inductance, and coupling capacitance, respectively; t_r is the rise time of the signal at the input of the CMOS circuit driving the interconnect. Under the conditions of equation (5.41), inductance needs to be considered.

Equation (5.41) is the conjunction (AND) of two rules. The first rule $l < 2/R\sqrt{L/C}$ is introduced to ensure that the transmission line's equivalent RLC circuit is overdamped. The equivalent circuit, in Figure 5.38, is created by using single RLC section approximation, where $L_t = Ll$, $R_t = Rl$, and $C_t = Cl$. The

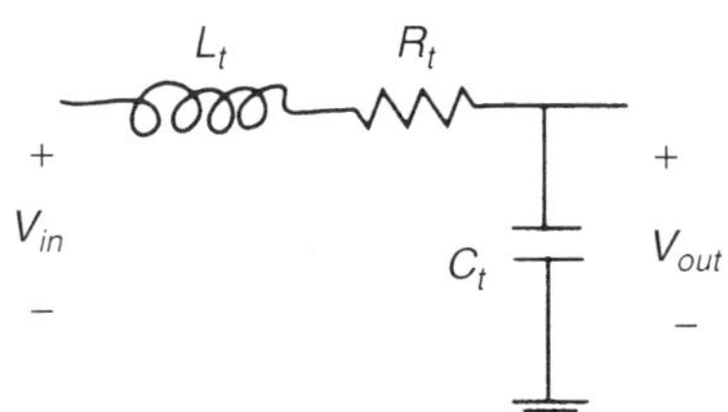

FIGURE 5.38 *Simple lumped RLC circuit model of an interconnect line.*

poles of this circuit are

$$p_{1,2} = \omega_0 \left[-\xi \pm \sqrt{\xi^2 - 1} \right] \tag{5.42}$$

and the damping factor ξ is

$$\xi = \frac{Rl}{2} \sqrt{\frac{C}{L}} \tag{5.43}$$

Also see equations (2.33), (2.38), and (2.39) for an alternative representation. If ξ is greater than one, the equivalent circuit is overdamped and has small inductance effects. The greater the value of ξ, the more accurate the RC model becomes. On the other hand, as ξ becomes less than one, the circuit becomes underdamped and the poles become complex and oscillations occur. In that case, inductance cannot be neglected. Therefore, we have the condition $l < 2/R\sqrt{L/C}$, under which inductance needs to be considered.

The second rule composed in (5.41), $t_r/2\sqrt{LC} < l$, is introduced to ensure the waveform agreement between the analytical solutions of the transmission line's characteristic impedance approximation and its RC approximation. Figure 5.39 shows these two approximations and the driving circuit considered. The characteristic impedance of an RLC transmission line looks like a resistance in series with a capacitance as defined in equations (4.216) and (4.217), rewritten:

$$Z_0 = R_0 + \frac{1}{j\omega C_0} \tag{5.44}$$

where

$$R_0 = \sqrt{\frac{L}{C}} \sqrt{\frac{1}{2} \left(\sqrt{1 + \left(\frac{R}{\omega L} \right)^2} + 1 \right)} \tag{5.45}$$

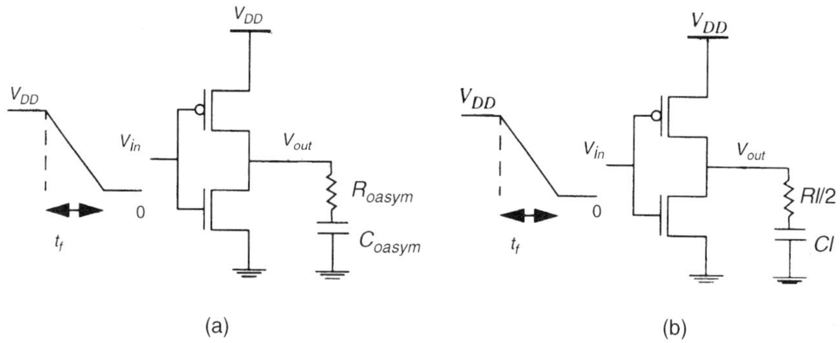

FIGURE 5.39 *(a) A CMOS inverter driving the equivalent characteristic impedance of an RLC transmission line. (b) A CMOS inverter driving an RC approximation of a transmission line.*

$$C_0 = \frac{1}{\omega\sqrt{\frac{L}{C}}\sqrt{\frac{1}{2}\left(\sqrt{1+\left(\frac{R}{\omega L}\right)^2}-1\right)}} \tag{5.46}$$

Both R_0 and C_0 monotonically decrease with the increase of frequency and saturate to the asymptotic values given by

$$R_{oasym} = \sqrt{\frac{L}{C}} \tag{5.47}$$

$$C_{oasym} = \frac{2\sqrt{LC}}{R} \tag{5.48}$$

when the frequency is beyond R/L. They are used in the circuit of Figure 5.39(a). The authors of [14] derived the analytical solutions for the outputs of both circuits based on two assumptions: the driving resistance of the inverter in Figure 5.39(a) matches or is greater than the impedance of the line (or $Z_{DRV} \geq Z_0$), and the driving MOS transistors are in saturation mode for most of the transition. By matching the two solutions, the authors found that the condition $t_r > 2\sqrt{LC}l$ is required. Therefore, we have the second part of the figures of merit for inductance to be important, which is the inverse of that condition. The $2\sqrt{LC}l$ in the condition is equal to twice the time of flight of the line or twice the time required by the electromagnetic wave to travel from one end of the line to the other.

Although the first match-or-underdrive assumption is valid for today's typical IC designs, it may not be true for future advanced high-speed designs. Overdrive may be necessary to minimize delays. Furthermore, in the derivation, the authors tried to use the asymptotic values of the line's characteristic impedance at very high frequency but, on the other hand, relied on this saturation assumption, meaning the inverter's output has to rise or fall very slowly.

The figure of merit in the paper is loose. Many RC-behaved lines pass the test. Figure 5.40 shows the simulation of our 2.5-μm-wide, 2000-μm-long, 8-wire structures with and without considering inductance. The drivers have 145 Ohm driving resistance (or Z_{DRV} = 145 Ohm), making the lines well underdriven. The 2000-μm length is much smaller than $2/R\sqrt{L/C}$. The input ramp time is 10 ps, which is much smaller than twice the time of flight delay[15]. According to the paper in [14], we would see significant inductance effects. But, it is not true in Figure 5.40. The maximal difference between the two waveforms is only 60 mV. From our simulation results, we observed that the inductance effects had little dependence on the driver's input transition times but were heavily dependent on

[15] The time of flight delay for this interconnect structure is 13.3 ps, which is equal to 2000u divided by the speed of EM wave. Since the dielectric constant is 3.9, the speed of EM wave is the speed of light divided by $\sqrt{3.9}$.

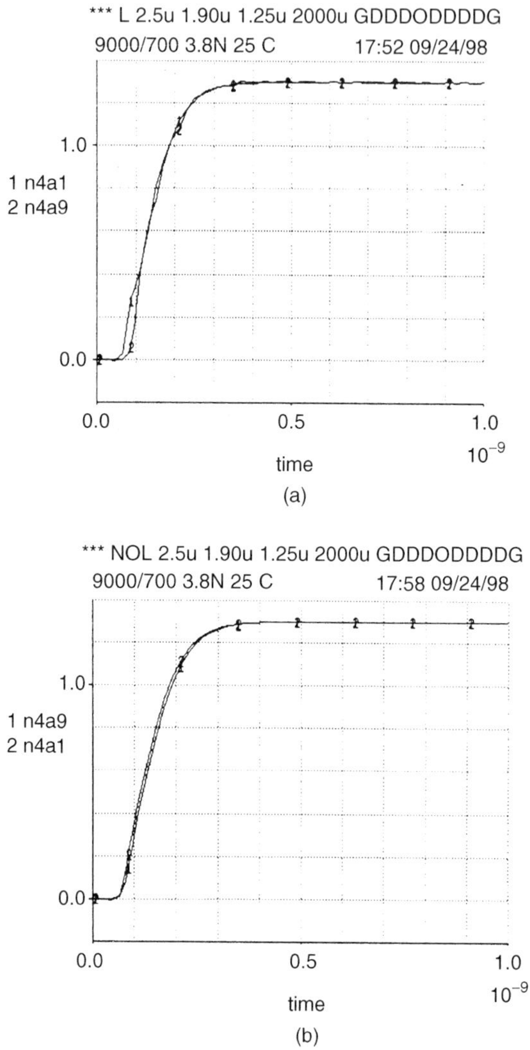

FIGURE 5.40 *Waveforms at the near and far ends (n4a1 and n4a9, respectively) of the fourth line of the 2.5 μm-wide, 2000 μm-long, 8-wire structure with and without considering inductance, (a) and (b), respectively. No other lines are switching.*

the driver's output transition times. A figure of merit should be related to the output transition times, which will be presented after we discuss the figures of merit introduced by Alina Deutsch in papers [15] and [16].

It is shown in [30] that the voltage on a long, uniform, lossy transmission line excited by a unit step $u(t)$ can be expressed as

$$v(l,t) = \left[e^{-(Rl/2Z_o)} + B(l,t)\right] u(t - l\sqrt{LC}) \tag{5.49}$$

where l is the length of the line and $B(l,t)$ is a slowly rising modified Bessel function [31]. For low resistance lines, Rl is small and the first term dominates. The line behaves like an LC circuit with fast-rising response As Rl exceeds Z_0, the second term starts dominating, and the rise time is slowed down, similar to the response of a slower RC circuit. Based on these relations and simulation results, Alina and other authors in papers [15] and [16] concluded that the error in prediction between RC and RLC representation will exceed 15% if

$$\begin{aligned} &C_L \ll Cl \\ &\frac{Rl}{2Z_0} \leq 1 \\ &Z_{DRV} < nZ_0 \end{aligned} \tag{5.50}$$

where n is between 0.5 and 1 and C_L is the loading at the far end of the line. The acceptable error limit will affect the number n. And the crosstalk prediction difference will be greater than 20% between RC and RLC modeling if

$$\begin{aligned} &C_L \ll Cl \\ &\frac{Rl}{2Z_0} \leq 1.5 \\ &Z_{DRV} < nZ_0 \end{aligned} \tag{5.51}$$

with n between 1 and 1.5. Therefore, they included the comparison between Z_{DRV} and Z_0 in addition to the figure of merit on length comparison between l and $2Z_0/R$ introduced previously.

Although these rules are important, they require the difficult calculation of the characteristic impedance Z_0. Because the penetrating nature of the magnetic field causes all on-chip nonorthogonal conductors to be magnetically coupled, the wire may not be modeled as a uniform RLC transmission line. Therefore, the solution of Z_0 is undetermined. We can only guess Z_0 to be a resistance between 50 and 75 Ohm. Furthermore, the value of n is dependent on the geometry, such as line width and line spacing, and the technology, such as the dielectric constant and resistivity of the metal This may also become a difficulty when applying the rules.

As discussed in section 5.1.2, the *significant frequency* f_s, defined as $0.34/t_r$ where t_r is the signal's ramp time at the near end of the line, is a very good representative frequency of a trapezoidal pulse signal. About 15% of the signal's frequency components lie above this frequency on the spectrum. Employing this frequency will determine a figure of merit. Here the ramp time is defined as the ratio of the time when the signal reaches its 80% of the pulse's final value minus the time when it reaches 20%, over 0.6 (or the signal's 20%-to-80% rise time divided by 0.6). The division of 0.6 is to determine the full-swing transition time and, due to that, the full-swing transition time is used in the definition of significant frequency in section 5.1.2. Inductance effects become important only when the inductive reactance, $j\omega Ll$, becomes a nonnegligible portion of the total reactance including driver's resistance, $(R + j\omega L)l + Z_{DRV}$. Therefore, as frequency gets higher and higher, the inductive reactance gets increasingly dominant, and larger and larger inductance effects show up. Here, R and L are the per-unit-length line-resistance and self-inductance, respectively. They are structure dependent and may be extracted by using a field-solver, such as Raphael R13 or MIT's FASTHENRY, at the estimated significant frequency.

From the simulations performed in the previous section, it was found that the results on a 0.8 μm-wide structure showed some inductance effects but they were minimal. Although the RC and RLC predictions on their crosstalk and delay results had about 20% difference, their absolute differences are small and, often, RC results are actually more conservative. Therefore, they are in the situation where either RLC or RC consideration is acceptable, depending on the design requirements. The structure's R and L are 36 Ohm/mm and 1.56 nH/mm, respectively. The ramp times at the driver's output for 2000 μm-long and 4000 μm-long structures are about 110 ps and 140 ps. The driver's resistance is 46 Ohm.

Based on the simulations performed and the fact that the inductive effects decrease with the increase of length confirmed by the first figure of merit introduced in [14], it was concluded that designers need to consider the inductance of an interconnect when

$$\begin{aligned} C_L &< \frac{1}{8}Cl \\ l &\leq \frac{2}{R}\sqrt{\frac{L}{C}} \\ 2\pi f_s Ll &> \frac{(Rl + Z_{DRV})}{2} \end{aligned} \tag{5.52}$$

Under these criteria, the inductive reactance occupies more than one-third of the total reactance, and the delay and crosstalk errors, without considering inductance, exceeded 25%. When the values of Cl and C_L were close, such as in the 1000 μm-long structures, the inductance effects (e.g., oscillations) tended to show up. Therefore, under the more stringent screening conditions

$$\begin{aligned} C_L &> \frac{1}{8}Cl \\ l &\leq \frac{2}{R}\sqrt{\frac{L}{C}} \\ 2\pi f_s Ll &> \frac{(Rl + Z_{DRV})}{4} \end{aligned} \tag{5.53}$$

the delay and crosstalk errors, without considering inductance, might exceed 25%.

The line resistance and self-inductance of a wire in the 2.5 μm-wide structure simulated in Figure 5.40 are 11 Ohm/mm and 1.42 nH/mm, respectively. The rise time t_r measured from the waveform at the near end of the fourth line is about 180 ps, giving $f_s = 0.34/180 \times 10^{-12} = 1.89$ GHz. Therefore, $2\pi f_s L =$ 16.8 Ohm/mm. Since l is 2mm and $Z_{DRV} = 145$ Ohm, the structure, whose Cl is greater than 8 C_L, fails the last condition of (5.52). Hence, the inductance for this structure does not need to be considered, which is proven by the waveforms shown in Figure 5.40.

It is difficult to estimate a priori the exact ramp rate at the near end of an RLC interconnect (or at the output of its driver). If we knew this ramp rate, the circuit including inductance would already have been simulated. That defeats the purpose of introducing the figures of merit. However, the ramp rate for the corresponding RC circuit may be easily estimated. That result may be used to determine the approximation of the significant frequency to be used in equation (5.53). The Asymptotic Waveform Evaluation (AWE) [32] and Effective Capacitance technique [33] may be used to quickly solve for the ramp times of an RC network.

5.3 ON-CHIP DESIGN SOLUTIONS TO COPE WITH INDUCTANCE EFFECTS

From section 5.2 we understand that the inductance effects, such as large oscillations and crosstalk noise, are mainly caused by oversized drivers and low-resistance interconnects. However, these solutions are used in order to achieve high performance. In this section, we are going to explore some on-chip design solutions to minimize the inductance effects in the advanced high-speed designs where oversized drivers and low-resistance wires are employed.

5.3.1 Dedicated Ground Wires

Although the partial inductance is determined only by the geometry, the loop inductance is determined not only by the geometry but also by where the current returns. Because the current has to return to the sources (the driver source or the noise source), what really matters to delays and crosstalk noise is the loop inductance. The smaller the loop inductance, the smaller the delay or noise. The example in Figure 5.41 shows a current loop composed of two parallel wires[16].

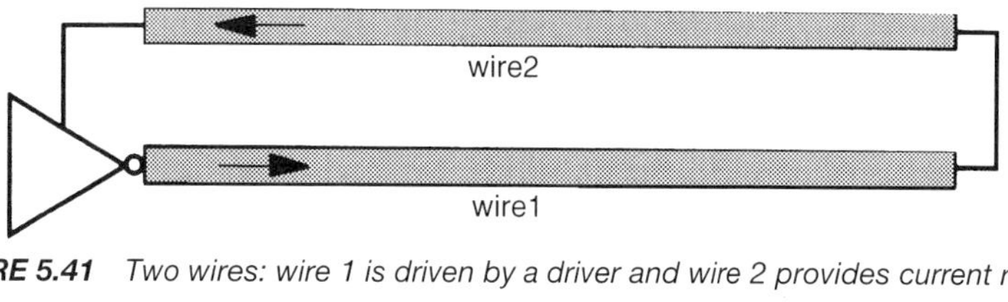

FIGURE 5.41 *Two wires: wire 1 is driven by a driver and wire 2 provides current return.*

[16] For the sake of simplicity, bypass capacitors between VDD and GND are not shown in Figure(5.41).

Wire 1 is driven by a driver and Wire 2 is the path for current return. The loop inductance can be calculated as follows:

$$L_{loop} = L_{11} - L_{12} + L_{22} - L_{21} \tag{5.54}$$

where L_{11} and L_{22} are self-partial inductances, and L_{12} and L_{21} are mutual partial inductances. Since the currents in the two wires flow in different directions, minus signs are in front of the mutual terms in (5.54). From these minus signs, we know that if the coupling between the signal wire and its return path is strong, the loop inductance is small; therefore, the return path should be as close as possible. It is also obvious from equation (5.54) that the loop inductance is small if each self-inductance is small, which can be achieved by having multiple return paths or by having wider return paths. Hence, the conclusion drawn from studying (5.54) is that we need to have as many return paths as possible and they should be as close as possible. In general, other signal wires may not be good return paths because they may switch in the same direction. Furthermore, the currents returning from signal wires have to go through devices, which may add a couple of hundreds Ohms of impedance. Therefore, the best return paths are nearby dedicated ground wires, such as those included in the structure of Figure 5.33.

Although we mentioned that wider return paths are desired due to their small self-inductance, the inductance does not decrease very fast with the increase of width. For our structures, the inductance decreases 10% by doubling the width. Using more dedicated ground wires is preferred. That is, if we have the wiring resource to afford twice the width of ground wires, we should still use the same width but add one more ground wire in between signal wires. Of course, the width cannot be too small; otherwise, the large line-resistance may overshadow its ground-return effect and the ground wire may not provide enough shielding for inductive crosstalks. The extreme case is that every signal wire is sandwiched (or guarded) by two ground wires. This approach actually has been used for clock lines and [34] in advanced high-speed designs because the clock lines have very stringent constraints on noise and oscillations. With the ground wires, capacitive crosstalk noise is also shielded and it is easy to control the parasitic environment, resulting in trustworthy delay predictions and skew estimations.

5.3.2 Differential Signals

In order to provide enough nearby return paths for fast signals, the differential signal technique has been used. The neighboring signals of a high-speed bus are arranged to switch in different directions. If designers know signals' probabilistic switching pattern, then the signals tending to switch in the same direction must be interleaved by signals tending to switch in the opposite direction. In some designs, differential signals are achieved by using redundant signals, in order to minimize the maximal number of bus signals switching in the same direction. For example, both a signal and its logic inversion (signal-bar) are used. However, we need to avoid placing the signal and its bar right next to each other, because doing so will slow down the signal when the capacitive couplings between them are strong. They should be placed with one or two other signals in between, which shields the capacitive coupling but still provides enough return paths through differential signals. From the worst-case discussion in section 5.2.2, we already know that the worst delay or crosstalk case happens when all aggressors are switching in the same direction. Therefore, employing differential signals may avoid this situation.

5.3.3 Buffer Insertion

Optimum buffer insertion is another effective technique to reduce inductive coupling (as well as delay and capacitive coupling) in an RC or RLC tree that can be formed by one or multiple sources and one or multiple sinks. The originally longer current return path is now shortened by guiding the current through the repeaters for a return, resulting in a smaller current loop and hence smaller inductive coupling.

In Figure 5.42, we illustrate the buffers evenly inserted into a uniform RLC line and the RLC in each segment. Since L is a nonscalable function of length, as mentioned in section 5.1, $L_{seg} = L_l$ when $n = 1$ and $L_{seg} = L_l/k^{(n/2)}$ when $n > 1$ instead of $L_{seg} = L_l/n$, where k is between 2 and 2.5, depending on the surrounding interconnect environment and its own length. When the total length is about a few thousand micrometers to 8 mm, k is approximately equal to 2.3. Therefore, when more buffers are inserted into the original RLC line, each segment has smaller L_{seg} as compared to L_l/n and therefore a smaller effect from inductance on propagation delay. The mutual coupling term, M_{seg}, has a similar dependence on length, $M_{seg} = M_l/k^{(n/2)}$ when $n > 1$, if there are coupled signal lines. Hence, with inserted buffers, inductive coupling decreases slightly faster than a linear rate. Each segment becomes more an RC line than an RLC line, because L per segment scales down at a rate faster than the R and C that are linearly scaled with length. However, when inductance effect has significant impact on delay and/or crosstalk, the RLC-induced crosstalk is higher than the RC-induced crosstalk before a certain length (we call it inductance-critical length) and smaller after the length as shown in Figure 5.44. Both RLC-induced and RC-induced crosstalk tend to saturate when the total interconnect length is long, and

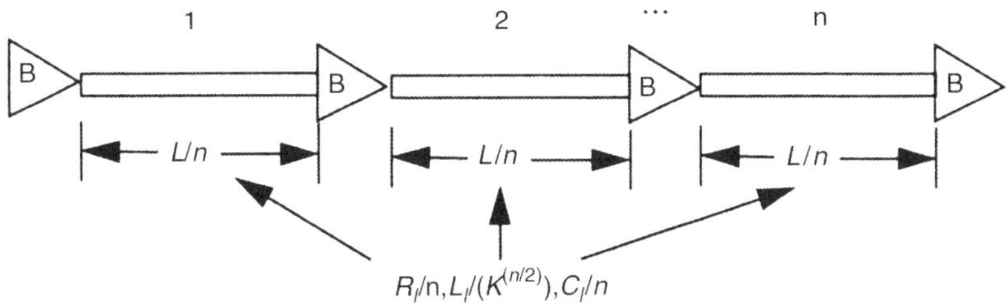

FIGURE 5.42 *Buffer insertion in an RLC line for the minimization of inductive coupling as well as capacitive coupling and propagation delay.*

in general the RLC-induced one is smaller than the RC-induced one at saturation. The basic reason for the above phenomenon is that the inductive crosstalk initially induced a larger opposite inductive voltage on the victim wire due to LdI/dt (with larger dI/dt component in the begining). The inductive crosstalk is in the opposite direction of the capacitive crosstalk under the assumption that all the aggressors switch in the same direction. Then the capacitive crosstalk (caused by CdV/dt) dominates on the victim signal with a larger dV/dt (due to victim signals pulled down by inductive crosstalk and therefore larger differential voltage with respect to the aggressors) compared to the RC-induced crosstalk alone. Therefore, the RLC-induced crosstalk is larger up to a certain length. However, due to faster charging during capacitive crosstalk (CdV/dt) in the RLC case, the slew rate of the signal is slower after the inductance-critical length and therefore the RLC-induced crosstalk has in general a smaller saturation voltage.

We illustrate this effect with the following example shown in Figure 5.43. Before repeater insertion, the original point-to-point (i.e., one source and one sink) length is 16,000 μm. If the signal length is broken up into N equal pieces with repeaters in between, the comparison of crosstalk between a single RLC and RC segment after repeater insertion is shown in Figure 5.44.

The comparison of the delay for these repeater-segmented lengths without the inclusion of buffer delays is shown in Figure 5.45.

With the same geometry configuration as in Figure 5.44, except the driver resistance decreases to about 40 Ohm using a larger driver (slightly in overdrive condition), we can see that the inductance-induced delay and crosstalk impact is sensitive to driver sizing, as shown in Figures 5.46 and 5.47.

Now if we keep the same driver, configuration, and switching patterns of Figures 5.46 and 5.47, except the width of the signals, Gnd, and Vdd becomes 5 μm, the impact of doubling the width is shown in Figures 5.48 and 5.49.

We can see that when the width is doubled, the impact of inductance on crosstalk and delay increases. The conclusion from the simulation runs above is that when the widths of the signals are relatively large (> 2 μm, for example) and the driver is strong (in overdrive situation), one would need to watch out for inductance effect on increased delay and crosstalk. In reality, these situations have to be simulated using RLC circuit formulation to determine the impact of inductance.

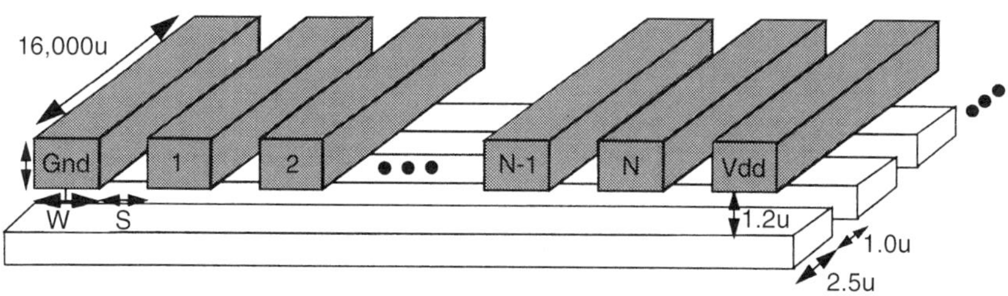

FIGURE 5.43 *An interconnect structure with N (N = 8) signal wires of 16,000 μ long, 1.9 μ thick, and 2.5 μ wide, sandwiched by Gnd and Vdd wires with the same thickness and width. The spacing between wires is S (S = 1.25 μ). The gap to the orthogonal layer below is 1.2 μ. The drivers of the signals have about 100 Ohm as source resistance.*

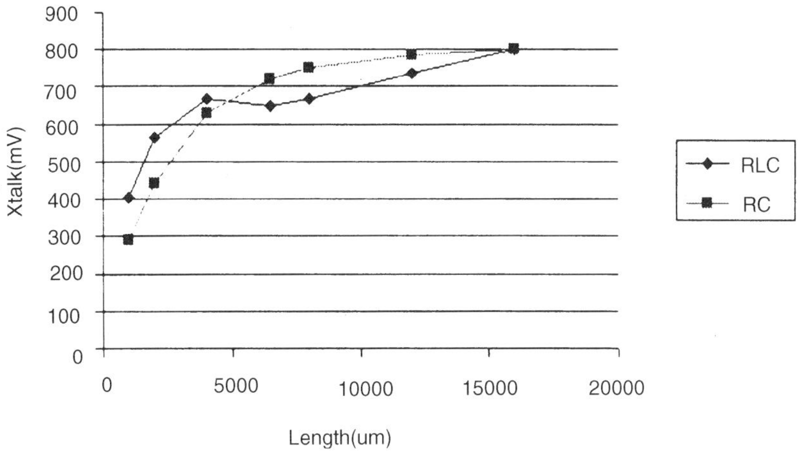

FIGURE 5.44 *Crosstalk versus length. The pattern of the signals in this case is GDDDODDDDV where G, D, O, and V represent ground (O V), transition from high to low (1.3 V to 0 V), staying at constant voltange (0 V in this case), and Vdd (1.3 V). RLC-induced cross-talk in absolute value is larger than the RC-induced when the segment length is less than 5000 μm in this case, and becomes smaller when the segment length is larger than 5000 μm.*

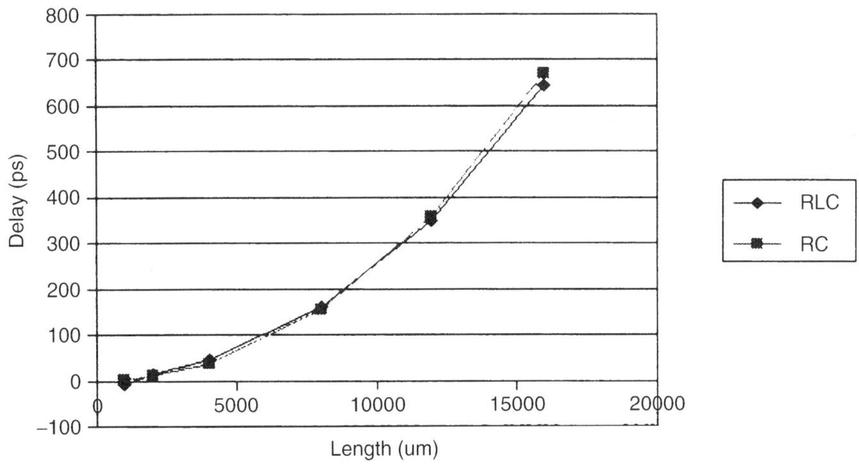

FIGURE 5.45 *Delay versus length. The pattern of the signals in this case is GDDDUDDDDV where G, D, U, and V represent ground (0 V), transition from high to low (1.3 V to 0 V), transition from low to high (0 V to 1.3 V), and Vdd (1.3 V). In this case, the difference of the delay between RLC and RC segments is small. The maximum difference is about 3.7% at 16,000 μm in length.*

5.3.4 Splitting Wires

The large oscillations caused by inductance effects prohibit us from using very wide wires. Although wire sizing for RC interconnects has been a successful technique to achieve higher performance [35], it may not be valid for future advanced high-speed designs where thick or low-resistive metals are used. In

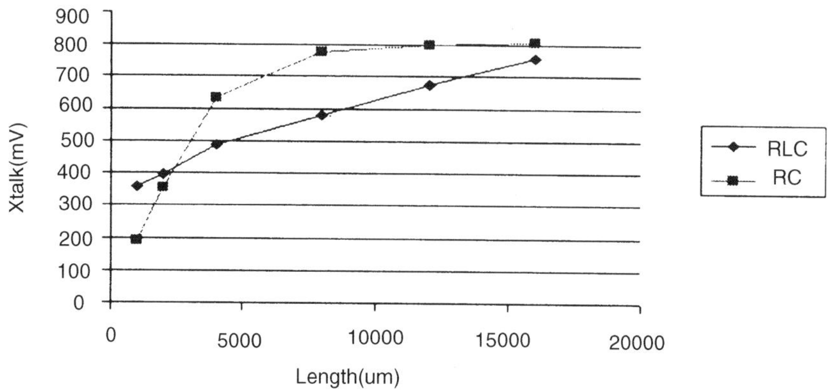

__FIGURE 5.46__ Crosstalk verus length. With the same geometry as in Figure 5.43 except a larger driver is used compared to a smaller one used in Figure 5.44. The pattern of the signals in this case is GDDDODDDDV where G, D, O, and V represent ground (0 V), transition from high to low (1.3 V to 0 V), staying at constant voltange (0 V in this case), and Vdd (1.3 V). RLC-induced cross-talk in absolute value is larger than the RC-induced crosstalk when the segment length is less than 2500 μm in this case and becomes smaller when the segment length is larger than 2500 μm.

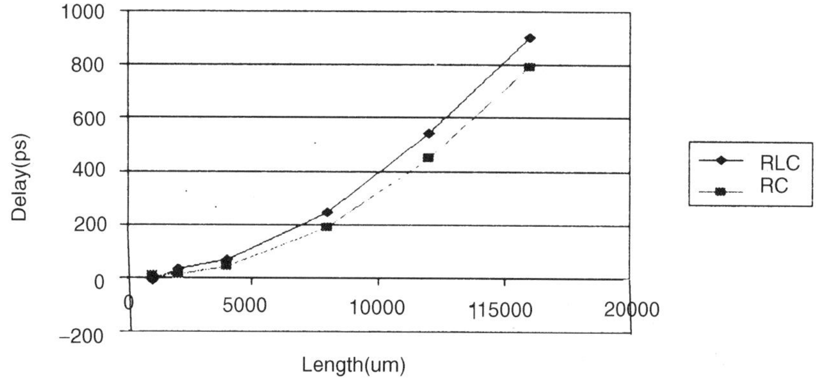

__FIGURE 5.47__ Delay versus length. With the same geometry as in Figure 5.43 except a larger driver is used compared to a smaller one used in Figure 5.44. The pattern of the signals in this case is GDDDODDDDV where G, D, O, and V represent ground (0 V), transition from high to low (1.3 V to 0 V), transition from low to high (0 V to 1.3 V), and Vdd (1.3 V). In this case, the difference of the delay between RLC and RC segments can be significant. The maximum difference is about 13.8% at 16,000 μm in length.

order to minimize the line reactances, wires wider than twice the skin depth (e.g., about 5 μm in our example structure), need to be split as shown in Figure 5.50(a) and (c). At high frequencies, the majority of the currents are conducted around the left and right corners of the wire (by assuming their return paths are on the same layer and their thickness is larger than twice the skin depth), resulting in about two-skin-depths conducting width. Wires wider than that do not achieve

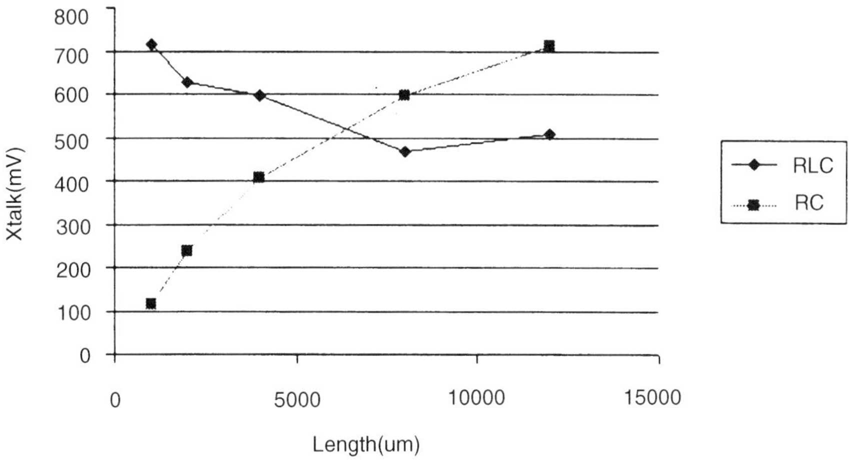

FIGURE 5.48 *Crosstalk versus length. The geometry pattern is the same as in Figure 5.46 except the width now is 5 μm. The pattern of the signals in this case is GDDDODDDDV where G, D, O, and V represent ground (0 V), transition from high to low (1.3 V to 0 V), staying at constant voltange (0 V in this case), and Vdd (1.3 V). RLC-induced cross-talk is larger in absolute value than the RC-induced crosstalk when the length is less than 6000 μm in this case and becomes smaller when the length is larger than 6000 μm. When the length is 1000 μm, the cross-talk is much larger for RLC circuit due to oscillation.*

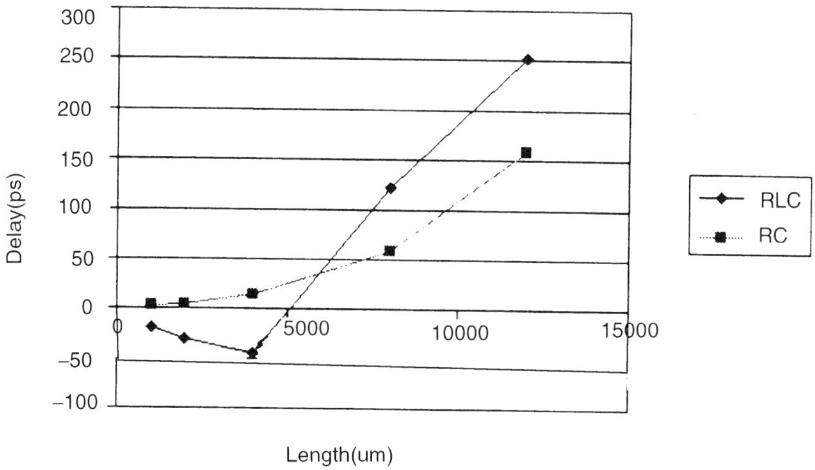

FIGURE 5.49 *Delay versus length. The geometry pattern is the same as in Figure 5.46 except the width now is 5 μ. The pattern of the signals in this case is GDDDUDDDDV where G, D, U, and V represent ground (0 V), transition from how to low (1.3 V to 0 V), transition from low to high (0 V to 1.3 V), and Vdd (1.3 V). The difference of the delay between RLC and RC segments is large. The maximum difference is about 57.8% at 12,000 μm in length. The negative delay under 5000 μm for the RLC segment is due to oscillation of the signals.*

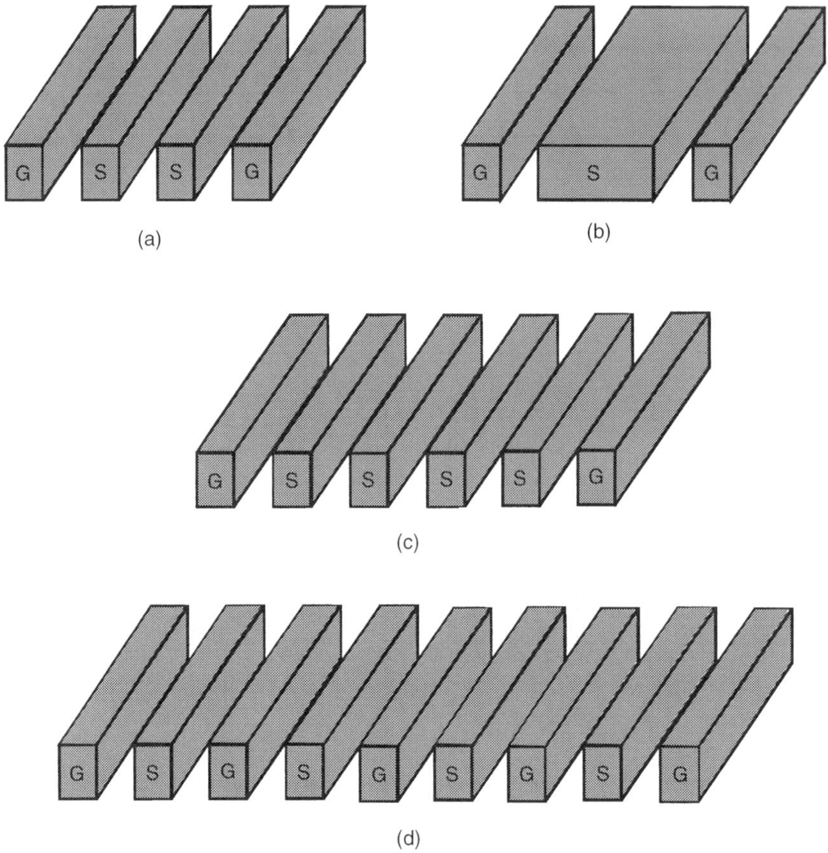

FIGURE 5.50 *(a) A signal wire has been split into two 2.5 μm-wide parallel wires. (b) A 7.5 μm-wide wire. (c) A signal wire has been split into four parallel wires. (d) A signal wire has been split into four parallel wires and, in between every two wires, a dedicated ground wire is inserted.*

much larger reduction on high-frequency resistance. Moreover, for thick wires, the low-frequency self-inductance does not decrease much with the increase of width. It only decreases about 10% with each doubling of the width. At high frequencies, the reduction on self-inductance will be even smaller due to the skin effect. Therefore, widening the wires does not seem to be effective in reducing line reactance. Furthermore, wider wires have stronger couplings between them, causing larger crosstalk noise and oscillation. On the other hand, splitting a wide wire into several, say N, parallel wires of about two- skin-depths width each may reduce the total reactance by a factor of N. That is a huge improvement. The width of each split wire is $(1/N)$th of the original width. The spacing between split wires also helps to reduce inductance.

We use the simulation of a 7.5 μm-wide and 2000 μm-long interconnect guarded by two 2.5 μm-wide ground wires as shown in Figure 5.50(b) to verify our conclusion. The simulation is compared with the simulation of two split wires of 2.5u width each and 2.5u spacing as shown in Figure 5.50(a). We use 2.5u width for the split wires to be more fair because there is a 2.5u spacing between them. In total, the wires and spacing occupy a 7.5 μm-wide channel. The split wires are guarded by two 2.5 μm-wide ground wires, too. The simulation results show that the delay of the 7.5 μm-wide wire is smaller by only 0.6 ps (out of 20.6 ps interconnect delay) and its 20%–80% rise-time is smaller by only 1 ps (out of 20.96 ps). However, the 7.5 μm-wide wire has the overshoot of 1.498 V (with 1.3 V Vdd) while the two 2.5 μm-wide split wires have the overshoot of 1.40 V. This proves that wire-splitting technique results in smaller inductance effects.

Combining this wire-splitting technique and the dedicated-ground-wire approach introduced earlier, we may use the interconnect structure as shown in Figure 5.50(d), where each split wire is guarded by ground wires. This technique may be used for noise-critical signals and clock signals. An interesting point to investigate is whether Figure 5.50(d) is faster than Figure 5.50(c) when the number of split wires is small. Figure 5.50(d) is definitely better in terms of noise immunity but it uses almost twice the wiring resource. Because of the skin effect, people might guess that there is not much current flowing in the center wires of Figure 5.50(c). On the other hand, Figure 5.50(c) might have the advantage of less capacitive loading at the center wires and, hence, is faster. We would like to find out whether either one of these possibilities is true.

We simulated two structures, as shown in Figure 5.50(c) and (d). An original 20-μm-wide signal wire is split into four parallel wires, each being 5.0 μm-wide, in the two structures. They are all 4000 μm-long. The ground wire width and the spacings are 8.0 μm and 2.5 μm, respectively for both. The metal thickness is 1.9 μm. All split signal wires are driven by CMOS drivers with impedance $Z_{DRV} = 145$ Ohm and 80 ps input rise time. Their SPICE simulation results showed that (d) indeed has smaller oscillations but is 24 ps slower than (c). About 45% of the total current flows through the center two wires of (c), contradicting people's intuition that there is much less current flowing in the center conductors. After considering (c)'s improvement on delay and almost twice the efficient usage of routing resource, it is much better. However, as the number of split wires increases, the performance of (c) is expected to decrease due to insufficient ground return. Therefore, it is advised to always use simulation to identify the best approach to adopt. Because of the efficient usage of routing resource of (c), designers can afford to use wider ground wires, which provides better crosstalk shielding.

5.3.5 Terminations

Termination technique has been used widely in PCB designs to minimize reflections [36] at the terminals of transmission lines. The technique is often called *matching* because terminations are chosen to match the complex-conjugate of the transmission line's characteristic impedance. Circuit theory [37] tells us that under this condition, the reflection coefficient at the matching node is zero. Although on-chip wires may not be treated as ideal transmission lines and their characteristic impedances are not well defined, we would still like to apply the termination technique to on-chip interconnects to investigate whether they may reduce inductance effects[17]. The purpose of the termination is *to reduce the coupling noise at the inputs of victim receivers with minimal sacrifice on performance.*

There are four different kinds of terminations available for on-chip designs, namely Shunt-RC termination, Series-R shunt-C termination, Series-R termination, and Diode termination, as shown in Figure 5.51. We did not show Shunt-R termination that uses a resistor (or resistors) connected between the terminal and power source (or both ground and vdd), because it will incur a power consumption problem due to constant current flowing. Furthermore, the terminations should not be inserted at the near end of the interconnect (or at driver's output) because they greatly slow down the transition, which defeats the purpose of termination.

Diode termination is not effective because with Vdd being 1.3 V, half Vdd is smaller than the turn-on voltage of a diode. That means the far-end noise may still be higher than half Vdd, enough to faultily trigger the receiver, even with diodes clipping the noise voltage. Although the clipping level may be adjusted by connecting the other end of the upper or lower diode to some voltage other than Vdd or ground, generating those voltages will increase power consumption

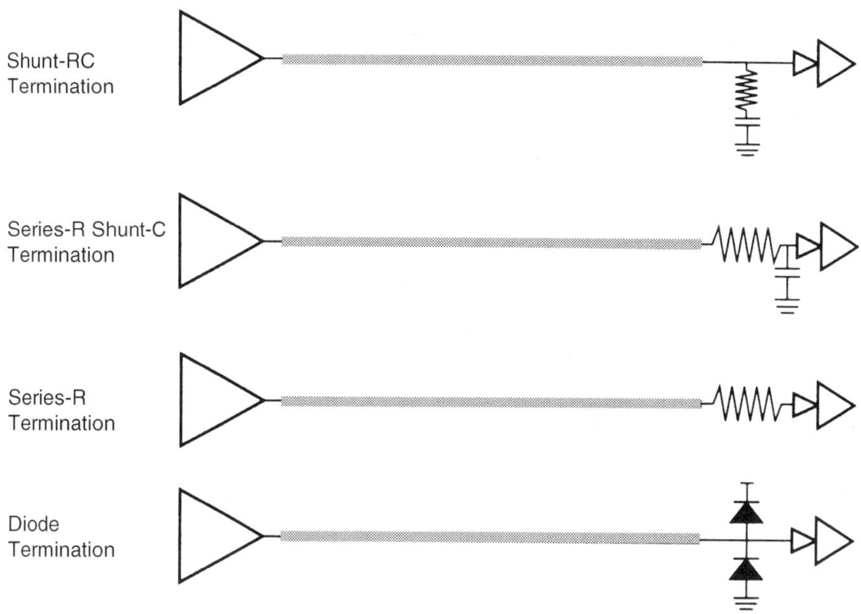

FIGURE 5.51 *Four different kinds of terminations for on-chip designs.*

[17] Because on-chip wires are not transmission lines, "matching" is not a correct term for the technique. It should be just denoted by termination.

and incur design difficulties. Therefore, for future deep-submicron designs, Diode termination is not preferred.

To determine the performance of each kind of termination, we ran many SPICE simulations of the 5 μm-wide, 2000 μm-long, 8-wire structures in Figure 5.33 with different terminations and different R or C. The worst-case aggressors' input switching directions are used in delay and crosstalk simulations and the ramp times at drivers' inputs are about 80 ps.

Shunt-RC Termination Table 5.7 summarizes the noise and delay results after connecting different Shunt-RC terminations. Both the maximum noise and performance sacrifice decrease with the increase of R and the decrease of C. However, we cannot completely eliminate C; otherwise, it becomes Shunt-R termination. With R = 190 Ohm and C = 0.1 pF, the noise is decreased by 75 mV, or 13%, while the wire delay is increased by only 5.5 ps, or 23%. The resistance of termination is also a design constraint. It is normally achieved by using poly line connecting to the device's gate. The poly resistance of 0.25u technology is about 6 Ohm per square. Therefore, 190 Ohm can be achieved by a 19 μm-long, 0.6 μm-wide poly line, which is reasonable. Resistance much larger than that will be difficult to insert.

Series-R Shunt-C Termination Table 5.8 summarizes the noise and delay results after connecting different Series-R Shunt-C terminations. Both the maximum noise and performance sacrifice decrease with the increase of R and the decrease of C, similar to Shunt-RC termination. Compared with Shunt-RC termination, it has better noise reduction but larger performance sacrifice. With R = 190 Ohm and C = 0.1 pF, the noise is decreased by 334 mV, or 60%, while the wire delay is increased by 24 ps, or almost 200%.

Series-R Termination Table 5.9 summarizes the noise and delay results after connecting different Series-R terminations and different drivers. Both the maximum noise and performance sacrifice decrease with the increase of R. Its noise

TABLE 5.7 Shunt-RC Termination

R	C	Max. Noise	Wire Delay
none	none	552 mV	21.6 ps
90	0.1 pF	542 mV	28.7 ps
90	0.4 pF	637 mV	34.1 ps
90	0.8 pF	647 mV	37.1 ps
140	0.1 pF	513 mV	28.1 ps
140	0.4 pF	528 mV	30.0 ps
140	0.8 pF	540 mV	31.1 ps
190	**0.1 pF**	**477 mV**	**27.1 ps**
190	0.4 pF	514 mV	27.9 ps
190	0.8 pF	520 mV	28.5 ps

TABLE 5.8 Series-R Shunt-C Termination

R	C	Max. Noise	Wire Delay
none	none	552 mV	21.6 ps
90	0.1 pF	457 mV	38.7 ps
90	0.4 pF	372 mV	65.1 ps
90	0.8 pF	307 mV	97.1 ps
140	0.1 pF	291 mV	39.3 ps
140	0.4 pF	246 mV	30.0 ps
140	0.8 pF	172 mV	117 ps
190	**0.1 pF**	**218 mV**	**45.3 ps**
190	0.4 pF	165 mV	85.9 ps
190	0.8 pF	111 mV	140 ps

TABLE 5.9 Series-R Termination.

Z_{DRV}	R	Max. Noise	Wire Delay
70	none	552 mV	21.6 ps
70	40	470 mV	22.4 ps
70	90	397 mV	23.6 ps
70	140	376 mV	24.3 ps
70	**190**	**350 mV**	**26.1 ps**
150	none	376 mV	18.9 ps
150	40	339 'mV	19.6 ps
150	90	321 mV	20.3 ps
150	140	308 mV	20.9 ps
150	**190**	**295 mV**	**21.5 ps**

reduction is better than Shunt-RC termination and worse than Series-R Shunt-C termination, and its performance sacrifice is the smallest. Therefore, it is preferred in high-speed designs. With R = 190 Ohm and 70 Ohm driver, the noise is decreased by 200 mV, or 36%, while the wire delay is increased by only 3.5 ps, or only 16%. With R = 190 Ohm and 150 Ohm driver, the noise is decreased by 80 mV, or 21%, while the wire delay is increased by only 2.6 ps or only 14%.

With the significant reduction and minimal increase of the delay, Series-R termination technique is very effective in controlling inductance and should be considered in advanced high-speed designs. Since only around 20 μm-long poly line needs to be used at each critical signal's receiver, it is easy to implement.

5.3.6 Continuous Power/Ground Planes

Inductive coupling can be greatly reduced if continuous power/ground planes are employed on-chip, as in Figure 5.52 [23]. This is due to the fact that the image current can flow in the opposite direction on the power/ground planes directly underneath the current flows on the signal lines. This greatly reduces the

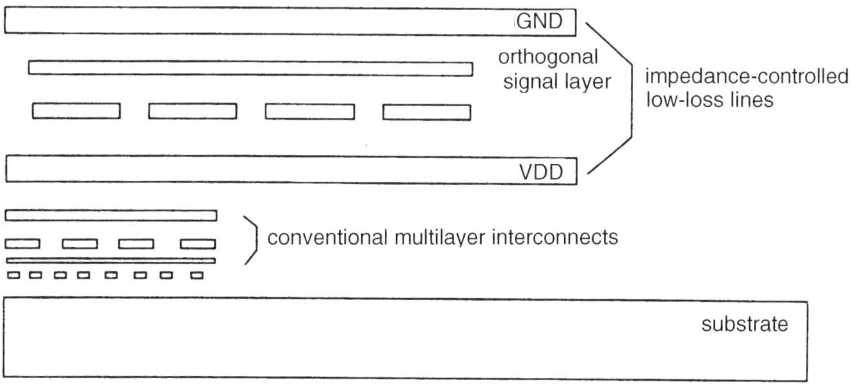

FIGURE 5.52 *Continuous power/ground planes on-chip provide impedance-controlled low-loss signal lines.*

excessive inductive coupling that occurs when signals travel a long distance for current return without planes nearby. Also, this technique is often used in printed circuit boards or multichip modules (MCM) as dual-signal stripline configurations in an impedance-controlled low-loss line environment. The dual-signal lines are orthogonally routed signals. Therefore, the signals have a tight control on propagation delay and crosstalk, because the signals are referenced to the surrounding planes, not the signals in the other layers. Some modern CPU designs have already used continuous power/ground planes for power delivery and reduced coupling purposes [21].

A detailed design of the impedance-controlled low-loss lines of a chip carrier for intrachip as well as interchip signaling is described in [20]. Nevertheless, a more economical way of including impedance-controlled low-loss lines should be to place them right on the interconnects of the chip, but not use multichip modules (MCMs) or printed-circuit boards (PCBs). A design including impedance-controlled and relatively low-loss lines on the top metal layer is shown in Figure 5.52 [23].

An example of how effectively nearby ground planes can cut down the inductive coupling and improve delay was shown earlier in Figure 5.3.

5.4 SUMMARY

In this chapter, we discuss the difference between on-chip inductance consideration and off-chip inductance consideration. The on-chip inductance is not scalable and, due to the lack of a ground plane, nonorthogonal wires are almost all inductively coupled. Therefore, the on-chip interconnects may not be treated as ideal transmission lines. We show how to calculate inductance. We mention that partial inductance should be preferred and let the circuit simulator determine the return path. We discuss the skin effect and the proximity effect. We show the inductance impact on delay and crosstalk noise for several different interconnect structures, which are mainly caused by overdrive and low resistance. The simulation results lead to the answer to the question, "When do we need to consider inductance?" Most importantly, we present several important design solutions to cope with inductance effects. We show that, based on the correct understanding of on-chip inductance, the inductive coupling noise and oscillation caused by overshoot may be effectively controlled in future multi-GHz chips.

BIBLIOGRAPHY

[1] J. Silberman, N. Aoki, D. Boerstler, S. Dhong, A. Essbaum, U. Ghoshal, D. Heidel, P. Hofstee, K. Lee, D. Meltzer, H. Ngo, K. Nowka, S. Posluszny, O. Takahashi, I. Vo, and B. Zoric, A 1.0 GHz single-Isue 64b PowerPC integer processor, *Proc. ISSCC-98*, San Francisco, pp. 230–231, Feb. 5–7, 1998.

[2] Frederick W. Grover, *Inductance Calculations Working Formulas and Tables*, Dover Publications, 1946.

[3] New dual inlaid copper interconnect, Motorola Press, http://motserv.indirect.com/press/html/ PR970930A.html, Sep. 30, 1997.

[4] Zhijiang He, Mustafa Celik, and Lawrence T. Pileggi, SPIE: Sparse Partial Inductance Extraction, *Proc. 34th Design Automation Conference*, pp. 137–140, June. 1997.

[5] B. Krauter, and L. Pileggi, Generating sparse partial inductance matrices with guaranteed stability, *Proc. IEEE International Conference on Computer Aided Design*, pp. 45–52, Nov. 1995.

[6] J. A. Brandao Faria, *Multiconductor Transmission-Line Structures: Modal Analysis Technique*, Wiley Series in Microwave and Optical Engineering, Nov. 1993.

[7] Albert E. Ruehli, Inductance calculations in a complex integrated circuit environment, *IBM J. Reseach and Development*, pp. 470–481, Sept. 1972.

[8] Pierce A. Brennan, Norman Raver, and Albert E. Ruehli, Three-dimensional inductance computations with partial element equivalent circuits, *IBM J. Research and Development*, 23 (6): 661–668, Nov. 1979.

[9] W. T. Weeks, L. L. Wu, M. F. McAllister, and A. Singh, Resistive and inductive skin effect in rectangular conductors, *IBM J. Reseach and Development*, 23 (6): 652–660, Nov. 1979.

[10] Peter K. Wolff, and Albert E. Ruehli, Inductance computations for complex three-dimensional geometries, *Proc. IEEE Int. Symp. on Circuits and Systems*, pp. 16–19, Apr. 1981.

[11] Simon Ramo, John R. Whinnery, and Theodore Van Duzer, *Fields and Waves in Communication Electronics*, 3rd ed, Wiley, 1994.

[12] M. Kamon, M. J. Ttsuk, and J. White, FASTHENRY: A multipole-accelerated 3-D inductance extraction program, *IEEE Trans. Microwave Theory Tech.*, 42 (9), Sept. 1994.

[13] Bryon Krauter, and Sharad Mehrotra, Layout based frequency dependent inductance and resistance extraction for on-chip interconnect timing analysis, *Proc. 35th Design Automation Conference, San Francisco*, pp. 303–308, June 1998.

[14] Yehea I. Ismail, Eby G. Friedman, and Jose L. Neves, Figures of merit to characterize the importance of on-chip inductance, *Proc. 35th Design Automation Conference*, San Francisco, pp. 560–565, June 1998.

[15] A. Deutsch, G. V. Kopcsay, P. J. Restle, H. H. Smith, G. Katopis, W. D. Becker, P. W. Coteus, C. W. Surovic, R. H. Dennard, G. A. Sai-Halasz, B. L. Krauter, and D. R. Knebel, When are transmission-line effects important for on-chip interconnects?, *IEEE Trans. Microwave Theory Tech.*, 45; 1836–1846, Oct. 1997.

[16] A. Deutsch, H. Smith, G. A. Katopis, W. D. Becker, P. W. Goteus, C. W. Surovic, G. V. Kopcsay, B. J. Rubin, R. P. Dunne, T. Gallo, D. R. Knebel, B. L. Karuter, L. M. Terman, G. A. Sai-Halasz, and P. J. Restle, The importance of inductance and inductive coupling for on-chip wiring, *Proc. IEEE 6-th Electrical Performance of Electronic Packaging*, pp. 53–56, Oct. 1997.

[17] A. C. Cangellaris, J. L. Prince, and O. A. Palusinski, Real inductance calculation for high-speed interconnect systems, *Proc. 8th IEPS*, Dallas, pp. 596–600, Nov. 7–10, 1988.

[18] R. Faraji-Dana, and Y. L. Chow, The current distribution and AC resistance of a microstrip structure, *IEEE Trans. Microwave Theory Tech.*, 38; 1268–1277, Sept. 1990.

[19] G. I. Costache, Finite element method applied to skin effect problems in strip transmission lines, *IEEE Trans. Microwave Theory Tech.*, MTT-35; 1009–1013, Oct. 1987.

[20] E. E. Davidson, B. D. McCredie, and W. V. Vilkelis, Long lossy lines (L^3) and their impact upon chip performance, *IEEE Trans. Comp. Packaging, Manuf. Technol. Part B*, 20 (4): 361–375, Nov. 1997.

[21] B. A. Gieseke, et al., A 600 MHz superscalar RISC microprocessor with out-of-order execution, *ISSCC97*, pp. 176–178, Feb. 1997.

[22] O. K. Kwon and R. F. W. Pease, Closely packed microstrip lines as very high-speed chip-to-chip interconnects, *Proc. IEMT*, San Francisco, pp. 34–39, Sept. 1986.

[23] K. Lee, On-chip interconnects: GigaHertz and beyond, *International Interconnect Conference*, 1998.

[24] B. D. McCredie, M. S. Veatch, and H. Merkelo, Issues and results in modeling dielectrically inhomogeneous propagation channels for high-speed digital applications, *Proc. 10th IEPS*, Marlborough, MA, pp. 719–755, Sept. 10–12,1990.

[25] L. P. Vakanas, A. C. Cangellaris, and J. L. Prince, A parametric study of the attenuation constant of lossy microstrip lines, *IEEE Trans. Microwave Theory Techn.*, 38; 1136–1139, Aug. 1990.

[26] P. Waldow and I. Wolff, The skin effect at high frequency, *IEEE Trans. Microwave Theory Tech.*, MTT-33; 1076–1082, Oct. 1985.

[27] W. T. Weeks, L. L. Wu, M. F. McAllister, and A. Singh, Resistive and inductive skin effect in rectangular conductors, *IBM J. Res. Develop.*, 23 no. 6: pp. 652–660, Nov. 1979.

[28] H. A. Wheeler, Formulas for the skin effect,*Proc. IRE*, pp. 412–424, Sept. 1942.

[29] R. B. Wu and J. C. Yang, Boundary integral equation formulation of skin effect problems in multiconductor transmission lines, *IEEE Trans. Magn.*, 25; 3013–3015, 1989.

[30] A. Deutsch, G. V. Kopcsay, V. A. Ranieri, J. K. Cataldo, E. A. Galligan, W. S. Graham, R. P. McGouey, S. L. Nunes, J. R. Paraszczak, J. J. Risko, R. J. Serino, D. Y. Shih, and J. S. Wilczynski, High-speed signal propagation on lossy transmission lines, *IBM J. Res. Develop.*, 34; 601–615, 1990.

[31] R. E. Collin, *Field Theory of Guided Waves*, New York; McGraw-Hill, pp. 124–125, 1960.

[32] L. T. Pillage and R. A. Rohrer, Asymptotic waveform evaluation for timing analysis, *IEEE Trans. Computer-Aided Design*, 9; 352–366, 1990.

[33] F. Dartu, N. Menezes, and L. T. Pileggi, Performance computation for precharacterized CMOS Gates with RC Loads, *IEEE Trans. Computer-Aided Design*, 15; 544–553, May 1996.

[34] Y. Massoud, S. Majors, T. Bustami, and J. White, Layout techniques for minimizing on-chip interconnect self inductance, *Proc. 35th Design Automation Conference*, pp. 566–571, June 1998.

[35] Jason Cong, and Lei He, An efficient approach to simultaneous transistor and interconnect sizing, *Proc. International Conference on Computer-Aided Design*, pp. 181–186, Nov. 1996.

[36] Mark I. Montrose, Printed circuit board design technique for EMC compliance, *IEEE Press Series on Electronics Technology*, 1995.

[37] R. Plonsey, and R. E. Collin, *Principles and Application of Electromagnetic Fields*, New York; McGraw-Hill, 1961.

[38] W. Wang, J. Fang, and Y. Chen, Efficient and accurate extraction of frequency-dependant resistance and inductance parameters of interconnects, submitted to *1999 Design Automation Conference*.

[39] Private communication Jiayuan Fang, 1999.

[40] J. Rubinstein, Jr., P. Penfield, and M. A. Horowitz, Signal delay in RC tree networks, *IEEE Trans. on Computer-Aided Design of ICAS*, 2; 202–212, Jul. 1983.

[41] T. Sakurai, Approximation of wiring delay in MOSFET LSI, *IEEE J. Of Solid-State Circuits*, 18; 418–426, Aug. 1983.

[42] S. Nakagawa, D. Sylvester, J. G. McBride, and S. Y. Oh, On-chip crosstalk noise model for deep-submicrometer ULSI interconnect, *HP Journal*, pp. 39–45, Aug. 1998.

[43] B. Troyanovsky, N. Chang, and D. Dowell, Integration of transient S-parameter simulation into HPSPICE, HPL-95-61, available at http://www.hpl.hp.com.

[44] N. Chang, L. Barford, and B. Troyanovsky, Fast time-domain simulation with frequency-domain data, *Proc. 47th Electronic Component Conference*, May 1997.

[45] N. Chang, V. Kanevsky, O. S. Nakagawa, K. Rahmat, and S. Y. Oh, Fast generation of statistically based worst-case modeling of on-chip interconnect, *International Conference on Computer Design*, Oct. 1997.

[46] P. Dowell, Effects of eddy currents in transformer windings, *Proc. of IEE*, 12, pp. 1387–1394, 1966.

[47] Raphael User Manual of Avant! Corp., 1998.

[48] HSPICE User Manual of Avant! Corp., 1998.

6

Synthesis: Overview and Static Topology Optimization

6.1 INTRODUCTION

Up to this point, this book has largely focused on analysis techniques for given interconnect structures. These techniques provide tools for assessing whether a design meets various interconnect-oriented design criteria: timing, noise immunity, and so forth. It is natural to ask what kind of CAD techniques can be used earlier in the design process to minimize the number of problems (timing or otherwise) that the analysis phase uncovers.

There are many degrees of freedom one might exploit during the design process to address interconnect-related issues. Among these are wire sizing, topology construction and buffer (repeater) insertion. Such synthesis techniques are the subject of this and subsequent chapters.

The focus throughout is on timing optimization with a strong emphasis placed on synthesis procedures and formulations that are of greatest value in practice. For example, there is almost always a cost (area, total switching capacitance) versus performance trade-off for these kinds of techniques and this trade-off must be explicitly considered to produce useful engineering solutions.

This chapter is presented in two parts: first, an overview of issues involved in interconnect synthesis and optimization is presented; critical issues such as optimization techniques, delay estimation, and relative merits of various problem formulations are discussed (the themes and issues introduced here will recur throughout this and subsequent chapters); second, algorithms for optimization of static routing topologies by buffer insertion and wire sizing. This presentation serves the dual purposes of detailing fundamental algorithms in interconnect synthesis and also introducing some of the algorithmic concepts that will be utilized in subsequent chapters for more complex optimization problems.

6.2 OVERVIEW OF INTERCONNECT SYNTHESIS

In interconnect synthesis and optimization, it is important to keep in mind system-level objectives. The performance (e.g., clock period) of digital systems is determined by the delay of signal paths through combinational logic from the outputs of memory elements to inputs of memory elements. The delay of such paths is determined by both logic gates and the interconnect between those gates.

It is important to keep this scenario in mind so as to ensure that the algorithms we design actually solve useful practical problems. As such, it is useful to consider interconnect synthesis in terms of three interrelated components: *delay estimators, design space* (e.g., the availability of buffers or the ability to determine the routing topology), and *problem formulations*.

6.2.1 Delay Estimators

Clearly, if optimization is performed using a grossly inaccurate delay estimation, the result is likely to be of little use. On the other hand, the delay estimator should be sufficiently easy to compute and well-behaved enough to enable the design of effective and practical synthesis algorithms. Striking an appropriate balance between these two issues is at the core of interconnect synthesis.[1]

While reseach in this area continues, the Elmore delay model introduced in section 4.5.8 appears to strike a good balance between these two objectives. Most of the algorithms presented herein adopt Elmore [5].[2]

We review the Elmore formulas here. For a given wire segment (u,v) of length l and width w (and fixed height), simple formulas for its distributed capacitance and resistance are

$$c = \alpha l \cdot w \qquad r = \beta l / w \tag{6.1}$$

where α and β are given technology constants.[3]

Under Elmore, the delay of a wire segment (u,v) is estimated by

$$r_{uv}\left(\frac{c_{uv}}{2} + c(T_v)\right) \tag{6.2}$$

[1] Further, since many optimization techniques must be carried out without complete information (e.g., detailed routing), complete accuracy is often unachievable.

[2] Section 6.2.5 gives a more in-depth discussion of the impact of error in delay estimators on the synthesis process as well as discussion of the advantages and disadvantages of Elmore itself and suggests that Elmore-based algorithms can often provide the foundation for algorithms incorporating more accurate estimators.

[3] The formula for capacitance accounts for area capacitance. A better measure includes fringe capacitance, which is independent of wire width: $c = \alpha l \cdot w + \gamma l$ where γ is the appropriate fringing coefficient. For ease of presentation, we utilize the simple formula throughout the book; however, the algorithms presented do not, in general, depend on this assumption.

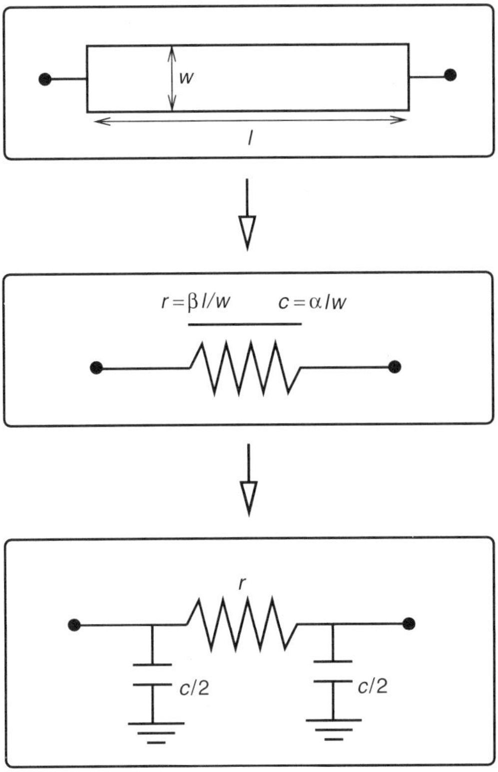

FIGURE 6.1 Wire model.

where r_{uv} and c_{uv} are respectively the resistance and capacitance of segment (u,v) and $c(T_v)$ is the total downstream capacitance at node v. The resulting wire model is illustrated in Figure 6.1.

A switch-level model for a signal buffer (e.g., an inverter) is shown in Figure 6.2; we can see that the delay of a buffer b of size w is determined by its output resistance $r_b = r0/w$ (where $r0$ is the on-resistance of a unit-sized buffer) and the total load on its output. In practice, typical cell libraries characterize a buffer b by three parameters: its output resistance r_b, its intrinsic delay d_b, and its input capacitance c_b. The intrinsic delay is defined to be the propagation delay when there is no load on the output. In this way, the effect of drain capacitance is usually absorbed into the intrinsic delay and also allows for some device properties neglected by the switch-level model to be accounted for in d_b. The resulting basic buffer delay formula is then

$$d_b + r_b \cdot c_L \tag{6.3}$$

where c_L is the total capacitive load driven by the buffer. More general and accurate models are frequently used; such a model incorporating the effect of signal slew on buffer delay is examined later in this chapter.

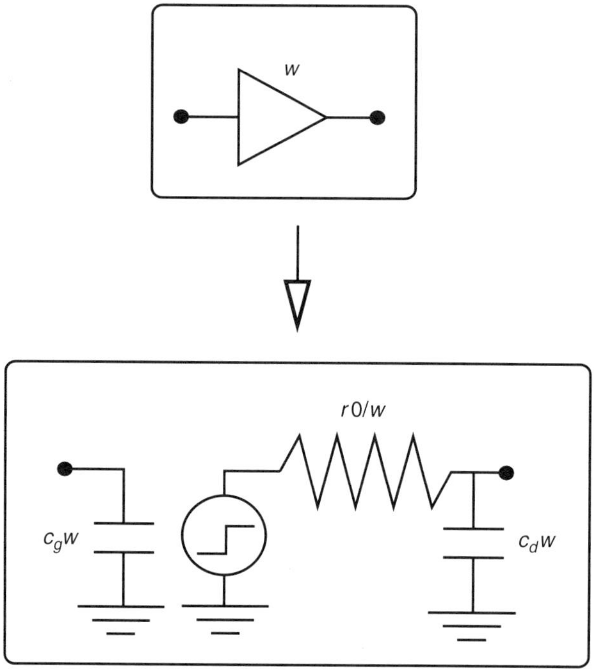

FIGURE 6.2 *Buffer model. The buffer size is given by w; r0 is the output resistance of a unit-sized buffer; c_g and c_d are the gate and drain capacitance of unit-sized buffers.*

6.2.2 Design Space

Many design decisions affect interconnect performance. A synthesis algorithm attempts to exploit these degrees of freedom to effectively address interconnect and more global issues. The particular design choices available may depend on the stage in the design flow and appropriate synthesis problems result. This section gives an overview of some of the more popular techniques available for net-level synthesis and optimization. We will usually focus on discrete design decisions rather than continuous. See section 6.4 for further discussion of this issue.

Buffer Insertion The decoupling effect of a buffer, as illustrated in Figure 6.2, is what makes buffer insertion a powerful tool for optimizing *RC* delay. Buffer insertion can optimize delay in two ways: (1) by partitioning long wires (e.g., acting as signal repeater) and (2) by decoupling off-path capacitance.

The first is illustrated in Figure 6.3. For simplicity, assume that terminal B is the input of yet another buffer. The total delay from A to B in the unbuffered case is

$$D_0 = d_b + r_b(\alpha l + c_b) + \beta l \left(\frac{\alpha l}{2} + c_b \right)$$

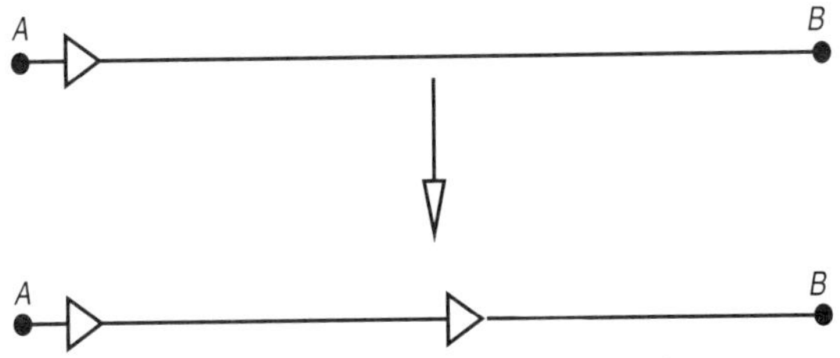

FIGURE 6.3 *Optimizing delay of long wires via buffer insertion.*

In the buffered case we have

$$D_1 = 2\left(d_b + r_b(\alpha l/2 + c_b) + \beta l\left(\frac{\alpha l}{4} + c_b\right)\right)$$

We leave it as an exercise for the reader to confirm that, under these circumstances, the optimal insertion point is indeed at $l/2$. It is also easy to derive that $D_1 < D_0$ when $l > 2\sqrt{(d_b + r_b c_b/\alpha\beta)}$; beyond this threshold, inserting a single buffer reduces signal delay. This simple analysis shows the value of buffer insertion for long lines particularly as geometries scale down: since gate capacitance is reduced, c_b decreases; intrinsic delay d_b decreases with each new technology; resistance per unit wire β increases; these effects are somewhat offset by the decreased per-unit wiring capacitance α, but the width-independent *fringe capacitance* reduces the impact of this trend. The net result is that more signal nets can benefit from buffer insertion.

Reducing delay by decoupling off-path capacitance is illustrated in Figure 6.4. By inserting a buffer at the root of the tree connecting sinks B, C, D, E, the load on the driver and interconnect up to the branch point is reduced, thereby reducing the overall delay to sink A (of course, depending on the parameters that the delay to the other sinks is also reduced).

Wire Sizing In deep-submicron technology, the resistance of minimum-width interconnect can be large enough that widening some wire segments near the

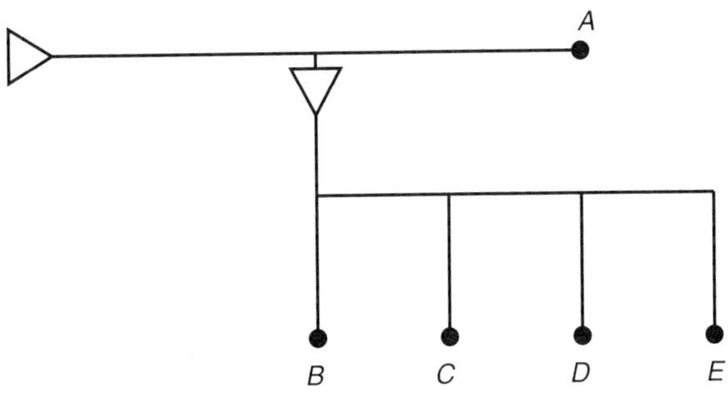

FIGURE 6.4 *Optimizing by decoupling off-path capacitance.*

driver of a routing topology can reduce delay. The trade-off between added capacitance and reduced resistance is clear from the simple equation (6.1).

Thus, the widths of individual wire segments is a natural design parameter to optimize. The works of Cong et al. [2, 3] represent early efforts in this area.

Topology For multipin nets (and even 2-pin nets depending on the situation), the design of the routing topology affects not only the routing resources consumed, but also the resulting performance.

As an example, consider Figure 6.5, showing three candidate topologies connecting sinks A, B and C to a driver. All three topologies have the same total wire length, but will yield different delay characteristics for each sink. For example, the first topology will have less delay to sink A than the second (since the segment from the branch point to A sees more downstream load); on the other hand, the second topology will have lower delay to sink C. Which topology is preferred depends on the particular context (see section 6.23). Methods for automatic synthesis of such topologies in a timing-driven scenario is the topic of Chapter 7.

6.2.3 Problem Formulations

Once delay estimators and the design space have been determined, the precise problem formulation must be decided to ensure that the net-level problem formulation is useful in achieving the system-level objectives.

For problems relating to multi-pin nets, we adopt formulations baseed on *required arrival times* or similar concepts. This allows a natural connection between the net being optimized and its surrounding context (e.g., delay of logical paths from the sinks of the net to the inputs of latches). Let q_t denote the

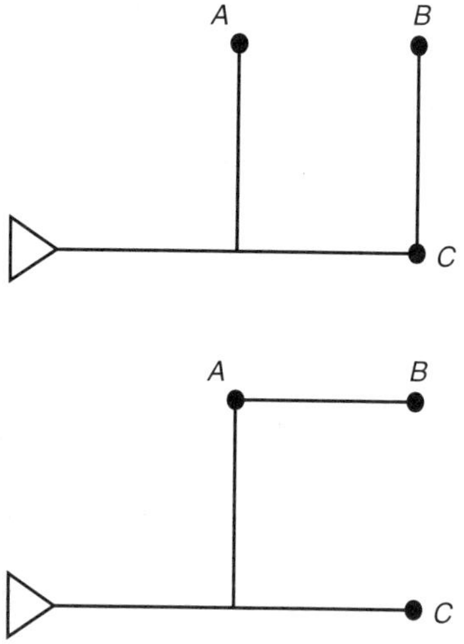

FIGURE 6.5 *Sample topologies.*

given required arrival time (or simply, required-time) at sink t of a net (i.e., the latest time at which the signal must be stable at t for the overall timing requirements to be satisfied). For a given topological configuration T, the required-times at internal nodes of the routing tree are determined recursively from sinks to source as

$$q_u = \min_{(u,v)\in T} (q_v - d(u, v))$$

where $d(u, v)$ is the signal delay from node u to v. This delay depends on the context; for example, u and v may be input and output pins, respectively, of a buffer or they may simply be ends of a wire. Note that the definitions apply for single-source nets. For the case of multisource nets (e.g., buses), a generalization of this concept—the *Augmented RC Diameter*—is applied (see Chapter 8).

Since we prefer larger required-times, one might wish to formulate a synthesis problem with the objective of simply maximizing the resulting required-time at the input of the net (e.g., find an assignment of buffers to insertion points that maximizes the resulting required-time at the input of the driver). Unfortunately, it is typically the case that such a solution is of little practical use. The reason is the trade-offs between performance and other design objectives (e.g., area, total capacitance). Thus, the system-level objectives are usually best viewed as *constrained optimization problems* (e.g., minimize area subject to a particular clock period being met). Further, optimization techniques at the net-level often exhibit a cost-versus-performance profile like that in Figure 6.6: after a certain investment (cost), we reach a point of diminishing returns. As a result, a solution that simply maximizes required-time is rarely the best engineering solution.

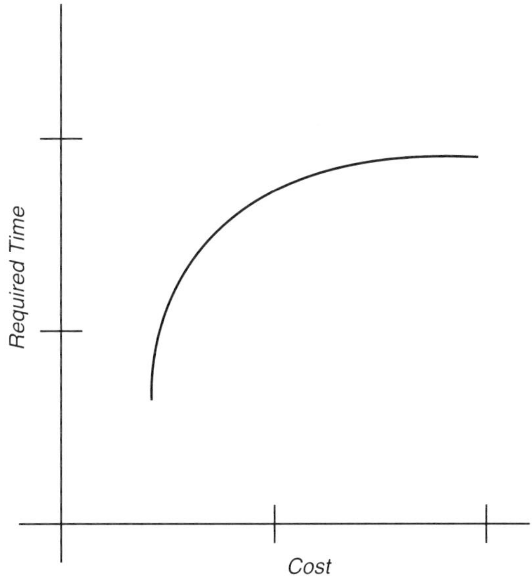

FIGURE 6.6 *Typical required-time versus cost trade-off.*

For these reasons—regardless of the optimization technique being applied—we focus on formulations that consider not only delay, but also cost overhead. This is done mainly via constrained optimization formulations or, more generally, formulations that uncover an entire family of candidate solutions that exhibit a performance-versus-cost trade-off.

Composite objectives are another, though less direct, way to consider cost overhead. For example, when optimizing two pin nets via buffer insertion or wire sizing, one might wish to minimize $A_1 \cdot \text{Delay} + A_2 \cdot \text{Cost}$, where A_1 and A_2 are weighting coefficients (in the case of required-times, the weighting coefficient could be negative). Such objectives are frequently adopted in the literature and can be of practical use; some of the formulations presented in Chapter 9 are of this type.

6.2.4 Scaling Coefficients

Depending on the precise problem formulation, it is usually very important to use an appropriate scaling coefficient δ for the delay estimator being used. Let d be a delay estimate under some model such as Elmore. We in turn estimate the true delay by δd. The value of δ may be chosen to center the error distribution of the given estimator (e.g., via empirical studies in which δ is chosen to minimize the resulting mean-squared error). In the case of Elmore, $\delta = \ln 2 = 0.69$ is a theoretically sound value, as mentioned in section 4.5.8.

Under a required-time formulation, an inappropriate scaling coefficient can cause a synthesis procedure to make incorrect decisions. The simple example in Figure 6.7 illustrates the issue. In the figure the delay estimate to sink A is 10 time units and to sink B is 20 time units; the required-time at A is 20 time units and the required-time at B is 25 time units. The graph then shows the resulting required-time with respect to each sink after subtracting the appropriate signal

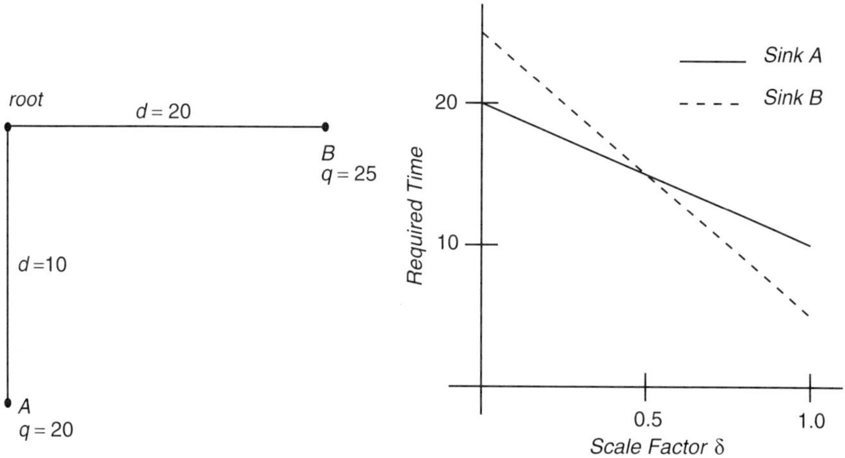

FIGURE 6.7 *Effect of scaling coefficient δ.*

delays as a function of δ. For $\delta < 0.5$, sink A is critical; for $\delta > 0.5$, sink B is critical. Thus, the decisions an optimizer makes are dependent not only on the delay estimator, but also on δ; if δ is not selected to maximize the chances that true critical paths are correctly identified (i.e., center the error distribution), an optimizer will more frequently make incorrect decisions and produce inferior solutions.

It is interesting that for formulations not based on required-time (e.g., minimization of maximum source-to-sink delay), the scaling coefficient is not an issue and we will get the same results regardless of δ (sink criticality being the same for any value of δ since there is no "offset" from given sink required-times). As a result, the *absolute accuracy* of a delay estimator is, unfortunately, often dismissed as unimportant, while clearly it is critical in required-time formulations.

6.2.5 Error in Delay Estimators

It is well known that Elmore can exhibit significant error versus simulated delay. In RC trees two main factors contribute to Elmore's error [7]:

- Its inability to account for resistive shielding (an increasingly important phenomenon in deep submicron)
- Its inability to account for the effect of signal slew rate τ

As a result, Elmore can grossly overestimate the delay to nearby sinks for which the resistive shielding effect is significant and the signal waveform is relatively sharp. Conversely, as the signal slew increases (i.e., as we travel to more distant sinks), Elmore becomes increasingly accurate.[4] The danger in a synthesis procedure is that this error will cause misidentification of critical paths and lead the algorithm to make incorrect decisions by optimizing delay to a false critical sink at the expense of the true critical sink.

Despite these drawbacks, Elmore and relatively simple buffer delay models have served as the delay estimation bases of many highly successful synthesis and optimization algorithms—including those presented in this book. Indeed, the use of Elmore has become ubiquitous in the literature. It can be argued that this success is largely the result of the following factors.

- *Balance of simplicity and accuracy*. Since Elmore is relatively easily computed and recursively decomposable, effective algorithms can be designed around it.
- *Evidence of fidelity of Elmore*. There have been several studies (e.g., [1]) than show Elmore to have a high degree of fidelity: an optimal solution under Elmore is typically not much worse than the true optimal. Such fidelity is, however, dependent on the precise problem formulation.

[4] Indeed, for ramp inputs Elmore approaches the true 50% delay as τ tends toward infinity [7].

- *Flexibility of Elmore-based synthesis algorithms.* This is perhaps the most compelling argument in favor of research and algorithm development based on Elmore: algorithms designed around Elmore seem to be flexible enough to systematically incorporate improved delay estimators. One such example appears in this chapter, in which an improved buffer delay model is used in an automatic buffer-insertion algorithm. Another example is [13], in which table-lookup methods were used in a topology synthesis procedure instead of Elmore (the methods accounted for the resistive shield effect and the effect of signal slew; improved solution quality was frequently achieved). The process of adapting Elmore-based algorithms to improved delay estimators is an increasingly popular industry practice.

6.3 OPTIMIZATION OF STATIC ROUTING TOPOLOGIES

This section examines synthesis problems for static (i.e., given) topologies. The basic tool used is that of Dynamic Programming, which has proved to give a remarkably robust framework in which many timing optimization–oriented problems can be solved. The use of Dynamic Programming for interconnect optimization was pioneered independently by van Ginneken [12] and Dhar [4] and subsequently refined by Lillis et al. [8]. The results presented herein can be viewed as generalizations and refinements of those seminal works. After developing the background and problem formulations, a general algorithmic framework is presented, into which several formulations fit naturally. An algorithm for simultaneous buffer insertion and wire sizing maximizing required-time is first presented. Subsequently, a solution to the more practical constrained optimization problem is addressed (minimization of total capacitance subject to timing constraints). Detailed implementation strategies for this formulation are given. Finally, the flexibility of the schemes is illustrated via the incorporation of a more sophisticated buffer delay model, which includes the effect of input signal slew.

6.3.1 Notational Conventions

We adopt the following notational conventions.

T_v	A routing tree rooted at node v
$l(v)$, $r(v)$	The left and right children of node v, respectively
e_v	Tree edge (wire) from node v to its parent
l_e	Length of edge e
c_e	Capacitance of edge e
c_v	Input capacitance of sink v
r_e	Resistance of edge e
c_b	Input capacitance of buffer b
r_b, r_g	Output resistance of buffer b or gate g
d_b, d_g	Intrinsic delay of buffer b or gate g
p	Polarity; usually referring to a signal, $p = 1$ meaning inverted

p_b	Buffer type; $p_b = 1$ indicates b is an inverter; $p_b = 0$ o.w.
q_v	Required arrival time of sink node v
W	Largest possible wire width ($1..W$ are possible)
B	Buffer library
leaves(T)	Set of leaves of tree T

We also use *elmore(e,w,c)* to indicate the elmore delay of an edge e at width w and having a downstream load capacitance of c. Similarly, *buf_delay(b,c)* is used to indicate the delay of a buffer b driving a load c.

6.3.2 Formulations

First, we present the cost-oblivious formulation.

Formulation 6.1

Given: A binary routing topology T, the technology parameters of the driver, buffer library B, candidate wire-widths $1..W$, and required-arrival times q_u for each sink u.

Objective: Find an assignment of buffers in B to branch points in T and widths $1..W$ to wire segments in T such that the resulting required arrival time at the input of the driver is maximized and signal polarity is correct at each sink.

An efficient algorithm for solving this problem is presented in section 6.3.4. The requirement regarding signal polarity enables us to use inverters in the buffer library provided that we ensure that each source-to-sink path traverses an even number of inverters. The assumptions that the topology is binary and that insertion points are at branch points are merely to ease the presentation of the algorithms. Simple transformations allow arbitrary insertion points and topologies (see Exercises).

Building on the algorithm for the cost-oblivious case, we then solve the more practical constrained-optimization problem.

Formulation 6.2

Given: A binary routing topology T, the technology parameters of the driver, buffer library B, candidate wire-widths $1..W$, required-arrival times q_u for each sink u.

Objective: Find an assignment of buffers in B to branch points in T and widths $1..W$ to wire segments in T which minimizes total capacitance subject to the required-time at the input of the driver being non-negative (i.e., the timing requirements of the sinks must be met).

This problem can be refined in a number of natural ways. The use of total capacitance is natural in this case because of its correlation with dynamic power consumption in CMOS technology and its less direct correlation with total area. However, virtually any monotone cost function could be used instead of total capacitance (e.g., the "cost" of inserting a buffer may be very high in some situations). Additionally, one may wish to have a family of solutions that exhibit a trade-off between cost and required-time rather than a single solution satisfying the timing requirements; the algorithm presented below provides such a suite of solutions.

The algorithms solving these problem formulations initially assume Elmore and the basic *RC* buffer delay models. Subsequently, a generalization of the algorithm is presented to take into account a buffer delay model that incorporates the effect of signal slew, illustrating the flexibility of the basic framework. For ease of notation, delay estimators are not scaled, as discussed in the previous section.

6.3.3 Algorithmic Framework

We first give the framework for a high-level dynamic programming algorithm into which all subsequent algorithms will fit. The framework covers all variants of interest. For instance, if we are only interested in wire sizing, we simply run the algorithm with an empty buffer library *B*. The specific algorithms will differ in their implementation of the basic routines invoked by the general algorithm and the characteristics of the solution sets they compute.

This General Dynamic Programming algorithm *GDP*() is given as pseudocode in Figure 6.8. The algorithm computes the solution sets $S_b(v)$ and $S_t(v)$. The set $S_b(v)$ can be thought of as the set of solutions for subtree T_v, including the possibility of inserting a buffer at v. Similarly, the set $S_t(v)$ can be thought of as the set of solutions for T_v *augmented* by the wire from its parent e_v and including possible sizing of e_v.

The four procedures *Base_Case*(), *Bottom_ Solutions*(), *Top_ Solutions*(), and *Optimal_Soln*() are routines that inductively compute solution sets from the solution sets of descendants. For each particular algorithm, these sets are parameterized differently. Intuitively, these routines can be thought of as follows.

```
Algorithm GDP (T,B,W )
  Foreach node v ∈ T in topological order from leaves to root
    If v is a leaf
      Compute Sb (v) ← Base_Case (v)
    Else
      Compute Sb (v) ← Bottom_Solutions (v,St (l (v)), St (r (v)))
    If v is not the root
      Compute St (v) ← Top _Solutions (v,Sb (v))
    Else /* v is the root */
      Compute Optimal_Soln (v,Sb (v))
```

FIGURE 6.8 *General algorithm structure.*

Base_Case(): Compute the singleton set giving relevant parameters at sink v.

Bottom_Solutions(): Given solution sets for left and right children, compute the solution set *at* v, including the possibility of inserting a buffer at v.

Top_Solutions(): Given the solution set *at* v, construct the solution set for T_v augmented by wire e_v.

Optimal_Soln(): Given the solution set at the root, select the best solution when combined with the driver.

The key to solving a particular problem formulation (e.g., minimizing total capacitance subject to timing constraints) is in the content of the solution sets, their efficient computation by the above routines, and limiting their size as much as possible to ensure computational efficiency.

In the next section we study the simplest scenario: maximizing the required arrival time $q(T)$ at the root of the tree under the basic *RC* delay model. This encompasses the basic algorithms of van Ginneken and Dhar. Subsequently, we generalize this to the problem of minimizing total capacitance subject to given timing constraints (again under the basic *RC* model). Finally, we further generalize the algorithm to solve either of the preceding problems with the generalized buffer delay model, taking into account the effect of signal slew. In each of these cases, the algorithms are sketched based on the framework of this section.

6.3.4 Maximizing Required Arrival Time

In this case we maximize $q(T)$, the required arrival time at the root of T, under the basic *RC* delay model. The solution sets $S_b(v)$ and $S_t(v)$ are each partitioned into two disjoint subsets as

$$S_b(v) = S_b^+(v) \cup S_b^-(v)$$
$$S_t(v) = S_t^+(v) \cup S_t^-(v)$$

The reason for this partitioning is to deal appropriately with the fact that, since we may be using inverters as buffers, the signal may be inverted for some portions of the net. The sets with superscript "+" contain solutions where we assume the incoming signal is *noninverted* and the sets with superscript "−" contain solutions where the incoming signal is assumed to be *inverted*. In other words, the superscript "+" ("−") ensures that each path to a sink passes an *even* (*odd*) number of inverters.

The solutions themselves are *load*, *required-time*, or (c,q) pairs. Intuitively, the English meaning of these sets is, for example,

$(c, q) \in S_b^+(v) \Longleftrightarrow$ "There exists an assignment to T_v with upward load c and required-time q at v when the incoming signal is not inverted."

An important initial observation made by van Ginneken [12] and Dhar [4] is the following.

Property 6.3.1 For (c, q), $(c', q') \in S$, if $c' \geq c$ and $q' < q$ then (c', q) is sub-optimal.

This is clear since a larger load can only worsen delay of ancestor components. In other words, we always prefer smaller load and larger required-time. Supposing that these sets are arranged in increasing order of load leads to the following property.

Property 6.3.2 Any *load-required_time* set S in increasing order of load may be replaced by $S' \subseteq S$ where S' is strictly increasing in required-time.

We maintain this sorted order as an invariant so that we may easily exploit this property.

In the context of our algorithmic framework, we fully specify the algorithm as follows. Recall that c_v and q_v are the input capacitance and required arrival time of sink v, respectively. The routine *Base_Case*(v) simply sets $S_b^+(v) \leftarrow (c_v, q_v)$ and $S_b^-(v) \leftarrow \emptyset$ since no valid solution will have an inverted signal at a sink.

Computation of *Bottom_Solutions*() is described in pseudocode in Figure 6.10; the algorithm first computes the optimal (c,q) pairs for unbuffered solutions by invoking the subroutine *join_cq_sets* in Figure 6.9. For each achievable arrival time q, we find the smallest load achieving q. This is done in a manner similar to the merging of two sorted lists and ensures that Property 6.3.2 holds. This is a linear time operation and the size of the resulting set is linear, as observed by van Ginneken [12]. This merging process is illustrated by the example in Figure 6.11. At each step we combine a pair of solutions from the left and right, after which we advance in the list that is currently critical to determine the next pair to merge.

Next we find the optimal buffer configurations in lines 5–10 by pairing buffers $b \in B$ with unbuffered solutions at v. We then perform merging and additional

```
Algorithm: join_cq_sets(S1, S2)
  S ← Ø
  i ← 1 ; j ← 1
  While (i ≤ |S1| and j ≤ |S2|)
    Let (c1, q1) = S1[i]
    Let (c2, q2) = S2[j]
   S ← S ∪ {(c1 + c2, min(q1, q2))}
    If (q1 ≤ q2)   /* S1 Critical */
      i ← i + 1
    If (q2 ≤ q1)   /* S2 Critical */
      j ← j + 1
  return S.
```

FIGURE 6.9 *Subroutine for merging cq-sets at a branch point.*

Algorithm: *Bottom_Solutions* $(v, S_t(l(v)), S_t(r(v)))$

1. /* First compute unbuffered solutions */
2. $S_b^+(v) \leftarrow$ *join_cq_sets* $(S_t^+(l(v)), S_t^+(r(v)))$
3. $S_b^-(v) \leftarrow$ *join_cq_sets* $(S_t^-(l(v)), S_t^-(r(v)))$
4. /* Now compute buffered solutions */
5. **Foreach** buffer $b \in B$
6.   **If** b is an inverter
7.   **Find** $(c,q) \in S_b^-(v)$ s.t.
            $q_b^+ = q - \text{buf\_delay}(b,c)$ is maximized
8.   **Else**
9.   **Find** $(c,q) \in S_b^+(v)$ s.t.
            $q_b^+ = q - \text{buf\_delay}(b,c)$ is maximized
10. Analogously compute q_b^-
11. $S_b^+(v) \leftarrow S_b^+(v) \cup \{(C_b, q_b^+) \mid \forall b \in B\}$
12. $S_b^-(v) \leftarrow S_b^-(v) \cup \{(C_b, q_b^-) \mid \forall b \in B\}$
13. Prune $S_b^+(v)$ and $S_b^-(v)$ by Property 6.3.2

FIGURE 6.10 *Bottom_Solutions( ) routine for max required-time.*

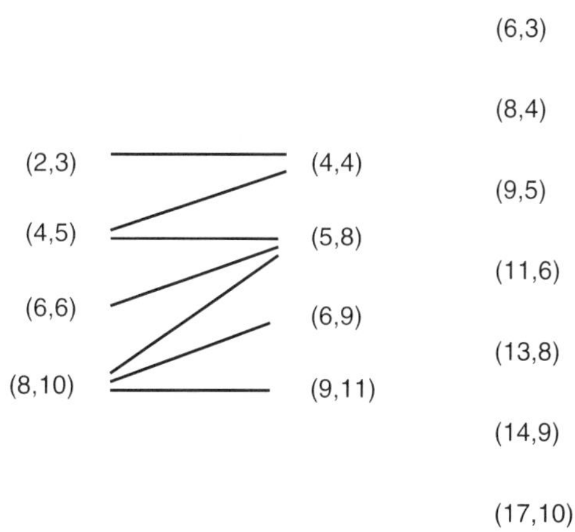

FIGURE 6.11 *Example of merge operation of the join_cq_sets routine in Figure 6.9. The two columns on the left are the sets of cq-pairs being merged; paris of solutions combined are denoted by black lines. The resulting set of solutions appears on the right.*

pruning in lines 11–13. Also note that the final pruning step is also linear since the sets are in sorted order.

We give the implementation of *Top_Solutions*() in Figure 6.12. We examine all pairings of widths w for wire e_v (having length l_{e_v}) with solutions $(c, q) \in S_b(v)$ at v. Since the loads c' are of the form $c + wZ$ where Z is fixed and $w \in \{1..W\}$, we can visit all c''s in order without explicitly sorting them. For each such pairing, we obtain a new required-time. In a final sweep we again apply Property 6.3.2 to ensure that the set is in strictly increasing order of both c and q.

Finally, we must implement *Optimal_Soln*$(v, S_b(v))$. This is done simple by pairing all of the previously computed solutions $(c, q) \in S_b^+(v))$ (since the signal

Algorithm: *Top_Solutions* $(v, S_b(v))$

1. $S_t^+(v) \leftarrow \emptyset$
2. **Foreach** $(c,q) \in S_b^+(v)$, $w \in \{1..W\}$ in increasing order of $c' = c + \alpha w l_{e_v}$
3. $S_t^+(v) \leftarrow S_t^+(v) \cup \{(c', q - \text{elmore}(e_v, w, c))\}$
4. **Compute** $S_t^-(v)$ analogously
5. Prune $S_t^-(v)$ and $S_t^-(v)$ by Property 6.3.2

FIGURE 6.12 *Top_Solutions( ) routine for max required time.*

leaving the driver is noninverted) with the properties (i.e., output resistance) of the driver and selecting the solution with the largest resulting required arrival time. Note that if we are not allowed to buffer the driver output, we can simply include only unbuffered solutions in $S_b^+(v)$.

To construct the solution achieving this timing we recursively revisit the tree to determine which choices of buffering and wire sizing led to the optimal solution. This is accomplished by storing with each (c,q) pair, local information indicating the choices that led to that solution.

Run Time We analyze the running time of the basic algorithm in three scenarios:

1. $|B| = 0$, $W > 1$ (wire sizing alone).
2. $|B| \geq 1$, $W = 1$ (buffer insertion alone).
3. $|B| \geq 1$, $W > 1$ (both methods).

In scenario 1 (wire sizing alone), we introduce the notion of "basic segment-length" to analyze the complexity. This parameter is derived from an underlying routing grid.

Property 6.3.3 In scenario 1 the size of each *load-required_time* set S is bounded by mW where m is the total number of basic grid lengths in the tree.

This can be seen by considering that the load at node v can be expressed as

$$c(T_v) = \text{Sink\_Load} + \text{Interconnect\_Load}$$

where *Sink_Load* is fixed and *Interconnect_Load* $= \delta \Sigma w_i$ where $w_i \in \{1..W\}$ is the width of the ith wire and δ is a constant derived from α, the basis grid length and the minimum width. Thus, the load is entirely determined by Σw_i, which can only take on integer values in range $\{m..mW\}$. This gives an upper bound of mW on the size of individual solution sets that the algorithm computes since it bounds the number of *distinct* load values. Thus, even though there are an exponential number of width assignments, there are only a polynomial number of distinct resulting loads. The resulting run time is $O(nW(mW)) = O(nmW^2)$.

In the case where every sizeable segment is of identical length, we have $O(n^2W^2)$ since, effectively, $n = m$ in this case.

Scenario 2 is a generalization of the situation for the algorithm of van Ginneken [12]. (It is a generalization only in the sense that is allows multiple buffers in its library). Since $W = 1$ in this case, computation of S_t sets is trivial. Thus the size of the S_b sets is the key factor in the run time. We first state the following properties alluded to earlier.

Property 6.3.4 For $S_b(v)$, let S_l and S_r be the S_t solution sets of v's left and right children, respectively, of the same polarity. The following inequality holds: $|S_b(v)| \leq |S_l| + |S_r| + |B|$.

Lemma 6.3.1 In scenario 2, for all *load-required_time* sets S, $|S| \leq n + n|B|$.

Proof: Let $M(n)$ be an upper bound on the number of solutions at the root of an n leaf tree. Suppose the tree has k and $n - k$ leaves in its left and right subtrees respectively. We then have the recurrence $M(n) = M(k) + M(n - k) + |B|$ where the first tow terms are for unbuffered solutions. By strong induction we have $M(n) = M(k) + M(n - k) + |B| = k + k|B| + (n - k) + (n - k)|B| + |B|$ or $M(n) = n + n|B|$. $\square$

These properties coupled with the fact that the merging operation is linear in $|S_l| + |S_r|$ gives an overall worst-case complexity of $O(n|B|(n + n|B|)) = O(n^2|B|^2)$.

Scenario 3 is complicated by the fact that the input capacitance of the buffers may not be simple multiples of the capacitance of a unit-length wire. However, in practice it is reasonable to assume that capacitive values can be linearly mapped onto a polynomially bounded integer domain with sufficient precision or are given as such. In such a situation we introduce another value c_{max}, which is the largest capacitance possible among the individual components of the tree (e.g., it may be the capacitance of the longest wire at the maximum width). Under this formulation we upper-bound the size of the solution sets by nc_{max} and the overall run time by $O(n^2c_{max}(\max(W, |B|)))$. In practice, observed run times are typically much less than this bound.

Comments For simplicity, we have presented the algorithm in terms of a binary tree, but note that the algorithm is easily applied to general trees. One straightforward method to achieve this is to convert a nonbinary tree to an equivalent binary tree simply by adding zero-length wires. For instance, suppose we have a node v with fanout 3 to nodes A, B, and C. We replace v with two nodes v' and v'' where node v' will have children A and B, and v'' will have children C and v' with the wire from v'' to v' having length 0. The algorithm can be modified to prohibit the placement of buffers at particular nodes—v' in this case.

Another issue is that, as described, the algorithm assumes exactly one sizeable wire segment between nodes and buffer insertion only at nodes in the tree.

However, the algorithm is easily implemented to accommodate multiple sizeable segments in a single wire and buffer insertion within a wire by introduction of intermediate nodes.

We further note that the algorithm can easily be extended to allow for optimal sizing of the driving gate. However, it should be realized that such sizing may have global effects by altering the input capacitance of the driver, thereby affecting the timing requirements and the system as a whole. These comments apply to all subsequent algorithms in this chapter.

6.3.5 Minimizing Total Capacitance Subject to Timing Constraints

We now extend the algorithm to minimize total capacitance subject to given timing constraints. The minimization of total capacitance is a natural choice because of its linear correlation with dynamic power consumption in CMOS technology. Further, it enables us to naturally unify the costs of wire sizing and buffer insertion. For clarity, we present this and subsequent algorithms without regard to signal polarity; application of the ideas of the previous section to deal with the presence of inverters is straightforward.

The first issue is how to parameterize the solution sets. We are not only concerned with the load and required-time of a subsolution, but its *total capacitance* (i.e., the total capacitance among all buffer stages in the topology). To avoid confusion with load capacitance as used in the previous section, we will refer to the total capacitance as "power" (again noting the relation with dynamic power consumption). Thus, a subsolution is now characterized by the parameters p (power), c (capacitance) and q (required-time).

It is convenient to organize solution sets by first partitioning them into subsets with identical power values p. In other words, solution sets $S_b(v)$ and $S_t(v)$ now contain pairs (p, S_p) where p is power consumption (as a capacitive value), and S_p is an ordered set of load, required-time pairs (c,q) as in the previous algorithm. For example, $(p, S) \in P_b(v)$ indicates that for power p and every $(c, q) \in S$, there exists an assignment for T_v consuming power p, presenting load c upward and yielding required time q at v.

We organize these sets first by sorting them in increasing order of power. Each set S_p is ordered by load c as in the basic algorithm.

One might think that the sets S_p are typically singleton sets; however, this is not the case. Because many different configurations may consume precisely the same power (by, for example, assigning an identical set of buffers to different locations), these sets can be quite dense.

Recalling that total capacitance is the correct measure for dynamic power dissipation, and thus, c_v is the "power" associated with sink v, *Base_Case*(v) simply sets $S_b(v) = \{(c_v, \{(c_v, q_v)\})\}$.

Pseudocode for *Bottom-Solutions*() is given in Figure 6.13. We visit all possible values of total power consumption at v. These values are from among the buffered and unbuffered configurations. The notion of a "nonbuffer" ϕ is introduced to unify the notation. In this case, we explicitly sort the values $p = p_l + p_r + p_b$.

```
Algorithm: Bottom_Solutions(v,S_t(l(v)), S_t(r(v)))
1. Let B' = B ∪ {φ}
   /* φ indicates "no buffer", c_φ = 0 */
2. S_b(v) ← ∅
3. Foreach triple (p_l,S_{p_l}) ∈ S_t(l(v)), (p_r,S_{p_r}) ∈ S_t(r(v)), b ∈ B'
              in increasing order of p = p_l + p_r + c_b
4.    S' ← join_cq_sets(S_{p_l}, S_{p_r})
5.    If (b ≠ φ)
6.       Find (c,q) ∈ S' s.t. q' = q − buf_delay(b,c) is maximized
7.       S' ← {(c_b,q')}
8.    Else
9.       S' ← S' \ {(c,q) ∈ S' | ∃(c',q') ∈ S_{p'}, p' < p, c' ≤ c, q' ≥ q}
         /* where (p',S_{p'}) ∈ S_b(v) */
10.   If (p,S_p) ∈ S_b(v)  (previous triple gave same p)
11.      S_p ← S_p ∪ S'
12.      Prune S_p by Property 6.3.2
13.   Else
14.      S_b(v) ← S_b(v) ∪ {(p,S')}
```

FIGURE 6.13 *Bottom_Solutions() routine for constrained optimization.*

However, we observe that the number of such *distinct* values p is often orders of magnitude less than the worst case (quadratic). Because of this observation we utilize a hash table to make an initial pass over all pairs to extract the *distinct* values, which we then sort. This avoids an overly expensive sorting operation.

Top_Solutions() is implemented in a similar manner; its pseudocode appears in Figure 6.14.

As described, these algorithms implement two types of pruning. First we prune solutions $(c, q) \in S_p$ for a power p in the same way as before by Property 6.3.2. However, an additional pruning condition is utilized in Figures 6.13 and 6.14 on lines 9 and 7 respectively. This pruning condition is given by the following property.

```
Algorithm: Top−Solutions(v, S_b(v))
1.  S_t(v) = { (p, ∅) | p is a possible power}
2.  Foreach pair w ∈ 1..W, (p_b, S_{p_b}) ∈ S_b
              in increasing order of p = p_b + αwl_{e_v}
3.     S_p ← ∅
4.     Foreach (c,q) ∈ S_{p_b}
5.        S_p ← S_p ∪ { (c + αwl_{e_v}, q − elmore(e_v, w, c)) }
          /* elmore delay evaluated at width w */
6.     Prune S_p by Property 6.3.2
7.     S_p ← S_p \ { (c,q) ∈ S_p | ∃(c', q') ∈ S_{p'}, p' < p, c' ≤ c, q' ≥ q }
       /* where (p', S_{p'}) ∈ S_t(v) */
8.     If (S_p ≠ ∅)
9.        S_t(v) ← S_t(v) ∪ {(p,S_p)}
```

FIGURE 6.14 *Top_Solutions() routine constrained optimization.*

Property 6.3.5 Consider solution (c, q) with total capacitance p. If $\exists$ solution (c', q') with total capacitance p', where $p' < p$, $c' \leq c$ and $q' \geq q$, then the solution (c, q) is suboptimal.

The efficient application of this property has proven essential in giving reasonable running times in practice. Efficient detection of Property 6.3.5 is addressed in the next subsection.

Optimal_Soln(v,$S_b(v)$) is implemented simply by selecting the lowest-power unbuffered solution at the root giving required-time $q(T) \geq 0$ when paired with the driver. Alternatively, this set of unbuffered solutions gives the full trade-off curve from which a solution may be selected. This is an interesting and potentially valuable side-effect of the algorithm. A suite of solutions giving a trade-off between cost and performance could very naturally be utilized as *sensitivity information* for a full-chip optimizer (i.e., it is unlikely that strict timing constraints must be enforced on a particular net — only for latch-to-latch paths). An example trade-off curve is given in Figure 6.15. The figure also illustrates a common phenomenon in many optimization problems: past a certain point, additional resources yield only incremental improvements in objective function. Further,

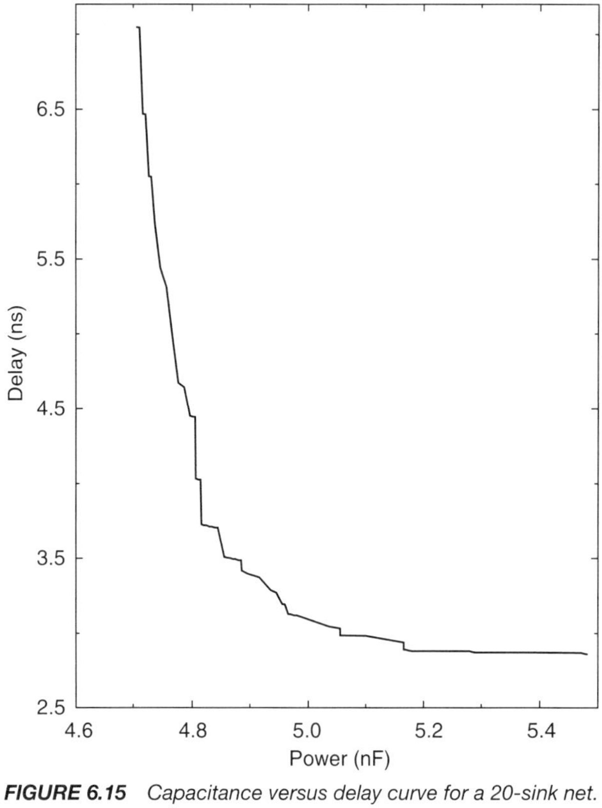

FIGURE 6.15 *Capacitance versus delay curve for a 20-sink net.*

the curve illustrates the impracticality of solutions produced by the "timing-only" version of the algorithm presented earlier — the "fastest" solution is likely to have unnecessarily high overhead.

Efficient Detection of Property 6.3.5 When constructing a solution set S_p consuming power p, we would like to efficiently determine if Property 6.3.5 holds for each $(c, q) \in S_p$. Since the solution sets can grow to substantial size, a linear scan of all other solutions to detect this property would likely be a disaster. We now outline a data structure to efficiently support this operation.[5]

Recall that we visit all power values in order and, therefore, we know that the entries in the data structure are for all power values $p' < p$. Thus the data structure need only concern itself with c and q values — any solution that can eliminate a solution consuming power p has previously been constructed and stored in the data structure; furthermore, all such previously visited solutions satisfy the $p' < p$ requirement by default.

We need a data structure that efficiently supports the following operations:

- *insert(c,q)*: updates the data structure to reflect solution (c,q).
- *sub_opt(c,q)*: returns TRUE if $\exists(c', q')$ previously inserted s.t. $c' \leq c$ and $q' \geq q$, FALSE otherwise.

The *sub_opt*() operation is illustrated in Figure 6.16. The data structure must store points in two-dimensional space and determine if a candidate point (c, q) is suboptimal (i.e., if the grey region in the figure contains any points). Such a data structure solves a special case of the *Orthogonal Range Query* problem from computational geometry (see e.g., [13]). Our problem is a special case for two

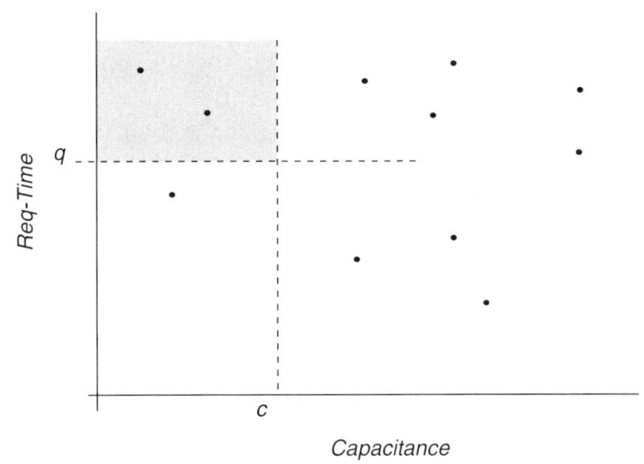

FIGURE 6.16 *Geometric illustration of detection of Property 6.3.5.*

[5] The help provided by Matthew Clegg of U.C. San Diego in devising this data structure must be acknowledged.

reasons: first we need not retrieve or count all (c', q')'s satisfying the property, and, second, the subspace we are interested in is defined by two inequalities, $c' \leq c$ and $q' \geq q$, rather than four; in other words, our subspace is the intersection of the half-planes formed by two inequalities rather than a rectangle formed by four such inequalities. These special properties of our problem allow us to support the operations above in $O(\log m)$ time and $O(m)$ space for m entries by use of an augmented binary search tree. In contrast, the fastest known approaches to the general two-dimensional orthogonal range query problem also run in $O(\log m)$ time but use $O(m \log m)$ space.

To support the operations we order a binary search tree tree by load values c. At each node t of the search tree we store the load value c *and the largest* q value among the entries in t's left subtree or *at* t itself. We refer to this value as $t.q_{l\_max}$. Given this data structure, *insert*() is easily implemented recursively and *sub_opt*() can be implemented by examining the four following cases with respect to the given c, q and $t.c$ and $t.q_{l\_max}$ stored at the current node in the tree (boundary conditions are not given for simplicity):

$c < t.c,\ q < t.q_{l\_max}$	Explore left subtree
$c < t.c,\ q > t.q_{l\_max}$	Return FALSE
$c > t.c,\ q < t.q_{l\_max}$	Return TRUE
$c > t.c,\ q > t.q_{l\_max}$	Explore right subtree

By following these guidelines recursively down the search tree, we can detect the property in time proportional to the depth of the tree.

Run Time With respect to wire sizing alone (i.e., $|B| = 0$), we notice that $p = c$ for every candidate subsolution since there is no decoupling by buffers. Thus the basic algorithm is sufficient to solve the low-power problem: we get power minimization "for free."

In the general case of simultaneous wire sizing and buffer insertion (or buffer insertion alone if $W = 1$), we have to take into account the quadratic nature of the merging steps of the algorithm. Since we examine all *pairs* of power values from the left and right children the solution sets are no longer assured to be linear in size. However, when the capacitive values are given as *polynomially bounded integers* or can be mapped to such, once again we can show the run time of the algorithm to be polynomial.

As in section 6.3.4, let c_{max} be the largest possible capacitive value among the components. Under this scenario, we bound the number of load-required_time pairs at a node by $(nc_{max})^2$. This gives an overall run time bound of $O(n(|B| + W)(nc_{max})^2 \log(nc_{max})) = O((|B| + W)(n^3 c_{max}^2 \log(nc_{max})))$. The log factor is an artifact of the sorting performed on the power values.

In practice, we observe much better run times as a result of the additional pruning described in the previous section (and not included in this analysis since we cannot prove it improves the worst-case performance).

6.3.6 Accounting for Signal Slew

A common generalization of this basic buffer delay model includes an additive term to account for the *slew* of the signal entering the buffer. One model for this delay is the product of a buffer dependent constant λ_b and the load delay of the previous stage, D_{Lprev} (i.e., the RC delay of the driving buffer). Thus we denote the augmented delay equation as

$$\text{buf\_delay}_{\text{slew}}(b, c_l) = \text{buf\_delay}(b, c_l) + \lambda_b D_{Lprev} \tag{6.4}$$

This and similar models have been proposed in various contexts.

This section gives a further generalization of the algorithm to account for the effect of signal slew on buffer delay. The techniques presented can be considered a case study in generalizing a known algorithm to incorporate an improved delay estimator. The key to the approach is the manipulation of piecewise linear functions to model the effect of signal slew.

Overview By equation (6.4), buffer delay is augmented by the term $\lambda_b D_{Lprev}$ (recall that λ_b is a characteristic constant of buffer b and that D_{Lprev} is the RC delay of the previous stage). Since our algorithm proceeds in bottom-up order, this is an unknown value when computing the delay associated with a buffer. Conceptually, we would like to support queries of the form "What is the optimal solution at v with capacitance c and $D_{Lprev} = x$?"

Since $\lambda_b D_{Lprev}$ is linear in D_{Lprev}, we utilize *piecewise linear* functions to model this effect. Where we previously had *load-required_time* pairs (c,q), we now have *load-required_time_func* pairs (c, f) where f is a piecewise linear function; $f(x) = q$ is the optimal required time q at v for load c and $D_{Lprev} = x$.

We illustrate the modeling of delay by a piecewise linear function in Figure 6.17. Figure 6.17(b) shows the piecewise linear delay function f at node v in Figure 6.17(a). The left and right subtrees have maximum delays of 5 and 4 units, respectively, when $D_{Lprev} = 0$. However, since the left and right subtrees are driven by different buffer types, they have different sensitivities λ_l and λ_r. The straight lines in Figure 6.17(b) are equal to the two delay functions contributed by the two subtrees with slopes corresponding to the sensitivities λ_l and λ_r. The resulting delay function f at node v is shown as a solid line, which is the max of the two. Thus, different values of D_{Lprev} can result in different critical paths.

A PWL function can be defined as follows.

DEFINITION 6.3.1

PWL: A *piecewise linear (PWL) function* f is a set of quadruples (y_o, *slope*, x_l, x_r) where each such quadruple represents a line segment and gives the y-intercept (y_0), *slope* and domain for which the segment is defined (i.e., all x where $x_l \leq x \leq x_r$). Additionally, no two segments may have overlapping domains.

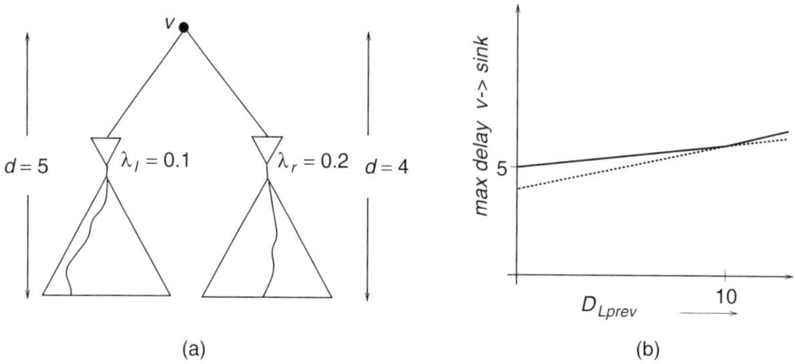

FIGURE 6.17 *Piecewise linear function modeling effect of signal slew.*

As is typical in most applications of PWL functions, we store line segments in a linked list ordered by domain.

Our manipulation of piecewise linear functions is founded on three basic operations:

$$f = \text{PWL\_Max}(f_1, f_2) \Longleftrightarrow f(x) = \max(f_1(x), f_2(x)) \quad \forall x$$

$$f = \text{PWL\_Min}(f_1, f_2) \Longleftrightarrow f(x) = \min(f_1(x), f_2(x)) \quad \forall x$$

$$f' = \text{PWL\_AddScalar}(f, d) \Longleftrightarrow f'(x) = f(x) + d \quad \forall x$$

The first two of these operations can be performed in a manner similar to the merging of two sorted lists in linear time by stepping through the lists and examining points of intersection. Further, they can be generalized to operate on *sets* of functions rather than pairs by repeated application, giving, for example, PWL_Max($f_1, \ldots, f_k$). The third operation is achieved simply by adding d to the starting y-coordinate (y_0) of each segment in the function.

With respect to the dynamic programming algorithm, we must also associate with each segment in the piecewise linear function f the relevant configuration information, which yields the solution (e.g., wire width, buffer type).

The algorithm modifications are summarized as follows:

1. Where we had *load-required_time* pairs (c,q), we now have *load-required_time_func* pairs (c,f).
2. Where we computed scalar *max* and *min* operations on arrival times q, we now compute *PWL_Max* and *PWL_Min* operations on piecewise linear functions.
3. Where we eliminated suboptimal solutions (c,q) by Properties 6.3.2 and 6.3.5, we now eliminate suboptimal *portions* of solutions (c,f) by analogous properties. These suboptimal solutions now give required arrival time of $-\infty$ for certain values of D_{Lprev}.

A detailed description of the generalized algorithm follows. For generality, we focus on the low-power formulation of the problem.

Algorithmic Details Before presenting *Bottom_Solutions*() and *Top_Solutions*(), we first discuss the basic pruning operations used in the algorithm.

Previously, a basic operation in our algorithms was the merging of two *load-required_time* sets. If we had two sets S_1 and S_2 where the solutions in each set consumed identical power p (or we were not concerned with power), we computed a new set $S_{1,2} = (S_1 \cup S_2) \backslash S^*$ where S^* was a set of provably suboptimal solutions by Property 6.3.2; that is, $(c, q) \in S^*$ implied that $\exists(c', q') \in S_{1,2}$ s.t. $c' \leq c$ and $q' \geq q$.

We generalize this concept for *load-required_time_func* sets. The property analogous to Property 6.3.2 is the following:

Property 6.3.6 Let load-required_time set $S = \{(c_1, f_1), \ldots, (c_m, f_m)\}$ be ordered by load; that is, $c_i < c_{i+1}$. For $(c_i, f_i) \in S$ let f_i' be defined for all x as:

$$f_i'(x) = \begin{cases} f_i(x) & \text{if } PWL\_Max(f_1(x), \ldots, f_{i-1}(x)) > f_i(x) \\ -\infty & \text{otherwise} \end{cases}$$

We may replace $(c_i, f_i) \in S$ with (c_i, f_i') while maintaining optimality. Further, if $f_i'(x) = -\infty \forall x$, then (c_i, f_i') may be eliminated altogether.

This property is implemented in Figure 6.18 as the routine *Merge_Load_Func_Sets*() and is used by both *Bottom_Solutions*() and *Top_Solutions*().

In addition, the previous low-power algorithm utilized a more general pruning method given by Property 6.3.5, which said that, for a pair of solutions (p, c, q) and (p', c', q'), if $p < p', c \leq c'$ and $q \geq q'$, then (p', c', q') is suboptimal and may be eliminated. The analogous property for the slew-sensitive generalization is as follows.

Property 6.3.7 For a solution (p, c, f), let S' be the set of all other solutions s.t. $(p_i, c_i, f_i) \in S'$ iff $p_i < p$ and $c_i \leq c$. Further, let $f_{max} = PWL\_Max(f_i)$ over

```
Algorithm: Merge_Load-Func_Sets (S1,S2)
  Let S' ← S1 ∪ S2
  fmax (x) = −∞  ∀x
  Foreach (c,f) ∈ S' in increasing order of c
     Let f'(x) = f(x)      if f(x) > fmax (x)
         f'(x) = −∞        otherwise
     S' ← S' \ { (c,f) }
     If ∃x  s.t. f'(x) ≠ −∞
        S' ← S' ∪ { (c,f') }
  /* Otherwise, (c, f') is useless */
  fmax ← PWL_Max (fmax, f )
Return S'
```

FIGURE 6.18 *Algorithm for merging load-func sets (presumably consuming the same power).*

all $(p_i, c_i, f_i) \in S'$. In a manner similar to Property 6.3.6, we may replace f with

$$f'(x) = \begin{cases} f(x) & \text{if } f(x) > f_{max}(x) \\ -\infty & \text{otherwise} \end{cases}$$

Again, if $f'(x) = -\infty \forall x$, then (p, c, f) may be eliminated entirely.

A generalization of the data structure in section 6.3.5 is be presented later in this section to efficiently implement the pruning of Property 6.3.7.

We give pseudocode for *Bottom_Solutions*() and *Top_Solutions*() in Figures 6.19 and 6.20 respectively. The routines follow the previous versions

```
Algorithm: Bottom_Solutions(v, S_t(l(v)), S_t(r(v)))
    B' ← B ∪ {φ} where φ indicates "no buffer" , c_φ = 0
    S_b(v) ← ∅
    Foreach  triple (p_l, S_{p_l}) ∈ S_t(l(v)) , (p_r, S_{p_r}) ∈ S_t(r(v)) , b ∈ B'
                 in increasing order of p = p_l + p_r + c_b
        /* Let (c_l,f_l) ∈ S_pt and (c_r,f_r) ∈ S_pr be indexed */
        /* E.g. (c_l[i] , f_l[i] is the i th smallest load in */
        /* and corresponding piecewise linear function in S_{p_l} */
        For i = 1..|S_{p_l}|
            If ( i = 1)
                j←1
            Else
                Let j be the smallest index s.t. f_l[i – 1](x) < f_r[j](x) for some x
            While ∃x s.t. f_r[j](x) < f_l[i](x)
                f_min ← PWL_Min (f_l[i] , f_r[j])
                If ( b = φ)
                    f'← f_min
                    c'← c_l[i] + c_r[j]
                Else
                    x← (c_l[i] + c_r[j]) · r_b
                    q← f_min(x)          /* Req. arrival time at output of buffer */
                    q← q – (c_l[i] + c_r[j] r_b – d_b
                    c'← c_l[i] + c_r[j]
                    Let f'(x) = q – λ_b x ∀x       /* One segment PWL func. */
                S'← S' ∪ {(c',f')}
                j←j + 1
            Prune solutions in S'per Property 6.3.7
            If (∃(p, S_p) ∈ S_b)
                S_b ← S_b – {(p, S_p)}
                S'_p ← Merge_Load – Func_Sets(S',S_p)
                S_b ← S_b ∪ {(p, S'_p)}
            Else     /* New power value */
                S_b ← S_b ∪ {(p, S')}
```

FIGURE 6.19 *Bottom_Solutions() routine for constrained optimization and signal slew.*

Algorithm: *Top-Solutions* $(v, S_b(v))$

$S_t(v) \leftarrow \emptyset$

Foreach pair $w \in 1..W$, $(p_b, S_{p_b}) \in S_b$ in increasing order of $p = p_b + \alpha w l_{e_v}$

  Foreach $(c, f) \in S_{p_b}$

    $d \leftarrow$ elmore (e_v) (at width w)

    $S_p \leftarrow S_p \cup \{(c + \alpha w l_{e_v}$, PWL_AddScalar $(f, -d)\}$ ($S_p = \emptyset$ initially)

  Prune suboptimal solutions from S_p by Property 6.3.6

  Prune suboptimal solutions from S_p by Property 6.3.7

  If $(S_p \neq \emptyset)$

    $S_t(v) \leftarrow S_t(v) \cup \{(p, S_p)\}$

FIGURE 6.20 *Top_Solutions() routine for constrained optimization and signal slew.*

quite closely. However, because Properties 6.3.6 and 6.3.7 introduce partially defined functions there is no total order on the (c, f) sets. Therefore, when combining solutions of children, we may look at all such pairs in the worst case rather than performing a simple "merge" as previously (i.e., invocation of the routine *join_cq_sets* in Figure 6.10). This can be seen in the *for loop* of *Bottom_Solutions*() in Figure 6.19—for a particular $f_l[i]$, we must find the appropriate $f_r[j]$'s to pair with $f_l[i]$. In the worst case, all j's may be candidates—and this may hold for all i's. However, in practice these sets tend to remain linear in size. A related issue is the complexity of the functions themselves. In principle, the size of the functions can grow exponentially. However, again we do not observe this phenomenon in practice.

Examples To give some intuition regarding the operation of the slew version of the algorithm, we now give some illustrations. Figure 6.21 shows function f having load c before and after buffer b is considered for insertion at v as is performed in *Bottom_Solutions*(). Since b will drive load c, D_{Lprev} for its descendants is cr_b. Therefore, the required arrival time *at the output* of the buffer

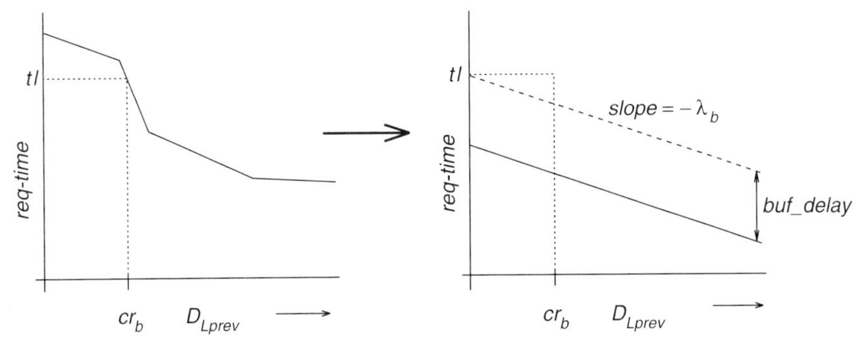

FIGURE 6.21 *Effect of inserting a buffer.*

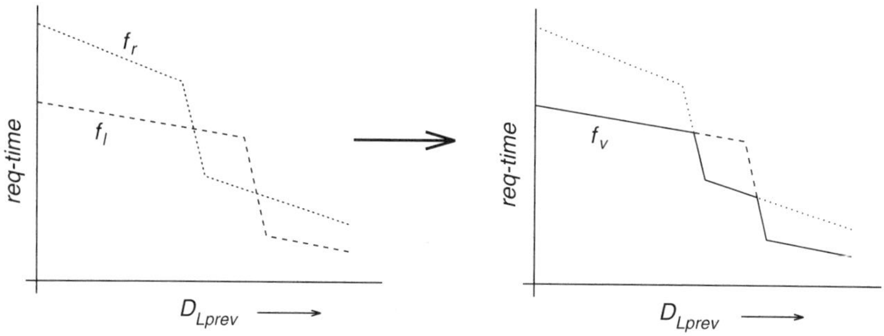

FIGURE 6.22 *Combination of left and right solutions.*

```
Algorithm: PWLPrune(t,c,f)
  if t = NULL return f
  else if c < t.c
     return PWLPrune(t.left,c,f)
  else if c ≥ t.c
     Define f' as:
        f'(x) = f(x) if f(x) > t.f_l_max(x)
        f'(x) = −∞ otherwise
     if (f'(x) = −∞ ∀ x or t.c = c)
        return f'
     else
        return PWLPrune(t.right,c,f')
```

FIGURE 6.23 *Implementation of Property 6.3.7.*

is $t1$. Subtracting the buffer delay *buf_delay*, we have the required arrival time function *at the input* of the buffer shown as a solid line in the right figure with slope $-\lambda_b$.

Figure 6.22 illustrates the operation of combining left and right ("top") solutions. For some values of D_{Lprev}, the left subtree is critical and for others, the right is. The solid line in the graph on the right shows the function resulting from the *PWL_Min* operation to capture the combination of these two solutions.

Implementation of Property 6.3.7 We now give a generalization of the data structure of section 6.3.5 to implement Property 6.3.7. Given a set of pairs (c_i, f_i) and a candidate pair (c, f), we want to efficiently compute f' as described in Property 6.3.7. (Recall that as the algorithm is organized, the power p associated with (c, f) is strictly greater than each p_i associated with each (c_i, f_i)).

To accomplish this we alter the previous augmented tree data structure to store the pair $(t.c, t.f_{l\_max})$ at each tree node t rather than $(t.c, t.q_{l\_max})$. As before, the search tree is ordered by c. The piecewise linear function $t.f_{l\_max}$ is $PWL\_Max(f_i)$ over all f_i where (c_i, f_i) resides either at t $(c_i = t.c)$ or in its left subtree.

Pseudocode implementing Property 6.3.7 is given in Figure 6.23—that is, given (c, f) we return the portions of f that are *not* suboptimal in the form of

f'. Updating the data structure can be done by recursively traversing the search tree and updating the f_{l_max} function at each node as we go.

The complexity of these operations is logarithmic in the number of entries in the data structure multiplied by the average complexity of the piecewise linear operations (which has been very small in experiments).

6.4 SUMMARY, DISCUSSION, AND FURTHER READING

In this chapter we have seen an introduction to automatic interconnect synthesis and optimization. An overview of important topics that practitioners and researchers alike should keep in mind was given:

- Relationships between net-level formulations and system-level design objectives (e.g., cost vs. performance trade-offs).
- The importance of required-time formulations.
- Delay estimators
  - Trade-offs between accuracy and computational ease.
  - Managing error distribution via scaling coefficients and connection to required-time formulations.
  - It was argued that algorithms based on Elmore often exhibit adequate fidelity of solution quality, but more importantly often can provide the basis for algorithms incorporating improved delay estimators. Nevertheless, this remains an area of active research.

In the second part of the chapter, algorithms for optimization of static routing topologies via buffer insertion and wire sizing were presented. The algorithms were introduced first with a "timing-only" version (of little use in practice). We then developed an algorithm for a constrained optimization version of the problem along with a detailed implementation. Many dynamic programming concepts that will be applied in subsequent chapters were introduced via this problem. Finally, the algorithm was generalized to consider a more sophisticated buffer delay estimator. This generalization can be considered a case study in reengineering an algorithm for one delay estimator to incorporate a better model.

The focus of this and subsequent chapters is largely on discrete optimization problems — for example, we have a fixed buffer library and set of wire widths (rather than being allowed to size arbitrarily). Continuous timing optimization problems (e.g., continuous transistor sizing in full custom design) have a longer history and more expansive literature and therefore are not covered here. Excellent resources in this area include the seminal work of Fishburn and Dunlop [6] and the work of Sapatnekar et. al. [10, 11]. Of course, some problems are fundamentally discrete (e.g., the insertion of a buffer is a 0/1 proposition). However, the practitioner should note that discrete and continuous optimization approaches can often complement one another. For instance, a continuous sizing of buffer sizes may follow a discrete buffer-insertion algorithm to fine tune performance (provided it is allowed by the design style).

6.5 EXERCISES

1. (Research Problem) For resistive interconnect, the lumped capacitance overestimates the "effective load" [9] on the output of a buffer or gate. As a result, simple *RC* delay estimators sometimes drastically overestimate buffer delay (depending on the structure of the interconnect they drive). Propose a scheme to take this into account; use simulation to assess the resulting improvement in accuracy; incorporate the scheme into a buffer-insertion algorithm and evaluate the effect on resulting solution quality (evaluated via simulation).
2. To simplify presentation, we assumed that insertion points were at branch points. Discuss an algorithm modification or topology transformation scheme which allows arbitrary insertion points.

BIBLIOGRAPHY

[1] K. D. Boese, A. B. Kahng, B. A. McCoy, and G. Robins, Near-optimal critical sink routing tree constructions, *IEEE Transactions on Computer-Aided Design of Integrated Circuits and Systems*, 14 (12): 1417–1436 Dec. 1995.

[2] J. Cong, K. S. Leung, and D. Zhou, Performance-driven interconnect design based on distributed *RC* delay model, *Proc. Design Automation Conf.*, pp. 606–611, 1993.

[3] J. Cong, and K. S. Leung, Optimal wiresizing under the distributed Elmore delay model, *Proc. Int. Conf. on Computer Aided Design*, pp. 634–639, 1993.

[4] S. Dhar, Delay optimization algorithms for tree models of MOS circuits, Ph.D dissertation, Dept. of Electrical Engineering, Washington University Stevens Institute of Technology, September 3, 1987.

[5] W. C. Elmore, The transient response of damped linear network with particular regard to wideband amplifiers, *J. Applied Physics*, 19 (1948), pp. 55–63.

[6] J. P. Fishburn, and A. E. Dunlop, TILOS: A posynomial programming approach to transistor sizing, *Proc. Int. Conf. on Computer Aided Design*, pp. 326–328, 1985.

[7] R. Gupta, B. Tutuianu, and L. T. Pileggi, The Elmore delay as a bound for *RC* trees with generalized input signals. *IEEE Transactions on Computer-Aided Design of Integrated Circuits and Systems*, 16 (1): 95–104 Jan. 1997.

[8] J. Lillis, C. -K. Cheng, and T. -T. Y. Lin, Optimal wire sizing and buffer insertion for low power and a generalized delay model, *IEEE Journal of Solid State Circuits*, 31 (3): pp. 437–447, March 1996.

[9] J. Qian, S. Pullela, and L. Pileggi, Modeling the "effective capacitance" for the *RC* interconnect of CMOS gates, *IEEE Trans on Computer-Aided Design of Integrated Circuits and Systems*, 13: 1526–1535.

[10] S. S. Sapatnekar, V. B. Rao, P. M. Vaidya, and S. M. Kang, An exact solution to the transistor sizing problem for CMOS circuits using convex optimization, *IEEE Transactions on Computer-Aided Design*, 12 (11): 1621–1632, Nov. 1993.

[11] S. S. Sapatnekar, Wire sizing as a convex optimization problem: Exploring the area delay tradeoff, *IEEE Transactions on Computer-Aided Design*, 15 (8): 1001–1011, Aug. 1996.

[12] L. P. P. P van Ginneken, Buffer placement in distributed *RC*-tree networks for minimal elmore delay, *Proc. International Symposium on Circuits and Systems*, pp. 865–868, 1990.

[13] F. F. Yao, Computational geometry, ch. 7 in *Handbook of Theoretical Computer Science, Vol A*, Elsevier Science Publishers, 1990.

7

Global Routing Topology Synthesis

7.1 INTRODUCTION

In Chapter 6, techniques for optimizing static routing topologies by wire sizing and buffer insertion were presented. In this chapter, we turn our attention to the problem of finding the routing topology itself for global routing applications. The technique is called the P-Tree algorithm and is also able to perform simultaneous wire sizing.

The focus is on constructing rectilinear Steiner trees. Thus, since even the wire-length version of the problem is NP-complete [6], the algorithms presented herein are heuristic in nature. However, the heuristic procedure is structured around finding an optimal solution in a constrained, but carefully designed solution space in polynomial time. As a result, the approach is quite systematic and is not an ad-hoc constructive heuristic; the general principles should be applicable to similar computationally difficult problems.

The proposed approach is a departure from previous timing-driven Steiner methods in that, as in the previous chapter, we are able to explicitly capture the notion of an area/delay trade-off. As a result, we not only improve on the performance achieved by previous approaches but, perhaps more importantly, we are able to produce high-performance routing trees with substantially less overhead in routing area versus previous approaches.

7.2 BACKGROUND AND OVERVIEW

7.2.1 Algorithm Overview

A key concept in the P-Tree approach is the idea of routing topologies being induced by a sink permutation. Given a permutation of the sinks of a net, consider

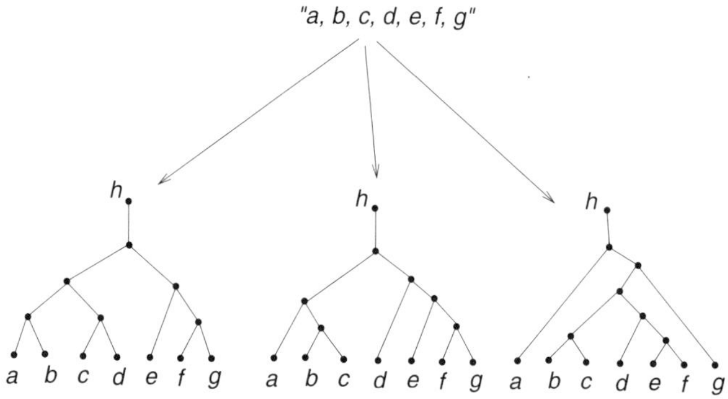

FIGURE 7.1 *Example trees induced by sink ordering "a,b,c,d,e,f,g".*

the (exponentially large) set of binary search trees with sinks identified with leaves that obey the order of the permutation (i.e., an in-order traversal of the tree visits the terminals in the order of the permutation). Figure 7.1 illustrates this concept by showing four trees which obey the permutation *"a b c d e f g"* (with the driver pin of the net labeled h). Now consider the possible mapping of the internal vertices of such a tree to a set of points in the plane or the vertices of a routing graph. Figure 7.2 shows several examples of such embeddings for a given set of pin locations. The P-Tree algorithms find routing solutions in this solution space (determining both topology and embedding simultaneously).

While the resulting solution space does not include every possible Steiner tree, the figure demonstrates the high degree of flexibility in determining the final routing solution. From the figure one can also see how various routing solutions in the space will exhibit different timing characteristics for each sink. The idea is that if care is taken in finding a high-quality permutation, the solution space will be sufficiently rich, containing many high-quality solutions while being sufficiently constrainted to allow efficient identification of such solutions. Furthermore, with respect to minimizing area, the proposed techniques for finding such high-quality permutations can guarantee a solution at least as good as that of a given structure (e.g., a minimum spanning tree or the result of another Steiner tree heuristic).

Thus, this work is a departure from the constructive greed heuristics proposed in the past (e.g., [1,8]). The approach bears some similarity to the alphabetic tree approach of [15] in that both approaches draw their solutions from the set of binary trees induced by a sink permutation. However, the approach differs in a number of fundamental ways. First placement of Steiner nodes in [15] is done heuristically while we go to great effort to optimally select their placement. Second, the objective function used in [15] is a weighted sum of source-to-sink delays while we minimize maximum delay or, more generally, maximize required arrival time; in most cases a required-time formulation is preferred because it deals directly with the timing requirements seen in practice. Third, area overhead was not taken into account in [15] while the P-Tree approach derives an explicit area/delay trade-off. Fourth, we use more sophisticated techniques for finding high-quality sink permutations. Finally, we incorporate simultaneous wire sizing.

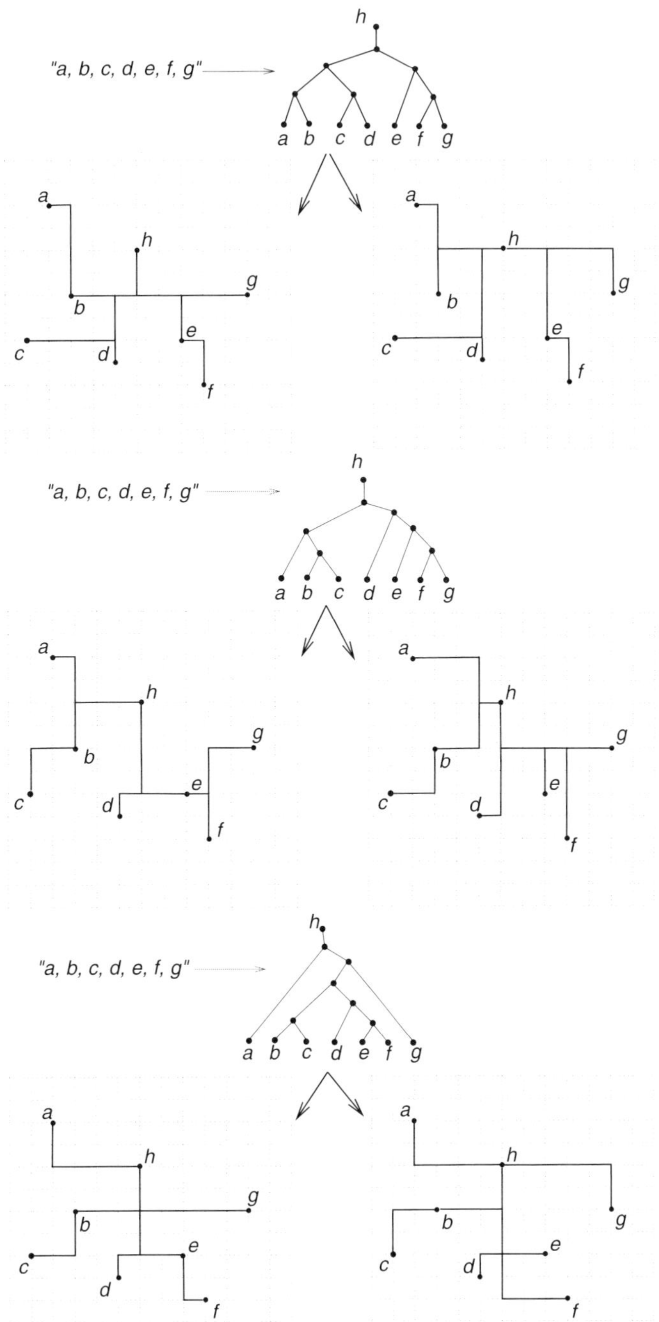

FIGURE 7.2 *Solution space illustration: From sink ordering to tree topologies to embedded topologies.*

The remainder of the chapter is organized as follows. Section 7.3 reviews delay models and gives some graph theoretic background; section 7.4 presents an algorithm for constructing good sink permutations; section 7.5 presents the P-Tree$_A$ and P-Tree$_{AT}$ routing algorithms for a given sink permutation; since the algorithms cannot guarantee optimality, section 7.6 presents experimental results supporting the effectiveness of the algorithms; we conclude in section 7.7.

7.3 PRELIMINARIES

7.3.1 Delay Models

As in Chapter 6, we adopt the Elmore delay model, which we review here. For a wire segment $e = (u, v)$, let r_e and c_e be the resistance and capacitance of e respectively (both of which are proportional to wire length l_e). Further, let $c(T_v)$ be the capacitive load at node v. The Elmore delay of the segment is expressed as

$$r_e \left(\frac{c_e}{2} + c(T_v) \right).$$

When wire sizing is taken into account, we need a model for resistance r_e, and capacitance c_e as a function of width w_e. A typical model is the following:

$$c_e = \alpha l_e \cdot w_e \qquad r_e = \beta l_e / w_e$$

where α and β are characteristic constants. For consistency with previous work, we adopt this model.

Delay of a driver g is defined similarly as

$$d_g + r_g \cdot c_L$$

where d_g is a constant known as intrinsic delay, r_g is the output resistance of the driver, and c_L is the capacitive load on g's output. Since d_g is constant, it may be neglected in relative comparisons of routing algorithms. We note also that we can consider the slew-based delay model presented in Chapter 6 to model driver delay. However, since it depends on the delay of the *previous* stage, it can be considered a constant and incorporated into the intrinsic delay.

Again, as in Chapter 6, we adopt a *required arrival time* formulation. Given a timing constraint on each sink u in the form of required arrival time q_u, the required arrival time for a routing tree T driven by gate g is

$$q(T) = \min_{sinks\ u} (q_u - delay(g \rightsquigarrow u))$$

where we include driver delay in the equation. If $q(T) \geq 0$, T is said to meet its timing requirements. A special case of required arrival time is maximum delay: letting $q_u = 0 \forall u$, the max source to sink delay is $-q(T)$.

7.3.2 Graphs, Trees, and Permutations

A classical result in the theory of Rectilinear Steiner Minimum Trees (RSMTs) is Hanan's theorem [7], which states that there always exists a RSMT for a terminal set N where all Steiner points are drawn from the n^2 points formed by the intersection of horizontal and vertical lines drawn through the terminals ($n = |N|$).

The grid formed by this set of points leads to a useful graph-theoretic formulation of the problem based on the *grid graph* of N.

DEFINITION 7.3.1

Grid Graph [9]: Given a set of terminals N, N's *grid graph* $GG(N) = (V, E)$ is defined by the following process:

Construct vertical and horizontal lines through each terminal.

Let V be identified with the set of intersection points of these lines.

There is an edge in E between two vertices iff the corresponding intersection points are connected by a single horizontal or vertical segment.

A grid graph is illustrated in Figure 7.3. The algorithms presented below produce routing topologies embedded in a graph structure such as *GG(N)*. We use $v_{\{r,c\}}$ to denote the vertex in *GG(N)* located at the rth row and cth column (position (1,1) being upper left). For convenience, we also use v_i and v_d to denote the vertex in *GG(N)* coincident with the ith pin of the permutation and the vertex coincident with the driving pin of the net, respectively.

For completeness, the following definitions relating pin permutations and routing topologies are provided.

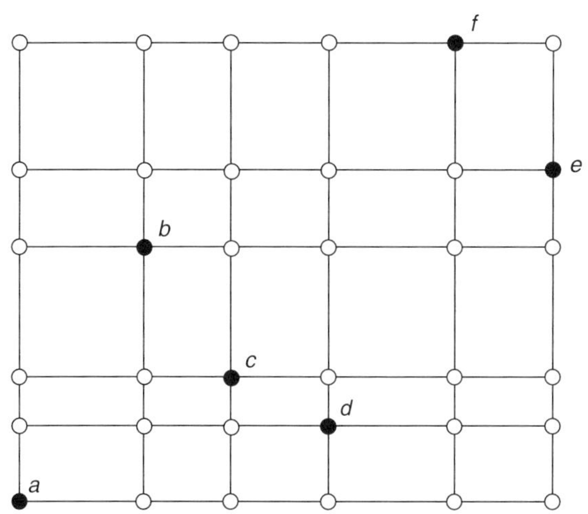

FIGURE 7.3 *Grid graph for pins a through f.*

DEFINITION 7.3.2

Permutation-Induced Abstract Topology: Consider a terminal set N and a permutation π on the sinks of N. A binary tree T is an abstract topology induced by π if its leaves are identified with the terminals in N and it obeys the ordering imposed by π when T is interpreted as a binary search tree (the driver being implicitly attached to the root).

DEFINITION 7.3.3

Abstract Topology Embedding: Given an abstract topology T (as a binary tree), leaves identified with the terminal set N and a target graph $G = (V, E)$ (e.g., $G = GG(N)$), an embedding of T into G is a mapping of the internal nodes of T to V.

To avoid confusion, we use the term "abstract topology" to emphasize that such a topology is not a true routing topology in that its internal nodes are not embedded. Only when such a topology is embedded in the plane or a routing graph such as *GG(N)* does a physical routing topology result.

The notion of "consistency" is a natural and convenient way of assessing the quality of a pin permutation. If we can show that a permutation π is consistent with a given routing topology T such as the Minimum Spanning Tree (MST), we ensure that the min-area solution induced by π is no worse than T. A similar notion is applied to a hierarchical decomposition of a terminal set (i.e., a binary tree with leaves identified with pins). Formally we have the following.

DEFINITION 7.3.4

Consistency: Given a set of terminals N and a tree structure T embedded in a routing graph and connecting N, a permutation π on N is *consistent* with T if π can induce a structure isomorphic to T (up to 0-length edges). Further, let H be a binary tree with all points $p \in N$ represented as leaves. We call H *a hierarchical decomposition* of N. H is *consistent* with T if *all possible depth-first traversals of* H visit the leaves in an order consistent with T.

Thus, if a permutation construction algorithm ensures consistency with the MST, we guarantee that the min area solution induced by π is no worse than $3/2$ times the optimal Steiner tree [10] while only relying on the MST indirectly in constructing the final topology. We present an algorithm for constructing permutations consistent with a given topology T such as the MST in the next section. Previous permutation-based approaches such as [15] provide no such assurance. Rather, they derive a sink permutation from a circular ordering of the sinks with the driver as the center of the circle, which can result in unbounded overhead versus the MST.

7.4 FINDING HIGH-QUALITY SINK PERMUTATIONS

The suggested method for constructing high-quality permutations is broken into three phases: (1) construction of a hierarchical decomposition of the terminals consistent with a given structure such as the MST, (2) reorienting the hierarchical structure such that the driver is attached to the root while maintaining consistency, and (3) application of a novel dynamic programming algorithm to further optimize the induced permutation.

Phase 3 is based on the following observation. Since the hierarchy from step (1) is consistent with the given topology, any depth-first traversal of the hierarchy visits the terminals in an order consistent with the given topology (e.g., MST). Can we perform such a traversal in a way to optimize the resulting permutation? In other words, we have a binary tree and we wish to assign a 0 or 1 to each internal node of the tree where 0 indicates that left subtree is explored first and 1 indicates that the right subtree is explored first. We propose that an appropriate metric for the quality of such an assignment is the tour length of the permutation induced by the resulting traversal in the traditional sense of the Traveling Salesman Problem (TSP). The intuition is that good TSP tours provide good clustering information that can identify good candidate sets for subtrees of a good Steiner tree. We present a dynamic programming algorithm for solving this subproblem optimally.

7.4.1 Hierarchy Construction and Reorientation

Consider the algorithm MST_to_hierarchy given in Figure 7.4, which takes a minimum spanning tree T_m and produces a hierarchy. The following consistency lemma holds by induction on the subtrees the algorithm computes.

Lemma 7.4.1 *The hierarchy formed by algorithm MST_to_hierarchy in Figure 7.4 is consistent with the minimum spanning tree of the given point set.*

Figure 7.5 illustrates the operation of *MST_to_hierarchy*, part (a) giving an example of a point set and its MST, and part (b) showing a hierarchical decomposition consistent with the MST produced by the algorithm. Labels on internal nodes of the hierarchy indicate the parent node from the merge step in the algorithm.

Algorithm: MST_to_hierarchy
Let T_m be a rectilinear Minimum Spanning Tree on terminal set N
For each $v \in T_m$, let $h(v)$ be a single node tree labeled v
Repeat $n - 1$ times
1. select a leaf node v and its parent u in T_m
2. delete edge (u,v) from T_m
3. replace $h(u)$ with a new tree t where $t.left = h(u)$ and $t.right = h(v)$

return last tree formed

FIGURE 7.4 *Pseudocode to construct hierarchy consistent with MST.*

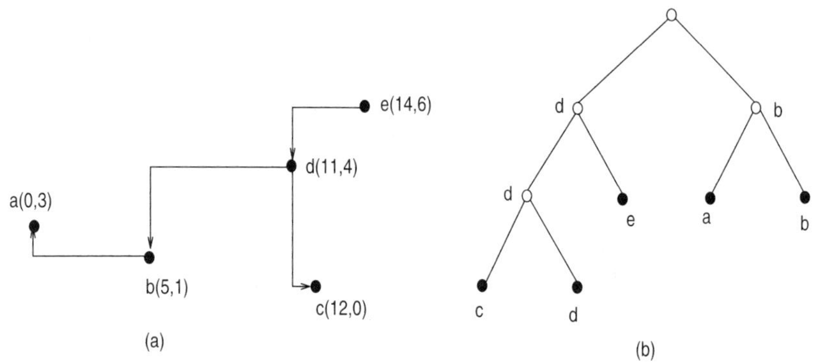

FIGURE 7.5 *Constructing a hierarchy consistent with MST.*

In the case where the given structure is a Steiner tree from another algorithm rather than a spanning tree, the algorithm may be modified as follows to deal with Steiner nodes. First, for each Steiner node u, $h(u)$ is initially NULL (for terminal nodes v, $h(v)$ is initialized as before). Second, when constructing a new tree in step 3 of the loop, if one of the subtrees is NULL, we simply copy the other tree. Finally, the number of times we execute the loop now equals the number of edges in the given Steiner tree.

The reader may have observed that in step 1 of the loop, there may be multiple candidate leaves v from which to choose and that the order in which they are chosen will impact the final hierarchy. We have examined several heuristics for selecting the next leaf. We have had success with the following: select v such that the depth of the tree resulting from step 3 is minimized; if there is a tie, select from among the candidates the node with minimum distance to its parent.

The intuition behind the heuristic is that a more balanced hierarchy is more likely to be better optimized by the subsequent tour length minimization phase and that nodes near each other should be clustered at the lower levels of the hierarchy.

After the hierarchy has been established, we must reorient the tree such that the driver is associated with the root. This is a necessary step since the signal direction is critical in performance driven routing: we want the signal to flow from root to leaves. Such reorientation does not disturb the consistency of the hierarchy and is illustrated in Figure 7.6 (note that the original root becomes a degree-2 internal node and therefore is redundant and eliminated).

7.4.2 Tour Length Minimization

In the third phase we solve the following problem.

Problem 7.4.1 Given a binary tree where leaves are identified with points in the plane, consider a labeling $s(v) \in \{0, 1\}$ of each internal node v with the following interpretation: if $s(v) = 0$, visit v's left child first; if $s(v) = 1$, visit v's right child first. Over all such labelings, find the one such that the tour length of the order in which the leaves are visited is minimized.

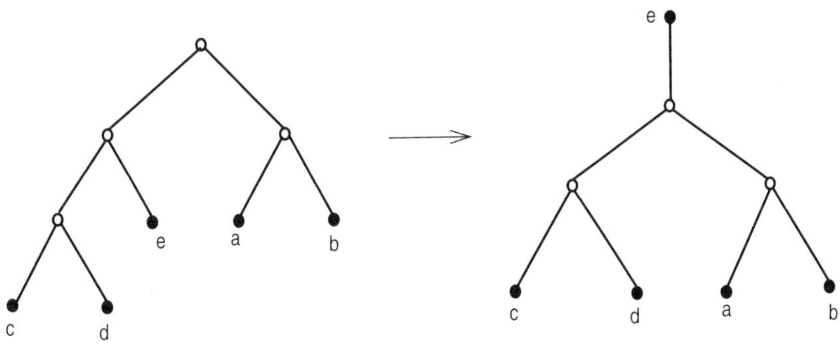

PERM	TOUR LEN
c d a b	24
d c a b	27
c d b a	21
d c b a	20

FIGURE 7.6 *Reorientation to reflect driver and possible permutations induced by reoriented tree.*

Referring again to Figure 7.6, we see the sink permutations that can be derived from the reoriented hierarchy and their (noncircular) tour length, "d c b a" being optimal. (There are actually 8 possible permutations corresponding to the 2^3 possible labelings of the internal nodes — the four shown and their reverse).

We solve Problem 7.4.1 by dynamic programming. For each node v in the tree let $P_l(v)$ and $P_r(v)$ be the sets of sink pins in v's left and right subtrees, respectively. For each pair $i \in P_l(v)$, $j \in P_r(v)$, we compute $D_v(i, j)$, the minimum length tour of the terminal set $P_l(v) \cup P_r(v)$ beginning at i and ending at j over all possible labelings of the internal nodes of the subtree rooted at v. Once we have computed $D_v(i, j)$ at the root for all possible i, j, we simply select the optimal solution. (This may be a circular tour or one that includes the driver itself, depending on implementation).

For a given v, i and j, let $l(v)$ and $r(v)$ be v's left and right children, respectively. We describe $D_v(i, j)$ by the following recurrence

$$D_v(i, j) = \min_{k_1 \in P_l(v) \setminus \{i\}, k_2 \in P_r(v) \setminus \{j\}} \{D_{l(v)}(i, k_1) + d(k_1, k_2) + D_{r(v)}(k_2, j)\} \qquad (7.1)$$

where $d(k_1, k_2)$ is the rectilinear distance between terminals k_1 and k_2. By symmetry we also know that $D_v(j, i) = D_v(i, j)$.

A naive algorithm for computing D_v has $O(n^4)$ complexity (i.e., we examine each i, j, k_1, k_2 quadruple). This gives an overall run time of $O(n^5)$ since this must be performed for each node in the tree.

However, a run time of $O(n^4)$ can be achieved by reducing the computation of D_v to $O(n^3)$. For a particular node v, consider the directed graph $G_{tsp} = (V, E)$ defined subsequently (all edges not explicitly defined have weight ∞) and Lemma 7.4.2.

For each terminal $i \in P_l(v) \cup P_r(v)$, we have nodes i and i' in V.
For each ordered pair $i, k \in P_l(v)$, introduce edge (i, k') with weight $w(i, k') = D_{l(v)}(i, k)$.
For each pair $k_1 \in P_l(v)$, $k_2 \in P_r(v)$, introduce edge (k'_1, k'_2) with weight $w(k'_1, k'_2) = d(k_1, k_2)$.
For each ordered pair $k_2, j \in P_r(v)$, introduce edge (k'_2, j) with weight $w(k'_2, j) = D_{r(v)}(k_2, j)$.

Lemma 7.4.2 *The shortest path from node $i \in P_l(v)$ to $j \in P_r(v)$ in G_{tsp} visits nodes k'_1 and k'_2 where k_1 and k_2 minimize expression (7.1) and the path length equals the tour length.*

The idea is that a path from a vertex $i \in P_l(v)$ to a vertex $j \in P_r(v)$ must be of the form i, k'_1, k'_2, j where k_1 and k_2 are possible intermediate nodes in the recurrence, equation (7.1), and the sum of the edge weights matches the sum in the recurrence. By Lemma 7.4.2 we can compute $D_v(i, j)$ for all i, j by the Floyd-Warshall all-pairs shortest-paths algorithm [3].

Since the Floyd-Warshall algorithm runs in $O(n^3)$ time and is applied at each of the internal nodes of the tree, we have the following.

Lemma 7.4.3 *Problem 7.4.2 is solved in $O(n^4)$ time by the above scheme.*

7.5 TREE CONSTRUCTION FOR A GIVEN PERMUTATION

This section describes the P-Tree algorithms, which take as input a sink permutation established by the techniques of the previous section. Again using the example point set begun in Figure 7.5, we give Figure 7.7, where an abstract topology is induced in 7.7(a) and that abstract topology is embedded in $GG(N)$ to form the routing topology in 7.7(b) (yielding, in this case, a min-area Steiner tree). It is this process that the P-Tree algorithms perform optimally.

We first give an overview and subsequently describe the P-Tree$_A$ algorithm for area minimization and then the P-Tree$_{AT}$ algorithm for deriving area/delay trade-off in detail.

7.5.1 Overview

Recall the definition of the grid graph $GG(N) = (V, E)$ of a terminal set $N(n = |N|)$; V contains n^2 vertices, $|E| = O(n^2)$ and all vertices have degree no greater than 4. Recall that $v_d \in V$ denotes the vertex associated with the driver of the

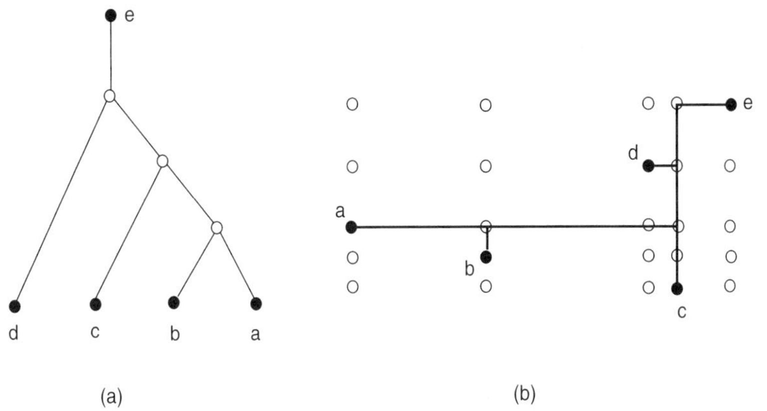

FIGURE 7.7. *A permutation-induced abstract topology and an embedding in GG(N).*

net. We also refer to the ith sink in the given permutation as sink i (i.e., without loss of generality we relabel the sinks based on the permutation). Also recall that $v_i \in V$ refers to the graph vertex coincident with sink i.

The algorithms construct trees rooted at vertices $v \in V$. These trees have the property that they drive a consecutive subsequence of the sink permutation; the subsequence is defined by its starting and ending indices i and j. The trees provide the subsolutions for a dynamic programming algorithm. More specifically, we inductively compute the following solutions or solution sets for a given sink permutation.

$S(v, i, j)$: the cost of the optimal solution(s) over all permutation induced (and embedded) routing topologies driving sinks $i..j$ and rooted at v.

$S_b(v, i, j)$: the cost of the optimal solution(s) over all permutation induced (and embedded) routing topologies driving sinks $i..j$ and rooted at v where v is additionally constrained to have degree 2—that is, v is a *branching point* $(i < j)$.

Notice that solutions represented in S may either have v as a branching point or be single-stem trees rooted at v. The idea is to use $S_b(v', i, j)$ to compute $S(v, i, j)$.

In the case of minimizing area alone, the "costs" $S(v, i, j)$ and $S_b(v, i, j)$ are scalars giving the area of the optimal solution. The min-area version of the algorithm can be thought of as a special case of the *Dreyfus-Wagner* (DW) algorithm [4] for the graph Steiner problem. The DW algorithm uses a similar decomposition into subsolutions; the difference is that instead of terminal subsets being determined by consecutive subsequences of a permutation, the DW algorithm considers *all* 2^n subsets of terminals.

For performance-driven routing, subsolutions are more complex since are we interested not only in the area of the subsolutions, but also in their timing properties and their capacitance since this will affect the delay of ancestor wires and

```
1. Compute S(v,i,i)          ∀v ∈ V, 1≤ i < n
2. For l = 1..n – 2
3.     For i = 1..n – 1 – l
4.         j = i + l
5.         Compute S_b(v,i,j)        ∀v ∈ V
6.         Compute S(v,i,j)      ∀v ∈ V using results of step 5
7. S (v_d, 1, n–1) now gives the subsolution(s) to be paired with the driver.
```

FIGURE 7.8 *Framework for permutation-constrained routing.*

the driver. Fortunately, area is proportional to capacitance. Thus, the capacitance c and required arrival time q are the key properties of a routing tree. In much the same manner as in Chapter 2, we maintain sets of (c, q) pairs ordered by c—this gives the notion of cost as a trade-off curve in the two-dimensional space of c and q. Again, as in Chapter 2, these sets have the property that they are strictly increasing in both c and q. Properties of these sets and primitives manipulating them are given later in this chapter.

While there are fundamental differences in the subsolutions computed by P-Tree$_A$ and P-Tree$_{AT}$, both algorithms follow the same basic structure given in Figure 7.8. The variable l indicates the length of the subsequences we are currently considering. This ensures that $S(v, i, k)$ and $S(v, k+1, j)$ have previously been computed for all $k \in \{i..j-1\}$ when we reach step 5 in the loop. Similarly, $S(v, i, j)$ is derived in step 6 from the result of step 5.

The details of the individual steps in Figure 7.8 for P-Tree$_A$ and P-Tree$_{AT}$ are given in the next two sections, respectively.

7.5.2 Routing for Min-Area: The P-Tree$_A$ Algorithm

The first and simplest step in P-Tree$_A$ is computation of base cases $S(v, i, i)$. We are just interested in the length of a wire from v to pin i as follows:

$$S(v, i, i) = d(v, v_i) \qquad \forall i \in \{1..n-1\}$$

where $d(v, v_i)$ is the rectilinear distance between vertices v and v_i in $GG(N)$.

Next we need to show how to compute $S_b(v, i, j)$. Since by definition of S_b, v must be a branching point we have the following.

$$S_b(v, i, j) = \min_{k \in \{i..j-1\}} \{S(v, i, k) + S(v, k+1, j)\}$$

Note that since $k - i < j - i$ and $j - k - 1 < j - i$, $S(v, i, k)$ and $S(v, k+1, j)$ have been previously computed.

Computation of $S(v, i, j)$ is more complex. A naive approach is given by implementing the following.

$$S(v, i, j) = \min_{v' \in V}\{d(v, v') + S_b(v', i, j)\}$$

In other words, any single-stem tree rooted at vertex v connects to some branching point v'; we find the optimal such branching point. Since there are n^2 candidates for v', sequentially computing $S(v, i, j)$ for all v by this procedure takes $O(n^4)$ time for given i, j. On the other hand, this procedure only computes $O(n^2)$ values. The $O(n^4)$ complexity for this step is unnecessary. In fact, $S(v, i, j)$ can be computed for each v in $O(n^2)$ time.

The key in achieving the $O(n^2)$ run time is the underlying graph $GG(N)$. The following simple lemma relating the optimal solution at a vertex v to the optimal solutions at v's neighbors in $GG(N)$ can be shown by contradiction. This gives the optimal substructure property required for dynamic programming.

Lemma 7.5.1 *For fixed i, j, the routing tree giving $S(v, i, j)$ either uses v as a branching point (i.e., $S(v, i, j) = S_b(v, i, j)$) or $S(v, i, j) = S(v'', i, j) + d(v, v'')$ where v'' is one of v's immediate neighbors in $GG(N)$.*

Referring back to Figure 7.7, the idea of the lemma is illustrated by the vertices labeled v and v'' where $S(v, b, a)$ is derived from $S(v'', b, a)$.

Without loss of generality, assume that the wire connecting v to its optimal branching point v' first travels along v's column to v''s row and then finishes the trip horizontally to v'. This leads to the four-phase strategy given in Figure 7.9 (for fixed i, j).

The algorithm computes intermediate solutions $LS(v)$, $RS(v)$, and $US(v)$, the meanings of which are summarized as follows.

$LS(v)$: the area (wire-length) of the optimal single-stem tree rooted at v with branching point v' constrained to be at v or to v's *left* in the same row as v.

```
/* Phase 1 */
LS(v{r,1}) ← Sb(v{r,1})          ∀r ∈ {1..n}
for r = 1 to n
    for c = 2 to n
        LS (v{r,c}) ← min {Sb(v{r,c}), LS(v{r,c−1}) + d(v{r,c}, v{r,c−1})}
/* Phase 2 */
RS(v{r,n}) ← LS(v{r,n})          ∀r ∈ {1..n}
for r = 1 to n
    for c = n−1 to 1
        RS(v{r,c}) ← min{LS(v{r,c}), RS(v{r,c+1}) + d(v{r,c}, v{r,c+1})}
/* Phase 3 */
US(v{1,c}) ← RS(v{1,c})          ∀c ∈ {1..n}
for c = 1 to n
    for r = 2 to n
        US(v{r,c}) ← min{RS(v{r,c}), US(v{r−1,c}) + d(v{r,c}, v{r−1,c})}
/* Phase 4 */
S(v{n,c},i, j) ← US(v{n,c})       ∀c ∈ {1..n}
for c = 1 to n
    for r = n−1 to 1
        S(v{r,c},i, j) ← min{US(v{r,c}), S(v{r+1,c},i, j) + d(v{r,c}, v{r+1,c})}
```

FIGURE 7.9 *Computation of $S(v, i, j)$ $\forall v \in V$ in P-Tree$_A$.*

$RS(v)$: the area of the optimal single-stem tree rooted at v with branching point v' constrainted to be in the same *row* as v.

$US(v)$: the area of the optimal single-stem tree rooted at v with branching point v' constrained to be in v's row or a row *above* v ("up" from v).

At the beginning of each phase, we have a bootstrap step. For instance, if v is in the bottom row then $US(v)$ effectively has no constraint on its branching position v' and therefore equals $S(v, i, j)$. With this a base case, we inductively compute $S(v, i, j)$ from bottom to top. Once $US(v)$ has been computed, we are then able to compute $S(v, i, j)$ $\forall v$ in a single additional pass. Each phase is implemented in a manner similar to the computation of the prefix sum or prefix max of a list of numbers, only examining an immediate neighbor to perform the computation (as provided by Lemma 7.5.1).

This process yields an $O(|V|) = O(n^2)$ complexity for computing, for all $v \in V$, $S(v, i, j)$ for a fixed i, j. Thus the overall complexity of *P-Tree*$_A$ is dominated by the computation of $S_b(v, i, j)$, which is $O(n)$ for each v,i,j triple. Since there are $O(n^4)$ such triples, this gives an overall complexity of $O(n^5)$.

Finally, we can show the optimality of the solution computed by P-Tree$_A$ inductively.

Lemma 7.5.2 *$S(v_d, 1, n-1)$ computed by the P-Tree$_A$ algorithm is the minimum area over all possible rectilinear Steiner trees induced by given sink ordering π. This solution is computed in $O(n^5)$ time.*

7.5.3 Routing for Performance: The P-Tree$_{AT}$ Algorithm

We extend the P-Tree$_A$ algorithm of the previous section to the timing-driven version of the algorithm, P-Tree$_{AT}$, which derives an area/delay trade-off. The basic structure of P-Tree$_{AT}$ is essentially identical to P-Tree$_A$ except in the primitives we use to manipulate solutions. Recall that both $S_b(v, i, j)$ and $S(v, i, j)$ were scalars in P-Tree$_A$. By contrast, P-Tree$_{AT}$ manipulates sets of *load, required-time* pairs. We refer to such sets as *cq-sets* and introduce new primitives for manipulating these sets. For instance, a load, required-time pair $(c, q) \in S(v, i, j)$ indicates that there exists a legal routing topology rooted at v and driving sinks $i..j$ that has lumped capacitance c at v (also proportional to routing area) and achieves required-arrival time q at v. Note that since we may be performing simultaneous wire sizing, we need to determine not only a routing topology but also the width of each wire segment between Steiner nodes.

As in the preceding chapter, we use the following pruning property.

Property 7.5.1 For $(c, q), (c', q') \in S$, if $c' \geq c$ and $q' < q$ then (c', q') is suboptimal.

Intuitively, this notion is clear since a larger capacitance can only slow down a path from the driver. Thus, given a cq-set S, all (c', q') pairs as described by

Property 7.5.1 may be eliminated. As a result we can organize cq-sets in strictly increasing order of c *and* q (recall that we prefer large required-times). This ordering captures the notion of an area/required-time (delay) trade-off and the effect of subsolutions on ancestors, thus giving capacitance values c the dual role of capturing area overhead and load presented to parents.

To manipulate and combine cq-sets, we need some primitive operations as follows.

join_cq_sets (S_1, S_2): This primitive was introduced in Chapter 6. It serves in computing $S_b(v, i, j)$. Its inputs are solution sets S_1 and S_2, both of which give solutions rooted at v, where S_1's solutions drive sinks $i..k$ and S_2's solutions drive sinks $k+1..j$ for some $k \in \{i..j-1\}$. From these sets it produces a set of solutions rooted at v driving sinks $i..j$, the size of which is proportional to $|S_1| + |S_2|$.

augment_cq_set (S, l, w): This primitive takes cq-set S and computes a new cq-set where each $(c, q) \in S$ is augmented by a wire of length l and width w. This is used in constructing single-stem solution sets $S(v, i, j)$.

prune_cq_set (S): This primitive takes a cq-set S and applies Property 7.5.1 to it, eliminating sub optimal solutions.

Recall from Chapter 6 the operation of the routine *join_cq_sets* in Figure 7.10. Sets S_1 and S_2 are in increasing order of c and q and indexed. The routine produces the new cq-set by first pairing the solutions with minimum c. We then advance through the sets noticing that we can only improve the current solution by improving the required-time of the critical branch—improving the required-time of a noncritical branch results in a suboptimal solution (i.e., one having more load than necessary).

The routine *augment_cq_set(S,l,w)* examines each $(c, q) \in S$ and adds to c the capacitance of a wire of length l and width w and subtracts the delay of such a wire from q to give (c', q') as

$$c' = c + \alpha wl \qquad q' = q - \frac{\beta l}{w}\left(\frac{\alpha l w}{2} + c\right)$$

```
Algorithm: join_cq_sets(S1,S2)
  S ← φ
  i ← 1; j ← 1
    While (i ≤ |S1| and j ≤ |S2|)
          Let (c1,q1) = S1[i]
          Let (c2,q2) = S2[j]
    S ← S ∪ {c1 + c2, min (q1,q2))}
     If (q1 ≤ q2)    /* S1 Critical */
      i ← i + 1
     If (q2 ≤ q1)   /* S2 Critical */
      j ← j + 1
    return S.
```

FIGURE 7.10 *Primitive join_cq_sets(S_1, S_2).*

Finally, we implement *prune_cq_set(S)* by making a linear pass over S in increasing order of c, eliminating suboptimal solutions as we go (by comparing each (c, q) pair with its predecessor).

To get started we need the base case

$$S(v_i, i, i) \longleftarrow \{(c_i, q_i)\} \qquad \forall i \in \{1..n-1\}.$$

where c_i and q_i are the input capacitance and required arrival time for sink i, respectively. Computation of $S(v, i, i)$ where $v \neq v_i$ is shown in Figure 7.11 (W is the largest allowable wire width as a multiple of the base wire width, which is normalized to 1).

Computation of $S_b(v, i, j)$ is illustrated in Figure 7.12. The algorithm visits the possible partition points k in the subsequence and takes the union of the resulting cq-sets and prunes the result per Property 7.5.1.

As in the min-area case we compute $S(v, i, j)$ for all $v \in V$ and fixed i, j in a four-phase process.

During the computation we partition the intermediate solution sets into W disjoint subsets to ensure that wire segments between Steiner points are of uniform width.[1] For instance, $LS(v, w)$ is a cq-set rooted at v where the wire to a branching point v' must have width w and v' is constrained to be at v or to v's left in the same row. In the fourth stage of the algorithm we collect in $S(v, i, j)$ all solutions regardless of wire width. Analogously to Lemma 7.5.1, we have the following, which gives the subsolution optimality property necessary for dynamic programming.

Lemma 7.5.3 *The routing tree giving $(c, q) \in S(v, i, j)$ either has v as a branching point (i.e., $(c, q) \in S_b(v, i, j)$), or $\exists (c', q') \in S(v', i, j)$ where*

```
S ← ∅
for w = 1..W
    S ← S ∪ augment_cq_set (S(v_i,i,i), d(v,v_i), w)
S(v_i,i,i) ← prune_cq_set (S)
```

FIGURE 7.11 *Computation of $S(v, i, i)$ in P-Tree$_{AT}$ ($v \neq v_i$).*

```
S ← ∅
for k = i..j−1
    S ← S ∪ join_cq_sets (S(v,i,k), S(v,k + 1,j))
    S ← prune_cq_set (S)
S_b(v,i,j) ← S
```

FIGURE 7.12 *Computation of $S_b(v, i, j)$ in P-Tree$_{AT}$.*

[1] The algorithm can be structured to allow multiple widths per wire segment. However, this is at the expense of higher computational complexity—such fine-tuning might be more appropriate at a later stage on a fixed topology.

```
/* Phase 1 */
LS(v_{r,n}, w) ← S_b(v_{r,n})    ∀r∈{1..n}, w∈{1..W}
for r = 1 to n
   for c = 2 to n
      for w = 1 to W
         LS(v_{r,c}, w) ← S_b(v_{r,c}) ∪
                        augment_cq_set(LS(v_{r,c−1}, w), d(v_{r,c−1}, v_{r,c}), w)
         LS(v_{r,c}, w) ← prune_cq_set(LS(v_{r,c}, w))
```

FIGURE 7.13 *Computation of* $LS(v)\ \forall v \in V$ *P-Tree*$_{AT}$.

v' is one of v's neighbors in $GG(N)$, $c = c' + \alpha w\, d(v, v')$, and $q = q' - (\beta d(v, v')/w)(\alpha d(v, v')w + c')$ for some $w \in \{1..W\}$.

Figure 7.13 gives the pseudocode for the first phase — that is, the computation of $LS(v, w)$. The other phases follow analogously as in the P-Tree$_A$ algorithm. Notice that instead of taking scalar minimums, we now apply the cq-set primitives and we now must deal with the parameterization of the sets by w.

Thus, we have shown how to compute cq-sets $S_b(v, i, j)$ and $S(v, i, j)$ for all $v \in V$ and $i \leq j$. Once this is complete, we must consider the cq-set seen at the driver position v_d. To obtain the final area/delay trade-off set S_{final} we augment each pair $(c, q) \in S(v_d, 1, n-1)$ by considering the resistance of the driver (much in the same way as we did with an augmenting wire) as follows.

$$S_{final} \leftarrow \{(c, q - r_d \cdot c) | (c, q) \in S(v_d, 1, n-1)\}$$

$$S_{final} \leftarrow prune\_cq\_set(S_{final})$$

Finally, we can show the optimality of the solutions derived by the P-Tree$_{AT}$ algorithm analogously to Lemma 7.5.2.

Lemma 7.5.4 *Consider $(c, q) \in S_{final}$ computed by the P-Tree$_{AT}$ algorithm. There exists no rectilinear Steiner tree induced by sink permutation π with Steiner points on the Hanan grid that yields required-time $q' \geq q$ and capacitance $c' \leq c$ (recall that load is proportional to area).*

Before proceeding, we would like to point out one of the properties of P-Tree$_A$. It is often the case that there are many distinct topologies induced by a permutation, all of which achieve min area. By incorporating the required-time q with the area (and capacitance) in P-Tree$_A$ we can derive the min-delay topology among the min-area topologies without increasing run time. This is done by using required-time q as the tie-breaker when multiple solutions derive identical area.

7.5.4 Complexity

As previously noted, the P-Tree$_A$ algorithm is dominated by the computation of the S_b sets and has $O(n^5)$ complexity. The complexity of P-Tree$_{AT}$, however, depends on the size of the cq-sets it manipulates, as reflected in the next lemma.

Lemma 7.5.5 *Algorithm P-Tree$_{AT}$ executes in time $O(\max(Wn^4m, n^5m)) = O(n^5m)$ where m is a bound on the maximum size of any cq-set and W is the number of allowed wire widths (W assumed to be $O(n)$).*

It can be argued that m is in fact polynomially bounded if the possible capacitives, are polynomially bounded integers or can be mapped to such with sufficient precision. This yields a *pseudo-polynomial* complexity.

The complexity of the P-Tree$_{AT}$ algorithm can be further formalized when the following assumptions are made.

1. The pin positions are drawn from a *uniform* $l \times l$ grid.
2. Let c_A be the cost (or total capacitance) of the min-area P-Tree solution. There exists a constant γ, such that solutions with cost greater than γc_A are considered infeasible.

The first assumption does not constrain the problem since l can be made arbitrarily large. It does, however, allow a complexity analysis in terms of l, which, in turn, gives some insight into the trade-off between precision and run time—that is, if a problem instance on an $l \times l$ is mapped to a coarser $l' \times l'$ grid, what is the impact on run time? The second assumption allows a tighter bound on the run time, and seems quite reasonable in practice.

We first bound the size of the routing trees in terms of the basic grid length in the following lemma.

Lemma 7.5.6 *Any set of n points drawn from an $l \times l$ grid has a rectilinear Steiner tree of length $O(l\sqrt{n})$.*

Proof: Uniformly divide the $l \times l$ grid into n subgrids of size $(l/\sqrt{n}) \times (l/\sqrt{n})$ or less. Let each point connect to the nearest boundary of the subgrid in which it lies; this connection has length $O(l/\sqrt{n})$ and thus, for all n points the cost is $O(l\sqrt{n})$. Clearly, the n resulting boundary points can be connected with $O(l\sqrt{n})$ additional cost since only segments on the $\sqrt{n}$ boundary rows (columns) need be considered. □

This is by no means a new observation (in fact, it can be shown that $O(l\sqrt{n})$ is the expected length of the MST—see Chapter 5 of [11] for related information). Since the RMST is within a constant factor of the RSMT, it follows that the topology produced by P-Tree$_A$ is $O(l\sqrt{n})$ in length.

From this analysis (and assumption 2), we can conclude that only trees of total length $O(\gamma l\sqrt{n}) = O(l\sqrt{n})$ are of interest and that this also bounds the number of *distinct* capacitive values c that must be considered for any cq-set. The overall run time follows directly from this discussion and is given by the following theorem.

Theorem 7.5.1 *Subject to assumption 2 above, P-Tree$_{AT}$ has run-time complexity $O(n^{5.5}l)$ or $O(n^{6.5})$ if $l = O(n)$. Space complexity is $O(n^{4.5}l)$ or $O(n^{5.5})$ if $l = O(n)$.*

7.5.5 Heuristic Limitation of cq-Set Size

As is clear from the preceding complexity analysis, cq-sets can grow to be quite large depending on the grid granularity, number of pins, and number of wire widths. To combat this we have implemented a simple heuristic to limit cq-sets to be no larger than a prespecified size. Given a maximum allowable cq-set size of m and a cq-set S with $|S| > m$, we first select $m/2$ entries in S uniformly by their ordering (always including the first entry) and select the remaining $m/2$ by their "marginal benefit"; that is, the marginal benefit of ith entry $(c_i, q_i) \in S$ is $(q_i - q_{i-1})/(c_i - c_{i-1})$. This strategy has proved reasonably effective in improving run time while maintaining high solution quality.

7.6 EXPERIMENTS

Since the P-Tree algorithm cannot guarantee optimality, we report experimental results in this section to demonstrate the quality and stability of the approach versus previous work.

The main experimental results appear in Table 7.2. Technology parameters are the same as those used in previous works such as [8] and [1] and appear in Table 7.1. For each net-size and technology pair 25 nets were routed (with terminal positions generated randomly and uniformly). We first ran the SERT and A-Tree algorithms and their wire-sizing variants on the nets and then used the delay figure achieved by these algorithms as timing requirements for the P-Tree$_{AT}$ algorithm. From this the min-area solutions produced by P-Tree$_{AT}$ achieving the same delay (or better) as SERT and A-Tree were derived.

As can be seen in both the unsized and wire-sized cases, SERT consumes substantially more area than a P-Tree solution with equal or better timing. For instance, for 12-pin nets in MCM technology without wire sizing, the SERT solutions used 66% more area on average than the P-Tree solution. In some instances, SERT consumes more than twice the area of the P-Tree solution. A-Tree solutions

TABLE 7.1 Technology Parameters for Experiments

Tech.	MCM	1.2 μm CMOS	0.5 μm CMOS
driver res.	25.0 Ω	212.1 Ω	270.0 Ω
wire res.	0.008 $\Omega/\mu m$	0.073 $\Omega/\mu m$	0.122 $\Omega/\mu m$
wire cap.	0.06 $fF/\mu m$	0.083 $fF/\mu m$	0.039 $fF/\mu m$
sink cap.	1000 fF	7.06 fF	1.0 fF

TABLE 7.2 Average Area Overhead versus P-Tree$_{AT}$ (area of P-Tree$_{AT}$ solution is normalized to 1.0)

		MCM		1.2 μm		0.5 μm	
		SERT	A-Tree	SERT	A-Tree	SERT	A-Tree
w/o	$n = 6$	1.672	1.055	1.353	1.122	1.381	1.076
wire	$n = 9$	1.406	1.060	1.608	1.160	1.533	1.081
sizing	$n = 12$	1.660	1.083	1.627	1.170	1.657	1.158
with	$n = 6$	1.289	1.116	1.461	1.171	1.638	1.171
wire	$n = 9$	1.300	1.105	1.570	1.193	1.700	1.184
sizing	$n = 12$	1.362	1.053	1.566	1.192	1.727	1.199

tend to produce smaller area solutions than SERT (delay figures appear to depend on the technology and net size — sometimes A-Tree outperforms SERT in delay also and vice-versa). As can be seen in the table, A-Tree consumes as much as 17% more area on average for unsized 12-pin nets in $1.2 - \mu m$ technology.

In the wire-sized cases the comparison is with the dynamically wire-sized variant SERT and the statically wire-sized variant of A-Tree described in [8]. We again see the substantial overhead of the SERT algorithm and significant overhead of A-Tree with SERT consuming more than 70% additional area for 12-pin $0.5 - \mu m$ nets and nearly 20% area overhead for A-Tree on those same nets[2]. We note that in the 12-pin cases with wire sizing (and only these cases), the heuristic described in the previous section to limit the size of the cq-sets with $m \approx 50$ was applied.

Another way of evaluating the quality of a routing topology is the *AT* metric (i.e., the product of area and delay). Results under this metric appear in Table 7.3. Results are normalized to the P-Tree$_{AT}$ solution with minimum area-delay product. Once again the P-Tree$_{AT}$ algorithm yields significant improvement.

The next set of experiments studied the min-delay solutions derived by P-Tree$_{AT}$ versus SERT and A-Tree. The results appear in Table 7.4. As can be seen, significant improvements can be made when absolute minimum delay is the goal. Of course, this performance comes at a price in area overhead (i.e., we are at the flattest portion of the trade-off curve).

Another set of topologies appears in Figure 7.14. The figures show a sampling of the trade-off curve computed by the P-Tree algorithm for a 12-pin 0.5 μm net; we begin with the minimum area topology and end with the minimum delay topology (maximum delay is the metric). The driver is indicated by a rectangle and in each case the critical sink is circled. The corresponding point in the

[2] In two of the 12-pin 1.2-μm trials and three of the 0.5-μm trials, the min-delay P-Tree$_{AT}$ solution was beaten by A-Tree. In these cases the difference in delay was less than 3% percent on average and never greater than 7%; area overhead was comparable to that in the table. We suspect that this is an artifact of the heuristic used for limiting cq-sets used in these cases. The figures in the table reflect the nets for which P-Tree$_{AT}$ equaled or bettered A-Tree.

TABLE 7.3 Comparison Under AT Metric (avg.)

		MCM		1.2 μm		0.5 μm	
		SERT	A-Tree	SERT	A-Tree	SERT	A-Tree
w/o	$n = 6$	1.406	1.130	1.536	1.174	1.568	1.119
wire	$n = 9$	1.695	1.161	2.008	1.231	1.751	1.163
sizing	$n = 12$	1.907	1.214	2.005	1.243	1.945	1.226
with	$n = 6$	1.527	1.368	1.780	1.259	1.865	1.293
wire	$n = 9$	1.504	1.278	1.927	1.286	1.875	1.257
sizing	$n = 12$	1.580	1.211	1.844	1.283	1.939	1.231

AT product of the P-Tree$_{AT}$ solution is normalized to 1.0.

TABLE 7.4 Average Min-Delay

		MCM		1.2 μm		0.5 μm	
		SERT	A-Tree	SERT	A-Tree	SERT	A-Tree
w/o	$n = 6$	1.018	1.301	1.146	1.051	1.102	1.056
wire	$n = 9$	1.045	1.421	1.271	1.077	1.179	1.065
sizing	$n = 12$	1.074	1.481	1.281	1.087	1.235	1.113
with	$n = 6$	1.054	1.048	1.206	1.059	1.372	1.057
wire	$n = 9$	1.109	1.145	1.265	1.074	1.396	1.049
sizing	$n = 12$	1.128	1.136	1.267	1.084	1.429	1.058

Delay of P-tree$_{AT}$ solution is normalized to 1.0.

trade-off curve is also circled for each topology. This sequence of topologies is a good example of the flexibility of the P-Tree approach.

The next set of experiments studied minimizing area alone. The P-Tree$_A$ algorithm was applied iteratively. In other words, we take a given topology such as the MST, induce a permutation consistent with that topology, and apply P-Tree$_A$ to that permutation. The process may be repeated on the new topology. The goal was to see how well P-Tree$_A$ compared with the state-of-the-art Batched 1-Steiner (B1S) algorithm of Kahng and Robins [12] and to see how much improvement could be obtained by using B1S to induce the initial permutation. The experiment was on fifty trials of 20-pin nets. The results appear in Table 7.5. The first column shows that the P-Tree$_A$ algorithm when applied to a permutation derived from

TABLE 7.5 Routing 20-pin Nets for Min Area

	MST + P-Tree$_A$	MST + 2P-Tree$_A$	B1S + P-Tree$_A$	B1S + 2P-Tree$_A$
Avg. Imp.	−0.418%	−0.164%	0.358%	0.380%

Figures are average percentage improvement versus BIS.

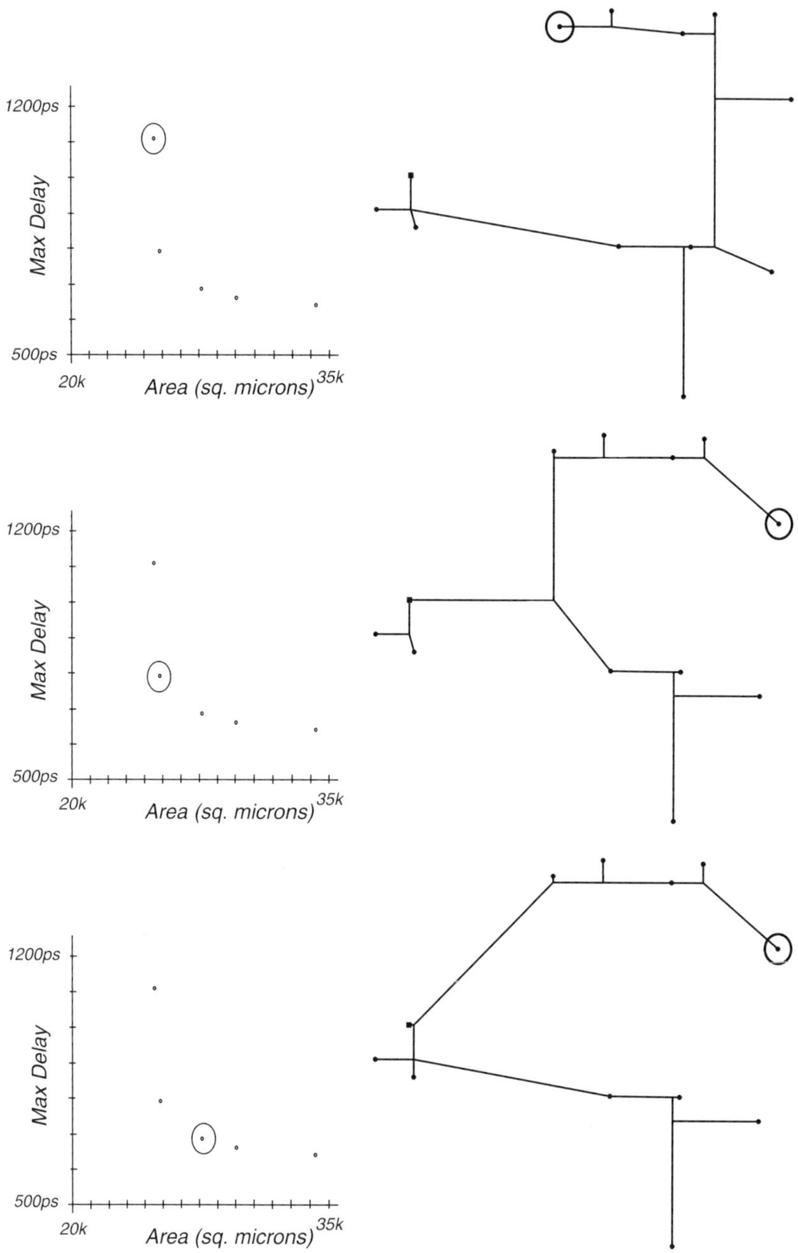

FIGURE 7.14 *P-Tree topologies for a 12-pin net in 0.5-μm technology. Driver is pin indicated by the rectangle. Corresponding point on trade-off curve and critical sink are circled. Topologies are rectilinear: Tree edges are drawn as Euclidean arcs for convenience.*

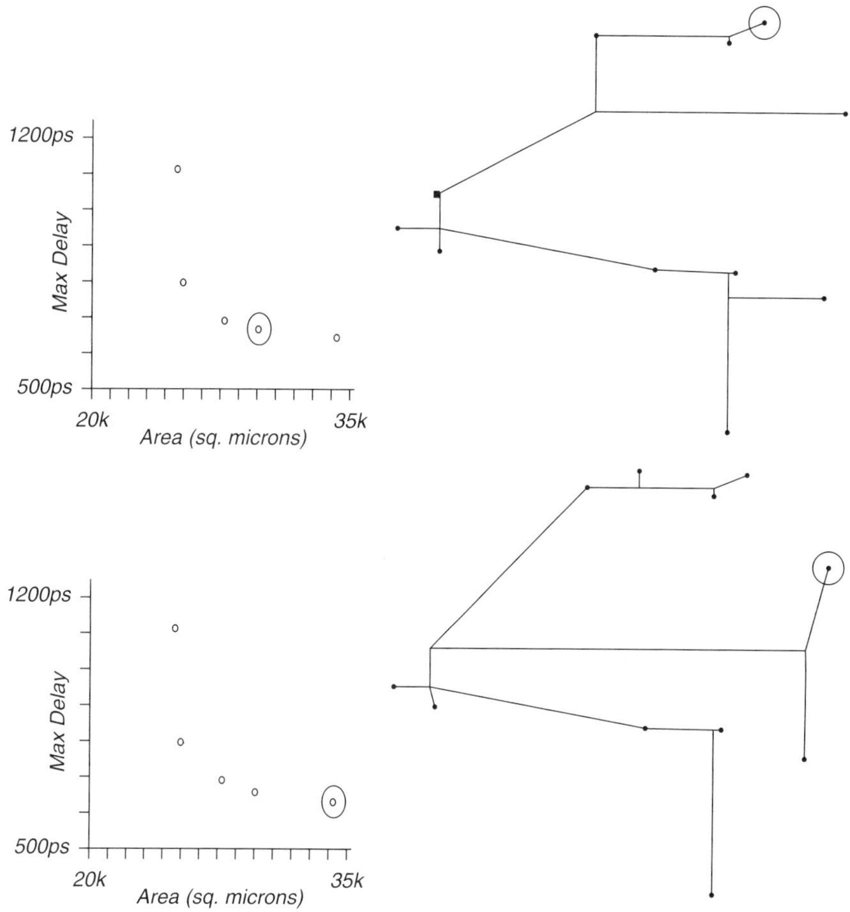

FIGURE 7.14 *(continued)*

the minimum spanning tree does not quite match the solution quality of B1S. However, when P-Tree$_A$ is applied to a permutation derived from the result in the first column, the solution quality comes much closer to B1S. Though not quite equaling B1S, these experiments confirm that MST-induced permutations are in fact of high quality. The third column shows the average area reduction achieved by applying P-Tree$_A$ to a permutation induced by the B1S topology itself and the fourth column shows the result of the next iteration. These results indicate that some improvement over B1S is possible (however, at the cost of $O(n^5)$ run time versus $O(n^3)$ for B1S). While the 0.38% average improvement appears quite small, it is interesting when put in the context of the results reported by Griffith et al. in [5]. In that paper, it was reported that for 20-pin nets, the *optimal* solution (derived through a branch-and-bound procedure) improved upon the batched 1-Steiner solution by only about 0.46% (this can be derived from Table III of [5]). Thus, it would appear that the 0.38% improvement achieved by applying

P-Tree is close to the best one could hope for. (This should be interpreted only as the trend since the experiments were on different data sets and optimal solutions were not derived for the instances in Table 7.5.)

The last experiment assessed the impact on delay by constraining topologies to Hanan grid points. We routed 9-pin nets first on the Hanan grid and then on a larger grid with $n - 1$ additional rows and columns. We then assessed the minimum delay solution of each run. We performed 50 such runs. In only 17 cases was delay improved and, of those cases, the largest improvement was 0.8%. The overall average improvement was 0.05%. We offer this as empirical evidence in support of the suggestion of [2]—that constraining Steiner points to the Hanan grid does not have significant impact on solution quality in performance-driven routing.

Finally, we give an idea of the run-time complexity of the algorithms in practice. We implemented the P-Tree algorithms in C on a Sun SPARC 20 workstation.

Representative run times for P-Tree$_A$ range from 0.05 cpu seconds for $n = 6$ to 1.8 cpu seconds for $n = 15$ and 49 cpu seconds for $n = 30$.

In the case of P-Tree$_{AT}$ without wire sizing, run times ranged from 0.1 cpu seconds for $n = 9$ to 7.3 cpu seconds for $n = 12$ and 120 cpu seconds for $n = 20$.

When running P-Tree$_{AT}$ *with* wire sizing, run times ranged from 0.8 cpu seconds for $n = 6$ to 70 cpu seconds for $n = 12$ and 150 cpu seconds for $n = 15$.

While the P-Tree algorithms have higher computational complexity than the other proposed algorithms, it appears that they remain quite practical for the vast majority of net sizes typically seen in VLSI systems (particularly when the distribution of those net sizes is taken into account). In addition, as shown experimentally, the payoff for this additional computational complexity in terms of reduced area and improved timing can be very high. Furthermore, as mentioned at the end of this chapter, there are many refinements of the algorithms to improve computational behavior in practice and a number of P-Tree-like algorithms that allow tradeoffs between solution quality and computational efficiency.

7.7 SUMMARY AND COMMENTS

This chapter introduced two algorithms, P-Tree$_A$ and P-Tree$_{AT}$, for finding high - quality rectilinear Steiner trees for global routing. The problems of area (wirelength) minimization and minimization of area subject to timing constraints were addressed. The algorithms find optimal solutions in the exponentially large solution space of routing topologies induced by a previously derived permutation on the sinks of the net. A technique was presented for finding high-quality sink permutations that are *consistent* with a given topology such as the minimum spanning tree or a previously computed Steiner tree. Such permutations are further optimized by a dynamic programming algorithm to minimize the tour length of the permutation.

Though they cannot guarantee optimality, the algorithms have been demonstrated to produce predictably high-quality solutions. It was shown that previously proposed performance-driven routing techniques produce routing topologies with substantially more routing overhead than the P-Tree algorithms. We also improve over previous works in minimum delay and under the AT metric. For area minimization, improvement on the state-of-the art Steiner tree heuristic was achieved by iteratively applying the algorithm.

In closing, we mention a few implementation issues relating to the P-Tree algorithm that are of practical interest. Additionally, we discuss a few natural P-Tree-like problem formulations that can be solved more quickly, but have less expansive solution spaces.

7.7.1 Speedup Techniques and P-Tree-like Algorithms

First, we point out that P-Tree$_{AT}$ can be made considerably faster in practice by applying bounding techniques. A particularly promising approach sets an a priori limit on the total cost of any topology it will consider; we gradually relax this limit as we explore the trade-off curve. By adopting such an approach and computing lower bounds on the size of the "rest of the tree" and upper bounds on required arrival time with respect to a particular subsolution, we can detect a vast number of subsolutions that either exceed the current cost limit or are dominated by a previously discovered solution (i.e., which was computed for a smaller limit on total cost) but cannot be discarded by the simple pruning properties presented in this chapter. The cumulative effect of early detection and removal of such subsolutions can be quite dramatic with respect to both run time and memory consumption. The idea of this iterative bounding process is illustrated in Figure 7.15. The figure shows the progression as we relax the cost limit. At each step, any subsolution that can be shown to contribute to only global solutions, which lie in the shaded area (as determined by bounding techniques), can be discarded.

Another implementation issue relates to memory usage. Clearly, it is not necessary to keep the solution sets $S_b(v, i, j)$ (the branching point solutions) once we have computed $S(v, i, j)$ (unconstrained solutions). Indeed, this can save some space, but it is not hard to see that we can instead store only the relevant branching point solutions ($S_b(v, i, j)$) and discard the unconstrained solutions ($S(v, i, j)$). For a particular i, j, we compute $S(v, i, j)$ for all v. Each unconstrained solution will have a corresponding branching point solution; we mark those branching point solutions in $S_b(v', i, j)$ for all v'. Once this is done we can discard $S(v, i, j)$ and the unmarked branching point solutions. In practice, the number of relevant branching point solutions is significantly smaller than the number of unconstrained solutions (i.e., many of the unconstrained solutions use precisely the same branching point solution). The approach now is to *recompute* $S(v, i, j)$ when necessary (e.g., when computing some $S_b(v, i, k)$ where $k > j$). It is easy to see that simple loop interchanging in the computation of $S_b(v, i, k)$ yields an algorithm with the same asymptotic complexity even though we perform such recomputation.

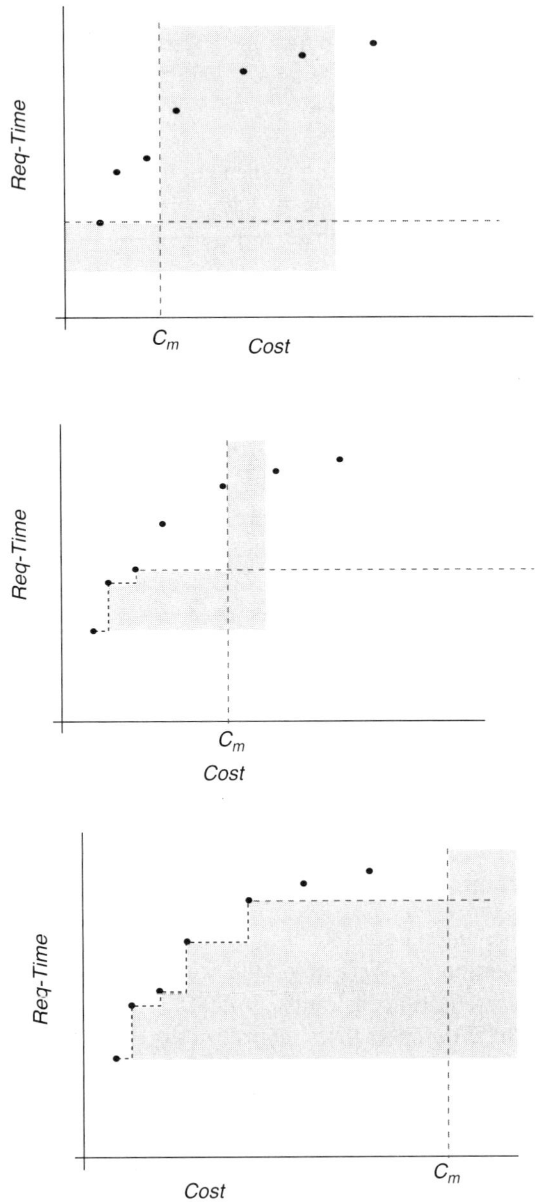

FIGURE 7.15 *Illustration of iterative bounding process.*

There are several natural P-Tree-like problem formulations that can be solved with essentially the same techniques as were used for P-Tree. For instance, computing the optimal trade-off curve for a *given* abstract topology routed on the Hanan grid can be done in $O(n^{3.5}l)$ time (see exercises below). When this algorithm is used on a routing graph of size $|G|$ (e.g., perhaps a finer-grain graph than the Hanan grid), we have a run time of $O(|G|n^{1.5}l)$. Thus, it is natural to use such an algorithm to *refine* a topology produced by the P-Tree algorithm by embedding it in a finer-grained grid.

While the previous approach speeds things up by assuming a *given* topology, it is also possible to accelerate things by considering a routing graph that is asymptotically *smaller* than the Hanan grid while maintaining the full set of candidate topologies induced by the given permutation. For instance, we may consider only the pin positions themselves as candidate Steiner points; such a topology can be thought of as a permutation-induced *spanning tree* (though it is important to realize that multiple links may connect to the same point, but not be electrically connected). This problem can be solved in $O(n^{4.5}l)$ time by P-Tree techniques. Note, however, that we do not have a regular grid as the underlying routing graph in this case; nevertheless, a naive algorithm for computing $S(v, i, j)$ that considers all possible branch points v' explicitly still yields the specified run time (with the mere computation of $S_b(v, i, j)$ producing the same run time). Note also that if we add some additional candidate Steiner points, the algorithm still has the same complexity provided we only add $O(n)$ additional points. For instance, we may add a uniform $\sqrt{n}$ by $\sqrt{n}$ grid to give additional flexibility.

Similarly, we can trade run time for solution quality (in fact, optimality) by combining the concepts of the P-Tree algorithm and the Dreyfus-Wagner optimal Steiner tree algorithm (see exercises below).

Thus, there are a wide variety of algorithms based on the P-Tree approach that can trade off solution space coverage for speed. There are clearly many interesting hybrids of these approaches that are worth exploring as "refinement" procedures. For instance, one may first compute permutation-constrained topologies with only $O(n)$ candidate Steiner points. From the resulting trade-off curve, it is natural to extract one or more abstract topology and route them on a grid with finer granularity.

7.7.2 Delay Model Generalizations

While the P-Tree algorithm has been presented under the Elmore delay model, some recent results demonstrate the flexibility of the framework by incorporating more sophisticated delay models. In [13], a table-lookup method was successfully employed to counteract Elmore's inability to account for the effects of signal slew and resistive shielding. It was shown that, at least under certain circumstances, significant improvement in real (SPICE-computed) performance and/or cost can be had. Similar results were reported in [14], though a direct delay computation method was employed rather than table lookup. In either case, optimality of the resulting solutions cannot be guaranteed for the given delay estimators. Nevertheless, they illustrate practical extensions of the basic framework.

7.8 FURTHER READING

An excellent resource on the Steiner tree problem is the book by Hwang, Richards, and Winter [9]. Kahng and Robins [11] present a wide variety of material on topology construction including SERT and related algorithms; additionally, their book provides very nice coverage of topology construction for clock nets, which is beyond the scope of this book.

7.9 EXERCISES

1. Devise a timing-driven algorithm for optimally embedding a given "abstract topology."
2. Incorporate simultaneous buffer insertion into the P-Tree algorithm.
3. The Dreyfus-Wagner algorithm [4] is an exponential time algorithm for the graph Steiner problem. In the P-Tree algorithm $O(n^2)$ candidate subtrees induced by the given permutation are considered; in the Dreyfus-Wagner algorithm, all 2^{n-1} subsets of the $n - 1$ sinks are considered to ensure optimality. Design and analyze a timing-driven version of the Dreyfus - Wagner algorithm.

BIBLIOGRAPHY

[1] K. D. Boese, A. B. Kahng and, G. Robins, High-performance routing trees with identified critical sinks, *Proc. ACM/IEEE Design Automation Conf.*, pp. 182–187, 1993.

[2] K. D. Boese, A. B. Kahng, B. A. McCoy, and G. Robins, Rectilinear steiner trees with minimum Elmore delay, *Proc. ACM/IEEE Design Automation Conf.*, pp. 381–386, 1994.

[3] T. H. Cormen, C. E. Leiserson, and R. A. Rivest, *Introduction to Algorithms*, MIT Press and McGraw-Hill, pp. 558–562, 1990.

[4] S. E. Dreyfus, and R. A. Wagner, The Steiner problem in graphs, *Networks*, 1: pp. 195–208, 1972.

[5] J. Griffith, G. Robins, J. S. Salowe, and T. Zhang, Closing the gap: Near-optimal Steiner trees in polynomial time, *IEEE Transactions on Computer Aided Design*, 13(11): 1351–1365, Nov. 1994.

[6] M. Garey, and D. S. Johnson, The rectilinear Steiner problem is NP-complete, *SIAM J. Applied Math.*, 32: 826–834, 1977.

[7] M. Hanan, On Steiner's problem with rectilinear distance, *SIAM J. Applied Math.*, 14: 255–265, 1966.

[8] T. D. Hodes, B. A. McCoy, and G. Robins, Dynamically-wiresized Elmore-based routing constructions, *Proc. IEEE Intl. Symp. Circuits and Systems*, 1994.

[9] F. K. Hwang, D. S. Richards, and P. Winter, *The Steiner Tree Problem*, Elsevier Science Publishers, pp. 213–214, 1992.

[10] F. K. Hwang, On Steiner minimal trees with rectilinear distance, *SIAM J. Applied Math.* 30: 104–114, 1976.

[11] A. B. Kahng, and G. Robins, *On Optimal Interconnections for VLSI*, Boston, Kluwer Academic Publishers, 1995.

[12] A. B. Kahng, and G. Robins, A new class of iterative Steiner tree heuristics with good performance, *IEEE Trans. Computer-Aided Design*, 11: 893–902, 1992.

[13] J. Lillis, and P. Buch, Table-lookup methods for improved performance-driven routing, *1998 ACM/IEEE Design Automation Conference*, San Francisco, pp. 368–373, June 15–19, 1998.

[14] F.-J. Liu, J. Lillis, and C.-K. Cheng, A new layout-driven timing model for incremental layout optimization, *Proc ASP-DAC '97. Asia and South Pacific Design Automation Conference 1997, IEEE*, pp. 127–131, 1997.

[15] A. Vittal, and M. Marek-Sadowska, Minimal delay interconnect design using alphabetic trees, *Proc. ACM/IEEE Design Automation Conf.*, pp. 392–396, 1994.

8

Optimization of Multisource Nets

8.1 INTRODUCTION

Chapters 6 and 7 focused on synthesis and optimization of single-source nets via buffer insertion, wire sizing, and topology design. However, after considering the various single-source formulations, optimization techniques, and algorithms, it is natural to ask what can be done in the case of multisource nets since buses are so prevalent in modern designs. This topic — synthesis and optimization techniques for buses — is the subject of this chapter.

The key points of the chapter are as follows.

1. The *Augmented RC-Diameter* (ARD) is introduced as a natural performance measure based on RC delay models, the arrival times at the inputs (sources) of a multisource, and downstream delays from the sinks (outputs) of the net. The ARD is a straightforward generalization of the typical single-source performance measures and is, we feel, more appropriate than measures such as weighted sums of pairwise delays (e.g., [3]) which have only an indirect relation to system objectives such as clock period.
2. A simple linear time algorithm for computing the ARD of a given topology is presented. Thus it is no harder than computing an RC-radius (see, e.g., [9, 7,13]; i.e., it is unnecessary to perform multiple single-source computations).
3. Algorithmic techniques for multisource optimization are developed through an algorithm for optimal repeater (i.e., bidirectional buffer) insertion for static routing topologies.[1] We adopt the objective of cost (e.g., area) minimization

[1] Throughout the chapter we use the term "repeater" to refer to a bidirectional buffer.

subject to a given timing spec. As in the preceding chapters, the technique produces a suite of solutions that exhibit a cost-versus-performance trade-off. The technique subsumes the discrete driver-sizing problem (subject to the drivers effectively being "two-stage").

In the process of developing the repeater insertion algorithm, general techniques for characterizing multisource timing optimization are illustrated. These techniques are based on manipulation of *piecewise linear* (PWL) functions to capture the complexities of multisource optimization. These fundamental operations have applications beyond the particular problem of repeater insertion, some of which are mentioned in exercises at the end of the chapter.

We focus on bidirectional repeater insertion in part because it has achieved a certain degree of maturity in the high-end design community and, in fact, is commonly used in the design of high-speed commercial microprocessors [8]. However, it was also selected because it exercises many of the issues involved in multisource optimization and thus serves as a good testbed in which to illustrate these techniques.

There are some important design and technology issues in repeater insertion that do not appear in the single-source problem. The primary issue is the choice of repeater design; should one use a pair of tristate buffers controlled by a bus arbitrator[2] or more sophisticated "direction-sensing" autonomous repeaters (e.g., [4,5])? The (perhaps) more typical choice — an arbitration-controlled repeater — is illustrated in Figure 8.1. Such decisions are largely dependent on designer preference [8] and are not addressed here; fortunately, such issues appear to have

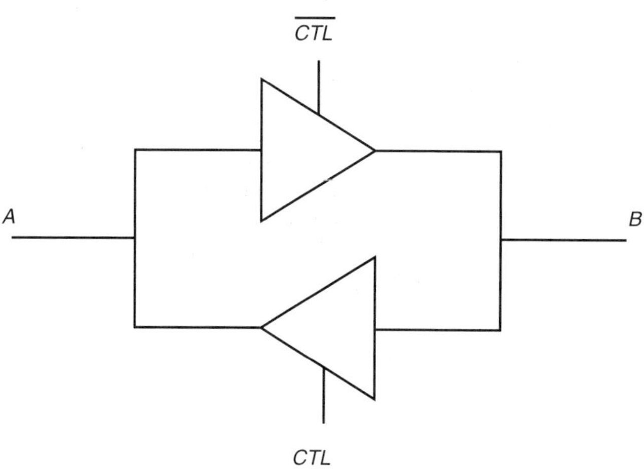

FIGURE 8.1 *Bidirectional repeater with control signals.*

[2] Note that a bus controller is typically necessary whether repeaters are inserted or not and thus the additional control overhead when repeaters are used can reasonably be expected to be small.

little impact on the fundamental nature of the optimization problem and, thus, the techniques presented here remain relevant independent of such decisions.

The remainder of this chapter is organized as follows. Section 8.2 introduces nomenclature, delay models, and problem formulations. Section 8.3 presents an algorithm for computing the augmented RC-diameter of a multisource topology. Section 8.4 presents an algorithm for optimal repeater insertion; the algorithm is based on functional abstraction and dynamic programming. Section 8.5 presents preliminary experimental results and we conclude in section 8.6.

8.2 PRELIMINARIES AND FORMULATIONS

Throughout the remainder of this chapter, the following are assumed to be given technology parameters.

- α and β, which are respectively resistance (in Ohms) and capacitance (in pF) per unit wire length (e.g., 1 micron) in the target technology.
- A library L_b of repeaters; each repeater has an *A-side* and a *B-side* (as shown in Figure 8.1), allowing us to refer to signal flow as *A-to-B* or *B-to-A* and thereby take into account the orientation of the repeater. The following parameters characterize a repeater where subscripts indicate signal direction.

$i_{AB}(b), i_{BA}(b)$:	intrinsic delay of b
$r_{AB}(b), r_{BA}(b)$:	output resistance of b
$c_{AB}(b), c_{BA}(b)$:	input capacitance of b
$s(b)$:	*cost* of b (e.g., area)

For completeness, we review the Elmore model for wire delay. The delay of a signal transmitted along a wire of length l from u to v is given as

$$\alpha l\left(\frac{l\beta}{2}+c_v\right)$$

where c_v is the lumped downstream capacitance at v.[3] Similarly, in a typical basic model, the delay of a buffer b is the sum of b's intrinsic delay and *RC* delay given as

$$i(b)+r(b)\cdot c_{load}$$

where c_{load} is the capacitive load on the output of b. Naturally, in the case of a repeater, this delay depends on the signal direction and the parameters are used accordingly. Subsequently when referring to the delay of a path, assume these models are applied.

In addition to the above technology parameters, the following net-specific parameters are also given.

[3] Note that for fixed-width wires as discussed here, fringe capacitance can easily be incorporated.

- Terminal set N in the plane; each terminal may act as an input or output and thus may have two buffers — an input buffer that drives the bus and an output buffer used when the terminal acts as a sink. The following parameters are associated with each terminal $v \in N$:

  $a(v)$: the maximum delay from a *primary input* (PI) of the circuit to the input buffer at v[4]

  $d(v)$: the maximum delay from output buffer at v to a *primary output* (PO) of the circuit[5]

  $c(v)$: the input capacitance associated with input buffer at v

  $r(v)$: the output resistance of the input buffer at v when acting as a source
- A rectilinear Steiner tree T connecting N with prescribed, degree-2, candidate buffer insertion points.[6]

The parameters associated with a terminal are illustrated in Figure 8.2. In the figure the terminal v serves as both a source and a sink by using the driver component enclosed in dashed lines. When acting as a source, its input arrives at time $a(v)$; conversely, when acting as a sink, the signal it produces will experience $d(v)$ additional delay to a *PO* of the circuit.

To assess the performance of a routing tree T spanning terminal set N with a driver assignment and a (perhaps null) buffer assignment to the candidate insertion points, we now formally define the augmented RC-diameter, *ARD(T)*.

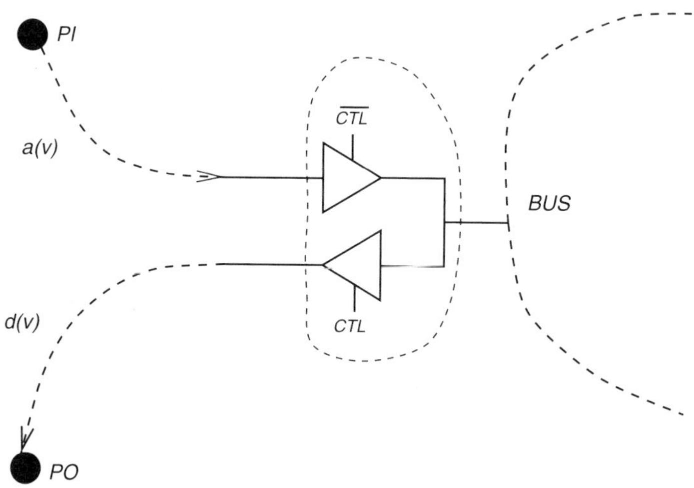

FIGURE 8.2 *Parameters associated with a terminal v.*

---

4 Excluding intrinsic and RC delay of the buffer.

5 In this case, intrinsic and RC delays of the output buffer are included since they are independent at the bus.

6 We assume that insertion points have degree two to avoid ambiguity with respect to which side of the repeater a branch connects.

DEFINITION 8.2.1

ARD(T): Let $PD(u, v)$ be the RC path delay from source pin u to sink pin v in topology T spanning N including delay of the driver (if it exists), buffers, and wires on the path. We define the *augmented RC diameter of T, ARD(T)* as

$$ARD(T) = \max_{u \in N} \max_{v \in N \setminus \{u\}} (a(u) + PD(u, v) + d(v))$$

The augmented RC diameter *ARD(T)* captures the maximum delay from the circuit's PIs to its POs among all paths traversing net N and, thus, captures the impact of the solution on, for example, system cycle time. In the next section, we present a linear time algorithm for computing *ARD(T)* where Elmore is used in the delay calculation.[7]

The definition of *ARD(T)* leads naturally to the *optimal repeater insertion problem* as follows.

Problem 8.2.1 Given routing topology T spanning multisource terminal set N, a set of technology parameters, and performance spec D_{spec}, find an assignment and orientation of repeaters to the insertion points of T such that the resulting cost is minimized subject to $ARD(T) \leq D_{spec}$.

Section 8.4 presents a solution to Problem 8.2.1.

We note that alternative formulations of the problem are possible. For instance, one may wish to minimize *ARD(T)* without regard to solution cost. However, we propose that Problem 8.2.1 is of more practical value (and in fact subsumes the cost-oblivious formulation). One may also consider a formulation in which arbitrary delay constraints are given for each source/sink pair. While Problem 8.2.1 implicitly imposes pairwise delay bounds, these bounds are not arbitrary as they are derived from the linear number of given parameters. We do not study the arbitrary pairwise constrained problem here but we note that this problem appears significantly more complex; fortunately, Problem 8.2.1 appears to capture the typical scenario in practice. Finally, we point out that no generality is lost by not explicitly specifying the sources and sinks of the net; for any nonsink v, we can simply set $d(v) = -\infty$ and for any nonsource u, we can set $a(v) = -\infty$.

8.3 LINEAR TIME COMPUTATION OF *ARD*(*T*) UNDER ELMORE

In the multisource formulation presented in the preceding section, the performance metric $ARD(T)$ is expressed as a maximum over a set of single-source delay computations. Clearly, by performing n single-source computations we can compute $ARD(T)$ in $O(n^2)$ time. However, we will show that $ARD(T)$—under

[7] Note, however that the concept of the ARD does not rely on the Elmore delay model; indeed the ARD is well-defined regardless of how $PD(u, v)$ is calculated.

Elmore — can be computed in linear time and thus is no harder (asymptotically) than computing the analogous performance metric for a single source net (e.g., maximum source-to-sink delay).[8]

Before proceeding, we clarify some additional assumptions and notation. First, without loss of generality, we assume that all terminals of the net are also leaves in the topology T; note that any nonleaf terminal can be made a leaf by adding a new vertex and a 0-length edge. For purposes of depth-first traversal, it will be useful to orient a topology T with respect to an (arbitrary) root vertex s. In such a rooted tree, we use $p(v)$ to indicate the parent of vertex v; we also note that edges $(u, v) \in T$ are now directed. We use $c(u, v)$ and $r(u, v)$ to respectively denote the capacitance and resistance of a wire $(u, v) \in T$. Finally, recall that we assume that repeaters are inserted along wires rather than at branch points.

The first step in computing $ARD(T)$ in linear time will be the computation of the capacitive values $c_L(u, v)$ and $c_L(v, u)$ for every edge $(u, v) \in T$ where $c_L(u, v)$ is the total downstream capacitance at v *as seen from* u and likewise, $c_L(v, u)$ is the load at u from v's perspective. Both of these load values are needed because a signal may traverse edge (u, v) in either direction. We first compute $c_L(u, v)$ where $(u, v) \in T$ by the following recurrence.[9]

$$c_L(u, v) = \begin{cases} c(v) & \text{if } v \text{ is a terminal} \\ c_{AB}(b) & \text{else if buffer } b \text{ placed at } v \\ \sum\limits_{(v,w)\in T} c_L(v, w) + c(v, w) & \text{o.w.} \end{cases} \tag{8.1}$$

After this bottom-up process, we compute $c_L(v, u)$ where $(u, v) \in T$ top-down as follows.

$$c_L(v, u) = \begin{cases} c(u) & \text{if } u \text{ is the root} \\ c_{BA}(b) & \text{else if buffer } b \text{ placed at } u \\ c_L(u, p(u)) + c(u, p(u)) + & \\ \quad \sum\limits_{\substack{(u,w)\in T \\ w\neq v}} c_L(u, w) + c(u, w) & \text{o.w.} \end{cases} \tag{8.2}$$

Finally, the overall algorithm for computing $ARD(T)$ appears in Figure 8.3. The algorithm recursively computes three values for each subtree T_v: a is the maximum augmented arrival time at v via sources in T_v; d is the maximum augmented delay from v to sinks in T_v; d_l is the maximum augmented RC-diameter among source/sink pairs in T_v. Once the root is reached, we can derive the ARD of the entire tree.

---

[8] This result gives some idea of the distinction between the formulation we propose and the "arbitrary pairwise constraint" formulation where an $\Omega(n^2)$ time is necessary simply to examine all constraints.

[9] This recurrence and the subsequent recurrence assume w.l.o.g. that the A-side of buffer b connects to the parent in the reoriented tree; simply reversing the subscripts of $c(b)$ accounts for the opposite orientation.

Routine: ARD(T)
Input: Multisource topology T spanning terminal set N
Output: The Augmented RC-Diameter of T

```
1. Orient T with an arbitrary pin s ∈ N as the root
2. Compute c_L(u,v) and c_L(v,u) for all (u,v) ∈ T
3. (a,d,d_l) ← ARD_R(s)        /* invoke recursive routine */
4. Let c_L(s) be the total load on pin s
5. RETURN max(a+d(s),d+a(s)+r(s)c_L(s),d_l)
```

Routine: ARD_R(v)
Input: vertex $v \in T$
Output: triple (a,d,d_l) where

$$a = \max_{u \in T_i \cap N} a(u) + PD(u,v)$$

$$d = \max_{u \in T_i \cap N} PD(v,u) + d(u)$$

$$d_l = \max_{u \in T_i \cap N} \max_{\substack{w \in T_i \cap N \\ w \neq u}} a(u) + PD(u,w) + d(w)$$

```
IF(v is a leaf)
   RETURN (a(v) + r(v)(c_L(v,p(v)) + c(v,p(v))),d(v),-∞)
ELSE
  a ← d ← d_l ← -∞
  FOREACH child u of v
    (a(u),d(u),d_l(u)) ← ARD_R(u)
    a'(u) ← a(u) + r(u,v) (c(u,v)/2 + c_L(u,v))
    d'(u) ← d(u) + r(v,u) (c(u,v)/2 + c_L(u,v))
    IF (buffer b placed at v)
      a'(u) ← a'(u)+r_AB(b)(c_L(v,p(v))+c(v,p(u)))    /* or use c_BA depending on orientation */
      d'(u) ← d'(u)+r_BA(b)(c_L(v,u)+c(v,u))           /* or use r_BA depending on orientation */
    d_l ← max(d_l,a + d'(u),d + a'(u),d_r(u))
    a ← max(a,a'(u))
    d ← max(d,d'(u))
  RETURN (a,d,d_l)
```

FIGURE 8.3 *Pseudo-code for combating ARD(T).*

The algorithm just presented applies to the Elmore delay measure. However, as previously stated, the ARD is well-defined for any delay measure. In any such delay measure, if the delays along wire segments are known, the resulting ARD can easily be computed in linear time also by using depth-first search. Details are left to the reader.

8.4 THE REPEATER INSERTION ALGORITHM

As in the previous section, assume that the routing topology T has been reoriented with an arbitrary terminal designated as the root. Our algorithm for optimal repeater insertion is based on bottom-up dynamic programming on this reoriented tree.

8.4.1 A Motivational Example

As in any dynamic programming approach, it is necessary to characterize the salient features of a subsolution and how they may affect a global solution. Consider a candidate repeater assignment s to a subtree rooted at vertex v; we refer to such a subtree as T_v. Intuitively, we must capture the following characteristics of the assignment.

1. Cost
2. Capacitance
3. Maximum augmented delay from v to a sink in T_v (i.e., $\max_{u \in N \cap T_v} (PD(v, u) + d(u)))$
4. Maximum arrival time at v from sources in T_v
5. Augmented RC-diameter of the subtree itself

The first three characteristics are inherited from the single-source problem in Chapter 6 and are scalars. Item 4 considers the case where one or more source is *internal* to T_v, and one or more sink is *external* to T_v; item 5 considers the case in which there are sources *and* sinks internal to T_v.

To motivate the issues involved in cases 4 and 5 above, consider the example in Figure 8.4. In the figure, we have arrived at vertex v and need to be able to assess the maximum arrival time at v from u and w and the internal augmented path delays u to w and w to u. The critical observation is that these values depend on the capacitance of the tree *outside* of T_v (i.e., $T \backslash T_v$). Of course, this *external capacitance* c_E is unknown at this stage of the algorithm. To assess the arrival time at v from terminal u, consider the case where $c_E = 0$; calculating the driver delay and interconnect delay of wire (u, v), we give the arrival time at v from u as

$$a(u, v) = a(u) + 2(2 + 4 + 1) + 5(1 + 4 + 1) = 204$$

Similarly, the arrival time at v via source w is given by $a(w, v) = 104$ (again, assuming $c_E = 0$). Thus, in general (i.e., for arbitrary c_E), we have

$$a(u, v) = 204 + 7c_E \quad \text{and} \quad a(w, v) = 104 + 12c_E$$

since the bottom-up accumulated resistance from u to v is 7 and that from w to v is 12. Looking at Figure 8.4(c), we see that not only does the arrival time depend on c_E, but so does the *critical source*—w dominating when $c_E < 20$ and u dominating when $c_E > 20$. Thus, the piecewise linear *maximum* of $a(u, v)$ and $a(w, v)$ shown in Figure 8.4(c) captures the arrival time at v from its descendants as a function of external capacitance c_E.

We must also consider the *internal* paths (u, v, w) and (w, v, u) as in Figure 8.4(d). Consider the path (u, v, w); the augmented RC delay of this path includes $a(u, v)$, the scalar delay from v to sink $w(PD(v, w))$, and $d(w)$. Thus,

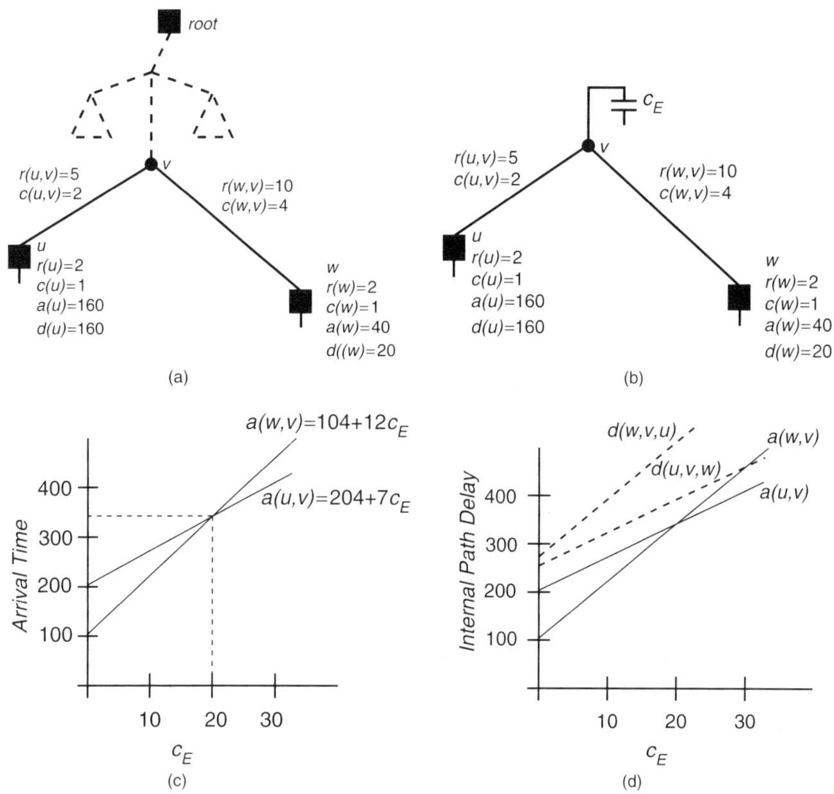

FIGURE 8.4 *Example bottom-up computation (a), abstraction of external capacitance* c_E*(b), corresponding arrival time functions at v from u and w(c), and augmented delay of internal paths (d).*

noting that $PD(v, w) = 30$ and $d(w) = 20$, we have

$$d(u, v, w) = a(u, v) + PD(v, w) + d(w) = 254 + 7c_E$$

Similarly, $d(w, v, u) = 274 + 12c_E$. These are shown by dashed lines in part (d) of the figure and amount to simply adding scalars to the y-intercepts of $a(u, v)$ and $a(w, v)$. Taking the piecewise maximum of $d(u, v, w)$ and $d(w, v, u)$ gives us the *internal augmented RC diameter* of T_v (notice in this case, however, that $d(w, v, u)$ dominates $d(u, v, w)$ *for all* values of c_E.)[10]

This discussion should give an idea of our approach: as we proceed bottom-up, we accumulate path resistance in the form of slopes in the PWL segments. As

[10] Note that this kind of decomposition is not possible in the case of the "arbitrary pairwise constraints" version of the problem where the critical source of the subtree cannot be so easily determined (e.g., it is not the case that every *external* sink will have the same critical source as in our problem).

we proceed and consider various options such as repeater insertion, appropriate PWL operators are applied.

8.4.2 Solution Characterization

The formal characterization of a solution s for subtree T_v is given by the following parameters (using C-like structure syntax to refer to the parameters associated with a solution s).

- $s.cost$ scalar cost
- $s.cap$ scalar capacitance
- $s.d$ scalar augmented delay to sinks in T_v
- $s.f_A$ PWL giving arrival time at v from sources in T_v as a function of c_E
- $s.f_I$ PWL giving augmented RC diameter for source/sink pairs internal to T_v

For each vertex v in the reoriented tree we compute a set $S(v)$ of *option* solutions for T_v where each $s \in S(v)$ is characterized as above.

8.4.3 PWL Primitives

Recall that piecewise linear functions were utilized in section 6.3.6 to take signal slew into consideration during buffer insertion. We review the basic concepts here and introduce various primitive operators. We first review the definition of a PWL itself.

DEFINITION 8.4.1

PWL: A *piecewise linear (PWL) function* f is a set of quadruples (y_0, slope, x_l, x_r) where each such quadruple represents a line segment and gives the y-intercept (y_0), *slope* and domain for which the segment is defined (i.e., all x where $x_l \leq x \leq x_r$). Additionally, no two segments may have overlapping domains.

As is typical in most applications of PWL functions, we store line segments in a linked list ordered by domain.

The following is a list of the PWL primitives we require.

$$f \leftarrow PWL\_Max(f_1, f_2) \iff f(x) = \max(f_1(x), f_2(x)) \quad \forall x$$

$$f \leftarrow PWL\_AddScalar(f_1, dy) \iff f(x) = f_1(x) + dy \quad \forall x$$

$$f \leftarrow PWL\_ShiftLeft(f_1, dx) \iff f(x) = f_1(x + dx) \quad \forall x$$

$$f \leftarrow PWL\_AddSlope(f_1, ds) \iff f(x) = f_1(x) + x \cdot ds \quad \forall x$$

All of these operations can be performed in time linear in the number of segments in the participating functions by appropriately stepping through the participating segment lists.

8.4.4 Solution Dominance

A key issue in virtually all multidimensional dynamic programming applications is the elimination of provably suboptimal solutions, and repeater insertion is no exception. The issue arises, for example, in single-source buffer insertion (as in Chapter 6) as well as other applications such as floorplanning. In the typical situation, a solution s is characterized by a set of k *scalars* $s[1], s[2], \ldots, s[k]$; assume without loss of generality that minimization in each of these k dimensions is preferred. This induces a partial order $\preceq$ on a set of solutions S; for $s_1, s_2 \in S$, we have the following

$$s_1 \preceq s_2 \Longleftrightarrow s_1[i] \leq s_2[i] \quad \forall i \in \{1.k\}$$

In the context of multidimensional dynamic programming, we need not consider any solution $s_2 \in S$ if there exists $s_1 \in S$ such that $s_1 \preceq s_2$ (assume w.l.o.g. all solutions are distinct). Thus, for a given solution set S, we compute the *minima* of S.

DEFINITION 8.4.2

Minimal Subset: The *minimal* or *dominant* subset of a set of k-dimensional points S is the largest subset $S' \subseteq S$ such that, for all $s_1, s_2 \in S'$, $s_1 \npreceq s_2$ and $s_2 \npreceq s_1$.

The problem of finding a minimal point set has a long history dating back to [6]. A variety of algorithms have been proposed; some achieve excellent asymptotic complexity (e.g. [6]) while others have been tuned for ease of implementation or fast *expected* run time (e.g., [1]). Note that in Chapter 6, we implicitly solved a three-dimensional minima problem.

The reader may have noted by now that because of the PWL functions $s.f_A$ and $s.f_I$ a subsolution s in the repeater insertion problem does not correspond to a single point in k-dimensional space; rather, it corresponds to an infinite set of five-dimensional points — one for every value of c_E. One way to deal with this complication would be to *discretize* the c_E domain and thus introduce a different point-dominance problem for each discrete value of c_E (and introduce a degree of error depending on the granularity of discretization). We have not adopted the discretization approach; instead, we introduce the notion of a *minimal functional subset* (MFS).[11]

DEFINITION 8.4.3

MFS: Let S be a set where $s \in S$ is a k-tuple of PWL functions of a variable x. For each $s \in S$, let s' be defined (for $i \in \{1..k\}$) as

$$s'[i](x) = \begin{cases} \infty & \text{if } \exists s_0 \in S s.t. s_0 \neq s \text{ and } s_0[j](x) \leq s[j](x) \forall j \in \{1..k\} \\ s[i](x) & \text{o.w.} \end{cases}$$

[11] We use the notion of *subset* somewhat loosely here; however, the meaning should be clear.

The *minimal functional subset* (MFS) of S is S':

$$S' = \{s' | \exists x \quad s.t. \quad s'[i](x) \neq \infty\}$$

Implicitly, if $s[i]$ is determined to be nonminimal for some value of x, all other $s[j]$'s must also be nonminimal for that same value of x. Note also that PWL functions in S' may no longer be "connected" since they may have suboptimal regions of the domain (i.e., ∞) separating useful regions of the domain. Finally, we point out that the fundamental pruning operation of the definition—for a given s and s_0, detect all ranges of x for which s_0 dominates s and alter the PWLs of s accordingly—can be implemented in time linear in k and the total number of line segments in question. This is done by first detecting the regions in which $s_0[i]$ dominates $s[i]$ (similar to a PWL_Max operation) for each i and then computing the intersection of these regions and finally updating the PWLs accordingly.

In our application, solutions are characterized by three scalars (degenerate PWL functions for the purpose of discussion) and two PWL functions. At each stage of the dynamic programming process, we would like to efficiently compute the FMS of a solution set S. We have devised an easy-to-implement algorithm for solving this problem that appears to works well in practice; it appears in pseudocode in Figure 8.5. The algorithm takes a divide-and-conquer approach motivated by the following intuition. Suppose that the (MFS) of S is much smaller than S itself (i.e., many suboptimal solutions can be eliminated).[12] The hope is that many of the suboptimal solutions will be discarded at relatively deep levels of the recursion and thus we can avoid pairwise comparisons at higher levels of the recursion. Thus, the approach targets fast performance in practice but retains an $\Omega(|S|^2)$ asymptotic worst-case complexity in terms of the number of pairwise comparisons.[13]

```
Routine: MinimalFS_DC(S)
Input:   Solution set S
Output:  The minimal functional subset of S
   IF (|S| = 1)
       RETURN S
   Divide S into equal sized sets S_l and S_r
   S'_l ← MinimalFS_DC(S_l)
   S'_r ← MinimalFS_DC(S_r)
   Compute S' = MFS(S'_l ∪ S'_r) by pairwise comparison between S'_l and S'_r
   RETURN S'
```

FIGURE 8.5 *Computation of MFS(S) by divide and conquer.*

[12] From experience, this is frequently the case; particularly when constructing solutions at a branch point from its children.

[13] Note however, that we haven't bounded the size of the PWL functions themselves (i.e., number of segments). In the worst case, the PWL functions can grow exponentially in the number of insertion points. However, such degenerate scenarios appear to occur infrequently in practice.

8.4.5 The Overall Algorithm

Finally, we present a high-level view of our algorithm for optimal repeater insertion in Figure 8.6. We recursively traverse the tree and inductively compute solutions sets $S(v)$ from the solution sets of v's children. The top-level code invokes different subroutines depending on whether the current vertex v is a leaf, Steiner (branch) node, insertion point, or the root itself; these routines appear in Figures 8.7, 8.8, 8.9, and 8.10, respectively. Additionally, a routine for extending a solution set for a tree rooted at vertex v to a solution set for that same tree augmented by the wire to v's parent appears in Figure 8.11.

The algorithm follows the intuition presented in the preceding sections. Consider, for example, the subroutine *RepeaterSolutions* in Figure 8.9. We visit each

```
Routine: MSRI(v)
Input:   Vertex v in oriented topology T
Purpose: Computes the minimal solution set S(v)
           among repeater assignments to T_v
  IF (v is a leaf)
    S(v) ← LeafSolutions(v)
  ELSE IF (v is a Steiner node)
    S(v) ← ∅
    FOREACH (child u of v)
       MSRI (u)
       S'(u) ← Augment (S(u), (v,u))
       IF (S(v) ≠∅ )
         S(v) ← JoinSets( S(v),S'(u))       /* Combine solutions from subtrees */
         S(v) ← MinimalFS_DC (S(v))
       ELSE
         S(v) ← S'(u)
  ELSE IF (v is a candidate insertion point (degree-2))
    LET u be the child of v
    S'(u) ← Augment (S(u), (v,u))
    S(v) ← S'(u) ∪ RepeaterSolutions(S'(u))
    S(v) ← MinimalFS_DC (S(v))
  ELSE       /* v is the root */
    LET u be the child of v          /* Recall: terminals are degrees-1 */
    S'(u) ← Augment (S(u), (v,u))
    S(v) ← RootSolutions(S'(u))
    S(v) ← MinimalFS_DC (S(v))
```

FIGURE 8.6 *Top-level algorithm for optimal multisource repeater insertion.*

```
Routine: LeafSolutions (v)
Input:   Leaf (terminal) vertex v
Output: The the (singleton) solution set for v
  s.cost ← 0
  s.cap ← c (v)
  s.d ← d (v)
  s.f_A ← a (v) + r (v) c_E   /* Single segment PWL with y-int at a(v) and slope r (v) */
  s.f_I ← − ∞                  /* No internal paths */
  RETURN {s}
```

FIGURE 8.7 *LeafSolutions() subroutine for optimal repeater insertion.*

Routine: JoinSets (S_1, S_2)
Input: Solution sets S_1 and S_2 from disjoint subtrees
Output: The minimal solution set S when the subtrees are joined at a common parent

$S \leftarrow \emptyset$
FOREACH (pair of solutions $s_1 \in S_1$, $s_2 \in S_2$)
    $s.cost \leftarrow s_1.cost + s_2.cost$ /* Scalar operations first */
    $s.cap \leftarrow s_1.cap + s_2.cap$
    $s.d \leftarrow \max(s_1.d, s_2.d)$
    $f_A^1 \leftarrow$ PWL_ShiftLeft ($s_1.f_A$, $s_2.cap$) /* Signals from left now see cap on right */
    $f_A^2 \leftarrow$ PWL_ShiftLeft ($s_2.f_A$, $s_1.cap$) /* Signals from right now see cap on left */
    $s.f_A \leftarrow$ PWL_Max (f_A^1, f_A^2) /* Combined arrival time functions */
    $f_I^{12} \leftarrow$ PWL_AddScalar (f_A^1, $s_2.d$) /* New left-to-right internal paths */
    $f_I^{21} \leftarrow$ PWL_AddScalar(f_A^2, $s_2.d$) /* New right-to-left internal paths */
    $f_I^v \leftarrow$ PWL_Max (f_I^{12}, f_I^{21}) /* v is the current root */
    $f_I^1 \leftarrow$ PWL_ShiftLeft ($s_1.f_I$, $s_2.cap$) /* Account for old internal paths on left */
    $f_I^2 \leftarrow$ PWL_ShiftLeft ($s_2.f_I$, $s_1.cap$) /* Account for old internal paths on right */
    $s.f_I \leftarrow$ PWL_Max(f_I^v, PWL_Max (f_I^1, f_I^2)) /* Put it all together */
    $S \leftarrow S \cup \{s\}$
RETURN S

FIGURE 8.8 *The JoinSets() subroutine for optimal repeater insertion.*

Routine: RepeaterSolutions (S)
Input: Unbuffered solution set S
Output: A solution set in which all solutions are buffered with an oriented repeater from the given library L_b

$S' \leftarrow$
FOREACH ($s \in S$ and $b \in L_b$)
    /* Assume that parent connects to the A-side of b; child to B-side */
    $s'.cost \leftarrow s.cost + s(b)$
    $s'.cap \leftarrow c_{AB}(b)$
    $s'.d \leftarrow s.d + i_{AB}(b) + r_{AB}\, s.cap$
    $a \leftarrow s.f_A(c_{BA}(b)) + i_{BA}(b)$
    $s'.f_A \leftarrow a + r_{BA}(b) \,.\, c_E$ /* accumulated resistance below b not affected by c_E */
    $s'.f_r \leftarrow s.fr(c_{BA}(b))$ /* internal paths independent of c_E */
    Compute S'' in the same way for opposite orientation of b
    $S' \leftarrow S' \cup \{s', s''\}$
RETURN S'

FIGURE 8.9 *The RepeaterSolutions() subroutine for optimal repeater insertion.*

unbuffered solution s and candidate repeater b and determine the characteristics of a solution s' in which b is placed at the root of the unbuffered solution (assume as in the figure that the A-side of b connects to the parent and the B-side connects to the unbuffered solution s). First, we must add the cost of the additional repeater b; next, b decouples downstream capacitance and thus, b determines the new capacitance; next $s'.d$ must take the delay of b from A to B

```
Routine: RootSolutions (S)
Input:   Solution set S which considers the entire tree
            except the root driver itself
Output: A solution set taking the root driver into account.
            Since these are global solutions, only the minimal
            solutions with respect to cost and final diameter ARD (T)
  S' ← Ø
  FOREACH (s ∈ S)
     s'.cost ← s.cost
     d ← s.d − r (v) s.cap + a(v)       /* root as source case */
     d' ← s.f_A (c(v)) + d(v)           /* root as sink case */
     d_I ← s.f_I (c(v))                 /* root not involved */
     s'.d − max (d, d', d_I)            /* store ARD(T) in s.d */
     /** Note: s'.cap, s'.f_A and s'.f_I do not matter **/
     s'.cap ← s'.f_A ← s'.f_I ← −∞
     S' ← S' ∪ {s}
  Delete from S' all nonminimal solutions w.r.t s.cost and s.d
       /* Easy in two dimensions */
  RETURN S'.
```

FIGURE 8.10 *The RootSolutions() subroutine for optimal repeater insertion.*

```
Routine: Augment(S, (u,v))
Input:   Solution set S and the wire (u,v) by which we are
            extending the subtree
Output: A solution set built from S set taking wire (u,v) into account

  S' ← Ø
  FOREACH (s ∈ S)
     s'.cost ← s.cost
     s'.cap ← s.cap+c(u,v)
     s'.d ← s.d + r(u,v) (c(u,v)/2 + s.cap)
     f'_A ← PWL_AddScalar (s.f_A, r (u,v) c(u,v)/2))
     s'.f_A ← PWL_AddSlope (f'_A , r(u,v))
     s'.f_I ← PWL_ShiftLeft (s.f_I, c(u,v))
     S' ← S' ∪ {s}
  RETURN S'
```

FIGURE 8.11 *The Augment() subroutine for optimal repeater insertion.*

into account. Again, because of the decoupling effect of b, the arrival time at the B-side of b is precisely known—that is, since we know that $c_E = c_{BA}(b)$, we simply evaluate $s.f_A(c_{BA}(b))$; $s'.f_A$ must now reflect the situation on the A-side of b, which is accomplished by adding the intrinsic delay and creating a new PWL with slope equal to $r_{BA}(b)$. Finally, we notice that $s'.f_I$ is now completely determined by $c_{BA}(b)$ and independent of c_E seen at the A-side of b. The other subroutines follow this same kind of reasoning.

Finally, optimality of the algorithm follows inductively from the bottom-up structure of the algorithm.

Theorem 8.4.1 *Consider $S(v)$ as computed by the MSRI algorithm in Figure 8.6. Let $s(c_E)$ be the 5-tuple of scalars (s.cost, s.cap, s.d, $s.f_A(c_E)$,*

$s.f_I(c_E))$. *The following property holds.*

$$s \in S(v) \Longleftrightarrow \exists c_E \text{ s.t. } \forall \textit{ repeater assignments}$$
$$s' \neq s \text{ to } T_v, \exists c_E \text{ s.t. } s'(c_E \not\preceq s(c_E)$$

8.4.6 Discussion

A number of issues with respect to the repeater insertion algorithm are worth mentioning:

- By considering a discretization of the various dimensions of the repeater insertion problem, it is possible to give a *pseudo-polynomial* bound on the algorithm's run time. The practical utility of such a bound is limited and, therefore, we defer to the experimental results of the next section for more meaningful evaluation of run time.
- The algorithm can also solve the driver-sizing problem subject to the assumption that drivers are single input (thus allowing us to easily take into account the effect a source driver has on its preceding stage).
- Organizational conventions such as maintaining solution sets in sorted order by cost and secondarily by capacitance can potentially improve performance by allowing us to easily avoid unnecessary comparisons when finding a minimal solution set.
- Many different strategies for finding minimal solution sets can be imagined and are perhaps worth exploring.
- An extension allowing the use of inverters as repeaters is possible and straightforward.

8.5 SUMMARY, DISCUSSION, AND FURTHER READING

This chapter focused on timing optimization for multisource nets (i.e., bus structures). A natural timing metric — the augmented RC diameter (ARD) — was introduced; a linear time algorithm for computing the ARD of a given topology configuration was presented.

We then turned our attention to timing optimization for multisource nets using the ARD concept. An optimal repeater insertion algorithm was presented that utilized piecewise linear functions to characterize subsolutions in a dynamic programming algorithm. This characterization is quite general and can be adapted to other scenarios (e.g., topology synthesis).

An open research problem in the area relates to more general problem formulations. While the ARD is natural and captures many practical situations, it is not the most general conceivable formulation. A formulation in which there is a delay constraint imposed for each pair of source and sink is more general than the ARD. Such a scenario is conceivable when, for example, the downstream delay from a sink vertex u is highly dependent on which pin is acting as the driver. Sometimes this can occur when a designer or high-level synthesis tool bundles a number of signals together onto a single bus for the purpose of conserving routing resources; in such a case data being put onto the bus from one clock cycle to the next may be logically unrelated and require completely different processing. In these situations, a more general problem formulation (i.e., pairwise delay constraints) may be desirable. Unfortunately, such a problem appears to be computationally much more difficult than an ARD-based problem. Nevertheless, an ARD-based algorithm may be useful in attacking the more general problem (see exercises below).

The literature on multisource optimization is not nearly as extensive as for the single-source case. In [2], Cong and Madden propose an algorithm for constructing low-cost *minimum-diameter* rectilinear Steiner trees ("diameter" being in the sense of geometric path length rather than RC delay). In [3], Cong and He studied the problem of minimizing a *weighted sum* of pairwise Elmore delays by wire sizing. Additionally, the convex programming techniques of [10] for *continuous* wire sizing can be generalized to accommodate the multisource case, including driver sizing [11]. A local search scheme for repeater insertion with pair-wise timing constraints was presented in [12].

8.6 EXERCISES

1. Adapt the techniques in this chapter and Chapter 7 to construct P-Tree global routing topologies for buses.
2. Suppose you are given arbitrary pairwise delay constraints $d_{\max}(u, v)$ for each pair of pins u and v. Let this be problem P_1. Derive an ARD formulation P_2 of the same problem in which a feasible solution to P_2 guarantees a feasible solution to P_1. Does such an approach guarantee optimality?

BIBLIOGRAPHY

[1] K. L. Clarkson, More output-sensitive geometric algorithms, *Proc. 35th Annual Symposium on Foundations of Computer Science*, pp. 695–702, 1994.

[2] J. J. Cong, and P. H. Madden, Performance-driven routing with multiple sources, *Proc. IEEE Symposium on Circuits and Systems*, pp. 203–206, 1995.

[3] J. J. Cong, and L. He, Optimal wiresizing for interconnects with multiple sources, *Proc. Proc. IEEE Intl. Conf. Computer-Aided Design*, pp. 568–574, 1995.

[4] T. B. Huang, Y. C. Jeng, et, al, Bidirectional bus repeater, U.S. Patent number 5,202,593, Apr. 13, 1993.

[5] D. Y. Kao, C. C. Tsai, C. K. Cheng, and T. T. Lin, New design and implementation for signal repeaters, *VLSI/CAD workshop*, Taiwan, pp. 173–176, Aug. 17–19, 1995.

[6] H. T. Kung, F. Luccio, and F. P. Preperata, On finding the maxima of a set of vectors, *J. ACM* 22: 469–476, Oct. 4, 1975.

[7] T. M. Lin, and C. A. Mead, Signal delay in general RC-networks, *IEEE Trans. Computer-Aided Design, CAD-3*, pp. 331–349, (1984).

[8] Noel Menezes, Intel Corp., *private communication*, Oct. 1996.

[9] J. Rubinstein, P. Penfield, and M. A. Horowitz, Signal delay in RC tree networks, *IEEE Trans. on CAD*, 2(3): 202–211, (1983).

[10] S. S. Sapatnekar, Wire sizing as a convex optimization problem: Exploring the area delay tradeoff, *IEEE Transactions on Computer-Aided Design*, 15(8): 1001–1011, Aug. 1996.

[11] S. S. Sapatnekar, Iowa State University, *private communication*, Oct. 1996.

[12] C. C. Tsai, D. Y. Kao, and C. K. Cheng, Performance driven bus buffer insertion, *IEEE Trans. on CAD*, pp. 429–437, Apr. 1996.

[13] R. S. Tsay, Exact zero skew, *Proc. IEEE Intl. Conf. Computer-Aided Design*, pp. 336–339, 1991.

9

Timing-Driven Maze Routing

9.1 INTRODUCTION

The classical maze-routing problem first studied by Lee and Moore [4, 5] has been a cornerstone of *design automation* for more than three decades. The fundamental problem in maze routing lies in finding a wiring connection between two pins s and t in a routing graph. Maze-routing algorithms have been successfully applied at both global and detailed routing stages.

In the deep-submicron era, natural generalizations of the maze-routing problem become important and are the topic of this chapter. The key issues addressed are the now-complex trade-offs between wiring cost and signal delay. Illustrations of this added complexity include the following.

- Interconnect technology parameters (e.g., resistance and capacitance) can vary significantly from one layer to the next.
- Vias now contribute significant resistance.
- It may be desirable to widen some wires if possible to reduce delay.

The net result of these trends is that determination of a "best" connecting path is now a complex multidimensional optimization problem for timing-critical signals.

Beyond traditional detailed routing, there are many other scenarios in which closely related timing-driven wiring problems are of practical interest. For example, in the early stages of layout it is now necessary to allocate some buffering resources for different signals. Further, depending on the floorplan, the availability of such buffering resources may not be uniformly distributed — for example, as a result of *intellectual property* (IP) blocks or precompiled cells, there may be regions of the layout where it is simply not feasible to insert a buffer (though we may be able to route an unbuffered signal over such regions). In the case of wire sizing, different routing regions may be more congested than others, disallowing wider wires; as a result the routes for nets that need to be wire sized are constrained. These issues complicate the wiring problem for timing-critical signals and lead to new maze-routing formulations. Algorithms for these problems can be used for what may be considered "resource-aware timing optimization" since optimization techniques such as buffer insertion and wire sizing are always performed in the context of the current chip layout or floorplan and the resulting constraints on resources.

This chapter presents formulations and algorithms for such practical problems. The algorithms are based on labeling techniques similar to techniques used in many shortest-paths algorithms.

We first review delay models, introduce the multigraph model, and propose a variety of natural problem formulations for a variety of situations (e.g., depending on whether buffer insertion is possible or the type of objective function we prefer). We then introduce algorithmic concepts with a simple label-correcting algorithm. This is then used as the basis of a more practical label-setting algorithm. The operation of the algorithm is illustrated with a computational example. We then discuss adaptations of the basic algorithmic framework for alternative problem formulations and provide a complexity analysis.

9.2 ASSUMPTIONS

As in previous chapters the algorithms are presented using the Elmore delay model and basic buffer delay models. However, the emphasis is on providing a clear and flexible algorithmic framework that can be generalized and refined for various specific situations.

Recall that, under Elmore, the propagation delay along a wire $e = (u, v)$ is approximated by

$$r_e(c_e/2 + c_L)$$

where r_e and c_e give the resistance and capacitance of the wire, respectively, and c_L gives the downstream capacitive load at vertex v.

Also recall that the delay through a buffer b is approximated by

$$d_b + r_b c_L$$

where r_b is the buffer's output resistance, c_L is the capacitive load that it drives, and d_b is the buffer's intrinsic delay.[1]

Finally recall that each such buffer has an input capacitance c_b seen by its ancestors.

[1] To simplify notation, algorithms will be presented without intrinsic delay. However, it can be trivially incorporated.

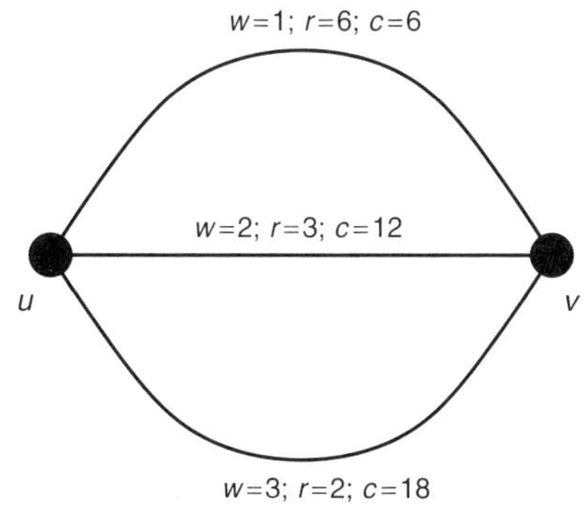

FIGURE 9.1 *Modeling sizing choices with a multigraph. Under a simple scaling model, the example shows three choices for connecting vertices u and v: a width 1 wire, a width 2, and a width 3 wire. Each such arc is labeled with the resulting resistance and capacitance. These parameters need not be derived from such a simple model however. For instance, context-sensitive parameters such as coupling could be incorporated when calculating the parasitics of a segment.*

In all formulations we model the routing target by a multigraph[2] — there may be multiple arcs between the same two vertices. Additionally, each arc in the multigraph has two labels: capacitance and resistance. This enables the natural modeling of various design choices such as wire sizing as shown in Figure 9.1. Further it allows the parameters of such choices to be determined by their context (e.g., neighboring wires). The parasitics of the routing segments can be determined by any desired method.[3]

9.3 FORMULATIONS

The first two formulations considered are wiring-only formulations; the first is a constrained optimization problem, defined as follows.

Formulation 9.1

Given: A multigraph $G = (V, E)$ in which each arc $e \in E$ is annotated with a resistance r_e and a capacitance c_e, a source terminal s with driving resistance r_s, a sink terminal t with load capacitance c_t and a delay constraint d_{spec}.

---

[2] Though traditional maze-routing assumes a k-layer grid (vias connecting one layer to the next), we assume an arbitrary routing target for generality. Routing blockages, of course, manifest themselves as absent arcs in the graph.

[3] For instance, the capacitive values may be adjusted depending on coupling with neighboring wires. Even the effect of "wire splitting" as mentioned in section 5.3.4 could be approximated and modeled as a single arc. Such modeling may be useful in high-frequency designs in which effective resistance is improved by splitting wires instead of widening them.

Objective: Find a path in G connecting s and t such that total wire capacitance is minimized subject to the delay not exceeding d_{spec}.

An alternative formulation seeks to optimize a single composite objective — the weighted sum of signal delay and cost (e.g., wiring area). Such a formulation is useful in many popular global optimization strategies (e.g., those based on Lagrange multiplier-like weighting and reweighting schemes). This formulation also has some interesting algorithmic properties, as follows.

Formulation 9.2

Given: A multigraph G in which each arc e is annotated with a resistance r_e, a capacitance c_e, and a cost COST(e), a source terminal s with driving impedance r_s, a sink terminal t with load capacitance c_t and weighting coefficients A_1 and A_2 where $A_1 + A_2 = 1$ and $A_1, A_2 \geq 0$.

Objective: Find a path in G connecting s and t such that A_1 (DELAY) $+A_2$ (COST) is minimized.

There is a great deal of flexibility in the interpretation of *COST*. Virtually any monotone cost measure can be incorporated into this formulation (e.g., total capacitance, some measure of area, or other heuristic measures based on congestion are all feasible). This is in contrast to Formulation 9.1 in which the use of total capacitance as the cost can be exploited for algorithmic efficiency (i.e., incorporating other cost functions would likely increase the complexity of the algorithm).

When buffer insertion is considered we have two very similar formulations. First we present a constrained optimization formulation.

Formulation 9.3

Given: A multigraph $G = (V, E)$ in which each arc $e \in E$ is annotated with a resistance r_e and a capacitance c_e and each vertex $v \in V$ is labeled with $v_b \in \{0, 1\}$ ($v_b = 1$ indicating that a buffer may be inserted at v), a buffer library B, a source terminal $s \in V$ with driving resistance r_s, a sink terminal $t \in V$ with load capacitance c_t and a delay constraint d_{spec}.

Objective: Find a path in G connecting s and t and an assignment of buffers in B to the vertices $v \in p$ (where $v_b = 1$ and some assignments may be NULL) such that total capacitance is minimized subject to the delay not exceeding d_{spec}.

Finally, we have the weighted sum formulation for the buffer insertion case.

Formulation 9.4

Given: A partial multigraph $G = (V, E)$ in which each arc $e \in E$ is annotated with a resistance r_e and a capacitance c_e and each vertex $v \in V$ is labeled with $v_b \in \{0, 1\}$ ($v_b = 1$ indicating that a buffer may be inserted at v), a buffer library B, a source terminal $s \in V$ with driving resistance r_s, a sink terminal $t \in V$ with load capacitance c_t and coefficients A_1, A_2, A_3 where $A_1 + A_2 + A_3 = 1$ and $A_1, A_2, A_3 \geq 0$.

Objective: Find a path in G connecting s and t and an assignment of buffers in B to the vertices $v \in p$ (where $v_b = 1$ and some assignments may be NULL) such that $A_1(\text{DELAY}) + A_2(\text{WIRING\_COST}) + A_3(\text{BUFFER\_COST})$ is minimized.

9.4 ALGORITHMS

Algorithms for solving the timing-driven maze-routing formulations are based on labeling schemes (see chapter 5 of [1] for background on labeling schemes). "Labels" of a vertex v take the form of lists of candidate (sub)solutions representing paths from a vertex v to the sink t or from the source s to vertex v. As in previous chapters, two or more parameters will characterize candidate subsolutions — in this case paths (note that in traditional maze-routing and shortest-paths algorithms a single parameter — distance — is sufficient).

9.4.1 Dominance

Since paths are characterized by two or more parameters (i.e., a vector) they must be compared by a partial order. If paths have k parameters, we use $(a_1, a_2, \ldots, a_k) \prec (a'_1, a'_2, \ldots, a'_k)$ to indicate that the first path dominates the second (i.e., $a_i \leq a'_i$ for all $i \in \{1..k\}$ and $a_i < a'_i$ for some $i \in \{1..k\}$). This notion of dominance assumes that smaller values are preferred in each dimension, which will indeed be the case for the algorithms presented.

Further, for notational convenience we use the dominance relation in the context of sets of vectors; let a be a k-dimensional vector and P be a set of k-dimensional vectors:

$$P \prec a \longleftrightarrow \exists a' \in P \quad \text{s.t.} \quad a' \prec a$$

9.4.2 A Sink-to-source Algorithm for Bufferless Constrained Maze Routing

A straightforward algorithm for Formulation 9.1 appears in Figure 9.2. This algorithm will form a basic framework for subsequent algorithms and illustrates the main issues involved.

```
    Algorithm: Sink-to-Source Bufferless Maze Routing
               Constrained Optimization Formulation
1   P(t) = {(c_t,0)}
2   P(u) = Ø  ∀u ≠ t
3   while (∃ arc e ∈ E between u and v and (c,d) ∈ P(v) such that
              P(u) ⊀ (c + c_e, d + r_e(c_e/2 + c) + r_s c_e))
          /* u is upstream; v downstream */
4         Let c_0 = c + c_e
5         Let d_0 = d + r_e(c_e/2 + c) + r_s c_e
6         Delete all (c',d') ∈ P(u) where (c_0,d_0) ≺ (c',d')
7         Insert (c_0,d_0) into P(u)
```

FIGURE 9.2 *Pseudocode for a basic label-correcting algorithm for Formulation 9.1. The algorithm is bottom-up (i.e., labels are computed from the sink toward the source).*

The algorithm maintains at each vertex u in the routing graph a list $P(u)$ of candidate paths from u to the sink t. Each such path is a candidate "tail" of a path $s \rightsquigarrow u \rightsquigarrow t$. These paths are characterized by two parameters:

- c: the total capacitance of the path (including the load capacitance of the destination pin t)
- d: the resulting Elmore delay from u to t (not including the delay of the driver at s)

Each time through the while loop, the algorithm finds a known solution at a vertex v and extends it to an upstream vertex u. We are only interested in such solutions that are not dominated by paths already known in $P(u)$. At each stage of the algorithm $P(u)$ is a set of *minimal* paths—that is, if $(c_1, d_1), (c_2, d_2) \in P(u)$, then $(c_1, d_1) \not\prec (c_2, d_2)$ and $(c_2, d_2) \not\prec (c_1, d_1)$. This property is preserved in line 7 of the algorithm. When we extend a path to the source vertex s, we then incorporate the driver delay.[4]

At the termination of the algorithm, we are left with a set of nondominated paths in $P(s)$. Each $(c, d) \in P(s)$ represents a path p connecting s and t with total capacitance c and Elmore delay d; further, there exists no alternative path p' with capacitance $c' < c$ and Elmore delay $d' < d$. The optimal solution is then given by some $(c, d) \in P(s)$ where $d \leq d_{\text{spec}}$ and c is minimized (assuming d_{spec} is feasible). Alternatively, the members of $P(s)$ can be considered a set of candidate solutions with different trade-offs between cost (total capacitance) and delay.

Actual paths can then be retraced by maintaining enough information (e.g., pointers) with each subsolution to determine how it was constructed (i.e., the edge used and the downstream subpath).

The algorithm can be considered "label correcting"[5] in that, at any given instant some $(c, d) \in P(u)$ (a "label" of vertex u) may not represent a nondominated path

[4] An alternative implementation would initially set $P(t) = \{(c_t, r_s c_t)\}$ and incrementally include additional driver delay due to each edge we extend solutions by.

[5] The Bellman-Ford shortest-paths algorithm is a classical example of a label-correcting algorithm.

from u to t. Although $P(u)$ is minimal during all phases of the algorithm — no path in $P(u)$ dominates another path in $P(u)$ — some path discovered subsequently may very well dominate such a path. This deletion of paths amounts to correcting the label of vertex u.

9.5 A LABEL-SETTING REFINEMENT

The algorithm of the previous section was "label correcting" — at any instant a path $(c, d) \in P(u)$ may be dominated by some as yet undiscovered path (c', d') — the label (c, d) is "corrected" (deleted) during the run of the algorithm. While the label correcting algorithm version illustrates key concepts, a "label-setting" version of the algorithm is a preferred implementation.[6] In a label-setting algorithm, we must ensure that all paths in $P(u)$ represent nondominated paths — not merely that $P(u)$ be minimal. This section presents this version by applying a simple queuing scheme.

In the label-setting version of the algorithm we maintain a priority queue of candidate solutions much in the way Dijkstra's algorithm maintains a priority queue for the shortest-paths fontier. The priority queue is ordered by the total capacitance c of the candidate solution and secondarily by d — the downstream delay of the configuration. This refinement ensures that in the case of a tie in capacitance, we visit the configuration with least delay first.

The overall algorithm appears in Figure 9.3. The algorithm uses a subroutine *Candidates*() that takes a candidate path and produces all augmentations of it (i.e., new frontier candidates by extending the path by one arc). This routine is used to initialize the priority queue and obtain new candidates to add to the priority queue once a new permanent label is given. The algorithm repeatedly removes the lowest capacitance candidate solution $(u, (c, d))$ from the queue (where u is the source of the suffix path) and tests if it should be added to the nondominated paths $P(u)$ in line m7 (i.e., the permanent labels). Once we have a new permanent label, new candidate solutions are added to the queue in lines m8–m11.

Because we visit candidates in nondecreasing order of c and d, we also construct lists $P(u)$ in increasing order of c. It is for this reason that the algorithm is label setting: when we add a solution (c, d) to $P(u)$, we know that all subsequent candidates will have higher capacitance (or equal capacitance and greater delay) and so cannot dominate (c, d). This property allows the tests of lines m6 and m10 to be done trivially. For example, in the case of m6, $P(u) \not\prec (c, d)$ if $\nexists (c_0, d_0) \in P(u)$ where (1) $c_0 \leq c$ and (2) $d_0 < d$; (1) holds by default for all $(c_0, d_0) \in P(u)$ and thus we need only test against the smallest d_0 among $P(u)$ — that is, the last element in $P(u)$. If the delay d of the candidate solution improves on the smallest delay we have in $P(u)$ so far, then we make (c, d) a permanent label. If $P(u)$ is stored simply in a linked list ordered in increasing

[6] Dijkstra's shortest-paths algorithm is a classical example of a label-setting algorithm.

Algorithm: Sink-to-Source Bufferless Maze Routing
Constrained Optimization Formulation
Label Setting Version

```
c1   Subroutine: Candidates(u, (c,d)) {
c2      L = ∅
c3      Foreach arc e = (u,v) ∈ E incident to u
c4         if (v = s) /* source */
c5            L = L ∪ {(v,(c + c_e, d + r_e(c_e/2 + c) + r_s(c_e + c))}
c6         else
c7            L = L ∪ {(v,(c + c_e, d + r_e(c_e/2 + c))}
c8         return L ordered by c and secondarily by d
     }

     /* Main routine */
m1   P(t) = {(c_t,0)}
m2   P(u) = ∅ ∀u ≠ t
m3   Q = Candidates(t, (c_t,0))  /* initialize priority queue */
m4   while (Q ≠ ∅) {
m5        (u, (c,d)) = dequeue(Q)
m6       if (P(u) ⊀ (c,d))
m7          P(u) = P(u) ∪ {(c,d)}  /* append */
m8          L = Candidates(u, (c,d))
m9          Foreach (v, (c',d')) ∈ L
m10            if (P(v) ⊀ (c',d'))
m11               enqueue(Q,(v,(c',d')))
     }
```

FIGURE 9.3 Label setting algorithm for bufferless maze routing (Formulation 9.1). Note that test on lines m6 and m10 can be performed in constant since solutions are visited in non-decreasing order of capacitance c (see text).

order of c and decreasing order of d, we simply compare the candidate delay with that of the last member of the list and append the new solution to $P(u)$. These observations allow a very straightforward and efficient implementation of the algorithm, which would not be possible in the label-correcting version. The most sophisticated data structure we need is a simple binary heap to make the priority queue operations efficient.

9.6 EXAMPLE

The behavior of the label-setting algorithm is illustrated using the simple example in Figure 9.4. The example is a 3×3 routing grid in which each of the 12 edges

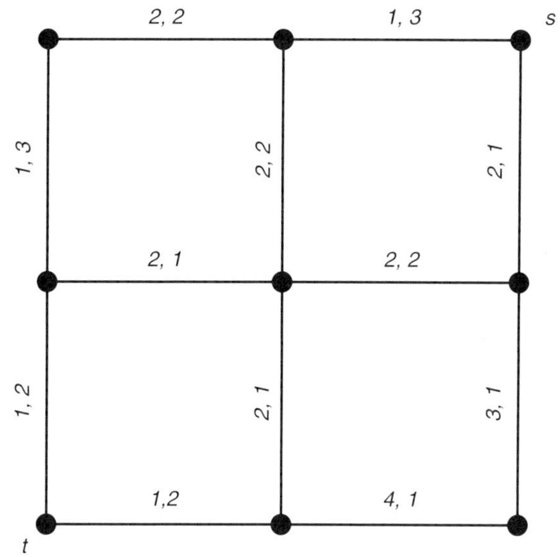

FIGURE 9.4 Example timing-driven maze-routing problem. Vertices are determined by their row/column coordinates. The upper left vertex is v_{11}, lower left is v_{31}, and so on. In this example, the source pin s is located at v_{13} and the sink pin t is at v_{31}. Further, the arcs are labeled with resistance, capacitance pairs. Source resistance $r_s = 3$ and sink capacitance $c_t = 3$.

are annotated with r, c pairs (no multi-arcs are included in this example). See the figure caption for labeling details and assumptions. The following sequence shows the behavior of the algorithm at each step. Each step 1–13 represents the act of removing a candidate solution from the priority queue and determining if it dominated by a previously discovered solution. If a solution is not determined to be suboptimal, it is added to the appropriate $P(v_{ij})$ and new configurations are added to the priority queue if they are not already dominated by preexisting solutions (note that, even if this is the case, by the time the candidate has reached the front of the queue it may no longer be true, at which point it will be discarded). At each stage configurations added to the queue at the end of the previous stage are underlined for clarity.

Each configuration in the queue is of the form $(u, v, (c, d))$ where u and v are the upstream and downstream vertices, respectively, and c and d give the characteristics of the solution: capacitance and delay. (The downstream vertex v is not strictly necessary, but clarifies the example in our opinion.)

Initialization:

$$P(v_{31}) = \{(3, 9.0)\}$$
$$P(v_{ij}) = \emptyset \forall v_{ij} \neq v_{31}$$

[1] $Q = \{\underline{(v_{21}, v_{31}, (5, 19.0))}, \underline{(v_{32}, v_{31}, (5, 19.0))}\}$
candidate $(v_{21}, v_{31}, (5, 19.0))$ added to $P(v_{21})$

[2] $Q = \{(v_{32}, v_{31}, (5, 19.0)), \underline{(v_{22}, v_{21}, (6, 33.0))}, \underline{(v_{11}, v_{21}, (8, 34.5))}\}$
candidate $(v_{32}, v_{31}, (5, 19.0))$ added to $P(v_{32})$

[3] $Q = \{\underline{(v_{22}, v_{32}, (6, 33.0))}, (v_{22}, v_{21}, (6, 33.0)), \underline{(v_{33}, v_{32}, (6, 44.0))},$
$(v_{11}, v_{21}, (8, 34.5))\}$
candidate $(v_{22}, v_{32}, (6, 33.0))$ added to $P(v_{22})$

[4] $Q = \{(v_{22}, v_{21}, (6, 33.0)), (v_{33}, v_{32}, (6, 44.0)), (v_{11}, v_{21}, (8, 34.5)),$
$\underline{(v_{12}, v_{22}, (8, 53.0))}, \underline{(v_{23}, v_{22}, (8, 53.0))}\}$
candidate $(v_{22}, v_{21}, (6, 33.0))$ suboptimal for $P(v_{22})$

[5] $Q = \{(v_{33}, v_{32}, (6, 44.0)), (v_{11}, v_{21}, (8, 34.5)), (v_{12}, v_{22}, (8, 53.0)),$
$(v_{23}, v_{22}, (8, 53.0))\}$
candidate $(v_{33}, v_{32}, (6, 44.0))$ added to $P(v_{33})$

[6] $Q = \{\underline{(v_{23}, v_{33}, (7, 66.5))}, (v_{11}, v_{21}, (8, 34.5)), (v_{12}, v_{22}, (8, 53.0)),$
$(v_{23}, v_{22}, (8, 53.0))\}$
candidate $(v_{23}, v_{33}, (7, 66.5))$ added to $P(v_{23})$

[7] $Q = \{(v_{11}, v_{21}, (8, 34.5)), (v_{12}, v_{22}, (8, 53.0)), (v_{23}, v_{22}, (8, 53.0)),$
$\underline{(v_{13}, v_{23}, (8, 84.5))}\}$
candidate $(v_{11}, v_{21}, (8, 34.5))$ added to $P(v_{11})$

[8] $Q = \{(v_{12}, v_{22}, (8, 53.0)), (v_{23}, v_{22}, (8, 53.0)), (v_{13}, v_{23}, (8, 84.5)),$
$\underline{(v_{12}, v_{11}, (10, 58.5))}\}$
canditate$(v_{12}, v_{22}, (8, 53.0))$ added to $P(v_{12})$

[9] $Q = \{(v_{23}, v_{22}, (8, 53.0)), (v_{13}, v_{23}, (8, 84.5)), (v_{12}, v_{11}, (10, 58.5)),$
$\underline{v_{13}, v_{12}, (11, 71.5))}\}$
candidate $(v_{23}, v_{22}, (8, 53.0))$ added to $P(v_{23})$

[10] $Q = \{(v_{13}, v_{23}, (8, 84.5)), \underline{(v_{13}, v_{23}, (9, 73.0))}, (v_{12}, v_{11}, (10, 58.5)),$
$(v_{13}, v_{12}, (11, 71.5))\}$
candidate $(v_{13}, v_{23}, (8, 84.5))$ added to $P(v_{13})$

[11] $Q = \{(v_{13}, v_{23}, (9, 73.0)), (v_{12}, v_{11}, (10, 58.5)), (v_{13}, v_{12}, (11, 71.5))\}$
candidate $(v_{13}, v_{23}, (9, 73.0))$ added to $P(v_{13})$

[12] $Q = \{(v_{12}, v_{11}, (10, 58.5)), (v_{13}, v_{12}, (11, 71.5))\}$
candidate $(v_{12}, v_{11}, (10, 58.5))$ suboptimal for $P(v_{12})$

[13] $Q = \{(v_{13}, v_{12}, (11, 71.5))\}$
candidate $(v_{13}, v_{12}, (11, 71.5))$ added to $P(v_{13})$

At termination of algorithm:

$$P(v_{11}) = \{(8, 34.5)\}$$
$$P(v_{12}) = \{(8, 53.0)\}$$
$$P(v_{13}) = \{(8, 84.5), (9, 73.0), (11, 71.5)\}$$
$$P(v_{21}) = \{(5, 19.0)\}$$
$$P(v_{22}) = \{(6, 33.0)\}$$
$$P(v_{23}) = \{(7, 66.5), (8, 44.0)\}$$
$$P(v_{31}) = \{(3, 9.0)\}$$
$$P(v_{32}) = \{(5, 19.0)\}$$
$$P(v_{33}) = \{(6, 44.0)\}$$

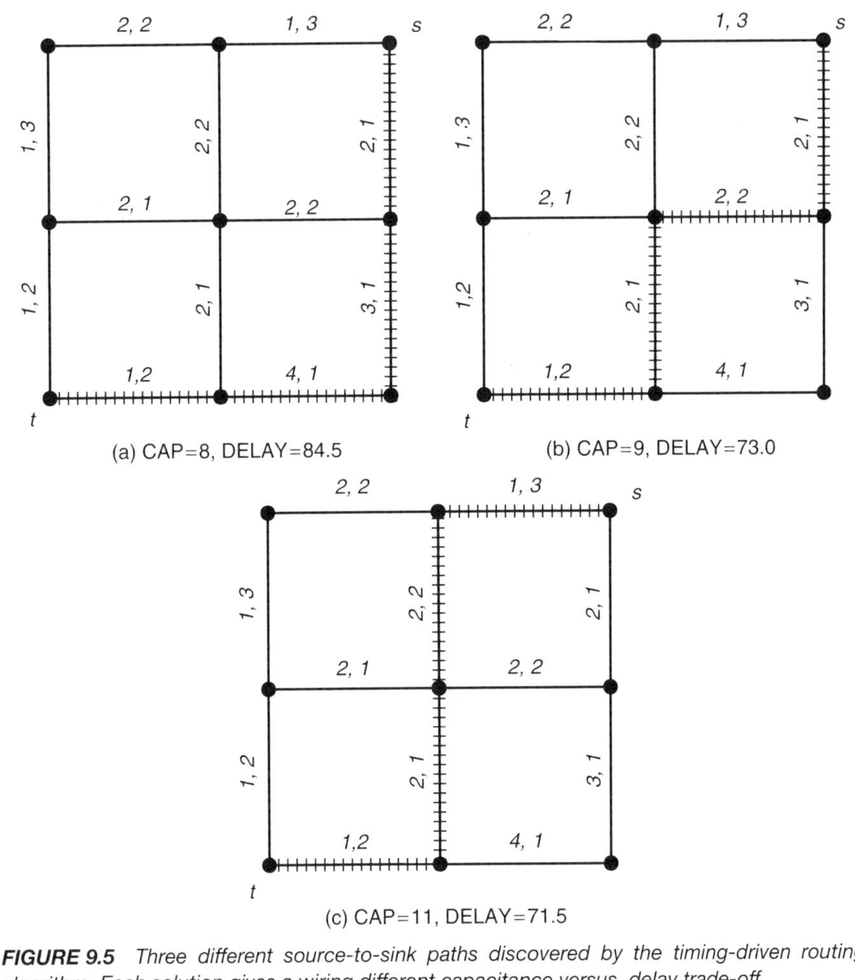

FIGURE 9.5 *Three different source-to-sink paths discovered by the timing-driven routing algorithm. Each solution gives a wiring different capacitance versus. delay trade-off.*

Actual paths in $P(v_{13})$:

$$(8, 84.5) : v_{13}, v_{23}, v_{33}, v_{32}, v_{31}$$
$$(9, 73.0) : v_{13}, v_{23}, v_{22}, v_{32}, v_{31}$$
$$(11, 71.5) : v_{13}, v_{12}, v_{22}, v_{32}, v_{31}$$

The algorithm discovers that there are three nondominated paths from s to t. These paths are illustrated in Figure 9.5. Which path is preferred of course depends on d_{spec}.

9.7 AN ALGORITHM FOR FORMULATION 9.2

In Formulation 9.2 of timing-driven maze routing, we minimize a single composite objective function $(A_1(\text{DELAY}) + A_2(\text{COST}))$ rather than solve a constrained optimization problem.

A sink-to-source solution to this problem is easily derived from algorithms for Formulation 9.1. Rather than maintaining load-delay pairs (c, d) to represent paths, we simply use load-objective pairs $(c, a) \in P(u)$ where a gives the weighted sum $A_1(\text{DELAY}) + A_2(\text{COST})$ of a particular path from vertex u to sink t. In the case of the label-setting algorithm we again order the priority queue first by c and secondarily by a.

However, there is one interesting difference between these two formulations. Formulation 9.2 permits an analogous source-to-sink algorithm, while such an algorithm for Formulation 9.1 would have increased complexity. We now describe a source-to-sink algorithm.

In the source-to-sink algorithm, we maintain pairs $(r, a) \in P(u)$ where r gives the cumulative resistance (including the driver resistance) of the candidate path from the driver at s to vertex u and a gives the corresponding cost of the path. (Recall that each are e has an associated cost, $\text{COST}(e)$.) The idea is to exploit the resistance-oriented interpretation of Elmore as discussed in section 4.5.8. The resulting source-to-sink algorithm appears in Figure 9.6. The refinements to the

```
Algorithm: Source-to-Sink Bufferless Maze Routing
           Weighted Sum Formulation
           Label Correcting Version
P(s) = {(r_s, A_1 r_s c_t + 0)}
P(u) = ∅ ∀u ≠ s

while (∃ arc e ∈ E between u and v and (r, b) ∈ P(u) such that
          P(v) ⊀ (r + r_e, b + A_1(r_e(c_e/2 + c_t) + rc_e) + A_2(COST(e))
      Let r_0 = r + r_e
      Let b_0 = b + A_1(r_e(c_e/2 + c_t) + rc_e) + A_2(COST(e))
      Delete all (r', b') ∈ P(v) where (r_0, b_0) ≺ (c', d')
      Insert (r_0, b_0) into P(v)
```

FIGURE 9.6 *Pseudocode for source-to-sink bufferless maze routing. Inside the while-loop we find a solution at u and extend it to vertex v (u is upstream, v downstream). We accumulate path resistance as we go from source to sink.*

sink-to-source algorithm for Formulation 9.1 in the previous section are easily adapted to this case.

The reason a similar algorithm will not work for Formulation 9.1 is that, in that algorithm, total capacitance served a dual role — it captured both the load (so that its incremental impact on delay can be computed as the path is extended) *and* the cost of the candidate path. In a source-to-sink approach, accumulating capacitance does not help in incrementally computing delay as we extend the path — rather, resistance does. Unfortunately, there does not seem to be a natural mapping of total path resistance to a reasonable cost measure; hence, resistance cannot play a dual role in source-to-sink computation as does that capacitance in sink-to-source computation. For a source-to-sink solution to Formulation 9.1 to work correctly, it appears necessary to carry along three parameters: accumulated capacitance (or some other measure of cost), accumulated resistance, and delay. We leave it as an exercise to adapt a label-setting version of this algorithm.

9.8 INCORPORATING BUFFER INSERTION

As a final illustration of the labeling techniques, we adapt the algorithm for Formulation 9.4 — maze routing with buffer insertion and a weighted sum objective function.[7] We present a sink-to-source algorithm in which paths are characterized by the following parameters.

c: total lumped capacitance of the path

a: the accumulated composite cost of the path

Pseudocode for a label-setting solution to Formulation 9.4 appears in Figure 9.7. The algorithm adopts the priority queuing scheme described previously; in this case the queue is ordered first by a and secondarily by c.[8] The subroutine *Candidates ()* takes a vertex u and a path from u to t characterized by (c, a) and produces all possible augmentations of this subpath:

- Extensions of the path to neighboring vertices
- Connecting the path by a single arc to the source s (if possible)
- Extending the path to a neighboring vertex v and then buffering the path at v

The main body of the algorithm follows the previously outlined strategy: remove a candidate from the queue; determine if it is dominated by a lower capacitance

[7] Formulation 9.3 is left as an exercise.

[8] This is an important if subtle point: load capacitance in the case of buffer insertion is not necessarily a monotone increasing function as we extend a solution. Therefore we visit candidate solutions in order of their total cost a, which is monotonically increasing as a path is extended, buffered, and so on.

Algorithm: Sink-to-Source Buffered Maze Routing
Weighted Sum Objective Function
Label Setting Version

Subroutine: Candidates$(u, (c, a))$ {
  $L = \emptyset$
  Foreach arc $e = (u, v) \in E$ incident to u
    if $(v = s)$ /* source */
      $L = L \cup \{(v, (c + c_e, a + A_1(r_e(\frac{c_e}{2} + c) + r_s(c + c_e)) + A_2\mathrm{COST}(e))\}$
    else
      $L = L \cup \{(v, (c + c_e, a + A_1(r_e(\frac{c_e}{2} + c) + A_2\mathrm{COST}(e))\}$
    if $(v_b = 1)$ /* valid insertion point */
    Foreach buffer $b \in B$
      $L = L \cup \{(v, (c_b, a + A_1(r_e(\frac{c_e}{2} + c) + r_b(c + c_e)) + A_2(\mathrm{COST}(b) + \mathrm{COST}(e)))\}$
  return L ordered by a
}

/* Main routine */
$P(t) = \{(c_t, 0)\}$
$P(u) = \emptyset \;\; \forall u \neq t$
Q = Candiates$(t, (c_t, 0))$ /* initialize priority queue */
while $(Q \neq \emptyset)$
  $(u, (c, a))$ = dequeue(Q)
  if $(P(u) \not\prec (c,a))$
    $P(u) = P(u) \cup \{(c, a)\}$
    L = Candiates$(u, (c, a))$
    Foreach $(v, (c', a')) \in L$
      if $(P(v) \not\prec (c', a'))$
        enqueue$(Q, (v, (c', a')))$
}

FIGURE 9.7 *Pseudo-code of a label-setting algorithm for Formulation 9.4 incorporating buffer insertion. The algorithm is bottom-up — ie, labels are computed from the sink toward the source. The subroutine Candidates() takes a vertex u and a path $(c, a) \in P(u)$; Three types of new candidate solutions are then constructed by augmenting (c, a): (1) candidates which augment the solution by a wire to some other vertex v not equal to the source, (2) candidates derived by connecting u to the source s and (3) candidates derived by extending the path to some vertex v and inserting a buffer provided v is a valid buffer insertion point ($v_b = 1$). Note that Q is a priority queue ordered first by a and secondarily by c (ordering by capacitance will not work because of the possible presence of buffers).*

solution; if not, update $P(u)$ and enqueue new candidate solutions derived from the just-dequeued solution.

9.9 COMPLEXITY

A worst-case analysis of the label-setting algorithm for Formulation 9.1 is derived in the following. With appropriate assumptions, similar bounds can be derived for Formulations 9.2 and 9.4. Let us make the following assumptions.

- The capacitive values c_e and c_b are given as integers. This is not a limitation since discretization is always possible.
- The node degree in the target routing graph G is bounded by a constant. This is certainly the case in most IC applications.

The analysis will give a pseudo-polynomial bound on the running time of the algorithm — the running time is not only a function of the size of the routing graph, but also of the values with which the graph is annotated (namely c's). Let $U = \Sigma_{e \in E} c_e$. The following observation follows from the minimality of $P(v)$.

Observation 9.9.1 In the worst case, $|P(v)|$ is limited by U.

We can then bound the size of the priority queue Q.

Observation 9.9.2 In the worst case $|Q| = O(U \cdot |V|)$. Since each vertex v has constant degree, each member of $P(v)$ can introduce only a constant number of candidates into Q, thus giving the bound. Further, over the lifetime of the algorithm, the total number of enqueues into Q is $O(U \cdot |V|)$. (Recall that in the label-setting algorithm all members of Q are derived from permanent labels of some $P(v)$).

From these observations an overall bound on the running time follows.

Theorem 9.9.1 *The running time of the label-setting algorithm for Formulations 9.1–9.3 is* $O(U \cdot |V| \log(U \cdot |V|))$.

Proof: The total running time is given by (number of dequeues) × (time to dequeue and test dominance of dequeued solution) + (number of enqueues) × (time to enqueue a candidate). The number of enqueues and dequeues is $O(U \cdot |V|)$, and the time to test the dominance of a dequeued solution at vertex v is a constant time operation. Further, the time to perform an enqueue/dequeue is $O(\log(U \cdot |V|))$. The resulting complexity is $O(U \cdot |V|(\log(U \cdot |V|) + 1)) = O(U \cdot |V|(\log(U.|V|))$. □

As is the case with previous pseudo-polynomial algorithms, this bound may be very pessimistic in practice, particularly since the upper bound of U on the number of nondominated solutions at a vertex is likely to be extremely pessimistic.

9.10 SUMMARY AND COMMENTS

This chapter examined several natural problem formulations of timing-driven maze routing relevant in deep-submicron technology. A multigraph model was introduced that naturally captures available design choices. Formulations incorporating buffer insertion were also discussed.

Algorithms for solving the problem were based on labeling techniques. In particular, a label-setting implementation, which is likely to be the most useful in practice, was presented and analyzed. Adapting the algorithms for various formulations was illustrated by example.

Finally, we note that a variety of techniques to improve running time in practice are possible. These techniques are similar in spirit to those for topology synthesis presented in section 7.7.1 where a trade-off curve is explored incrementally to provide bounding information. By using lower-bounding techniques, this solution can be used to discard candidate subpaths during the run of the algorithm that could not be discarded otherwise.

The main ideas are as follows.

- As a preprocess compute the minimum capacitance paths from s to all other vertices using Dijkstra's algorithm. Note that this also produces the min capacitance path from s to t and its resulting delay (i.e., the first solution in the final trade-off curve).
- Also as a preprocess compute the minimum *length* paths from s to all other vertices using Dijkstra's algorithm. This presumes that physical distance between routing nodes is known — a safe assumption in most cases.
- Impose an artificial upper bound on the total allowed capacitance among any s to t paths. This bound will be gradually relaxed.
- When running the algorithm and considering a candidate solution (c, d) at vertex u do the following:
  - If the minimum capacitance path from s to u plus c exceeds the current capacitance bound, discard the solution (or simply do not expand from it). *Any* s to t path using (c, d) as a suffix must exceed the capacitance bound.
  - Using the minimum path length from s to u and c, calculate a lower bound on the delay from s to u. If this delay plus d exceeds the delay of a previously known s to t path, we may discard (c, d).
- After a pass, relax the capacitance bound and continue using new delay bound information.

The process is illustrated in Figure 9.8. Experiments with such an approach show it to be remarkably effective — often resulting in one or two orders of magnitude speedup.

The question remains how to obtain effective lower bounds on the delay given only the minimum wiring distance and the capacitance c. One effective method is to use closed forms for optimal continuous tapering as in [3] and [2]. These results give closed forms for the optimal delay via wire tapering of a wire of given length and downstream load. Since the resulting delay is for the idealized case where arbitrary tapering can be performed, we obtain a legitimate lower bound

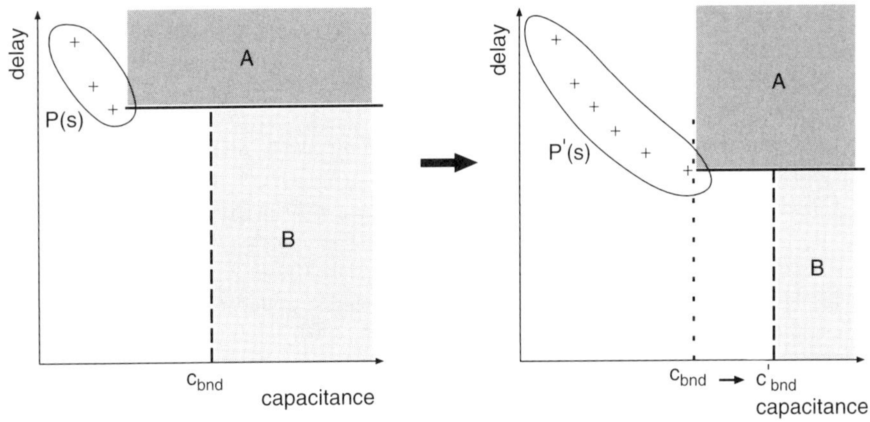

FIGURE 9.8 Iterative bounding process. If a subpath u⇝t can only contribute to s to t paths with capacitance in region B it violates the capacitance bound. Similarly if the resulting delay of all s to t paths is in the region A, the solution is discarded.

for our problem. Similar techniques such as table-lookup may also be viable. Note, however, that the maze routing formulations assumed nothing about the underlying technology while the results in [3] and [2] do. To give an appropriate lower bound, these parameters should be determined carefully.

9.11 EXERCISES

1. Devise a source-to-sink algorithm for the weighted sum buffered maze-routing Formulation 9.4.
2. Describe a label-setting algorithm for constrained buffered maze-routing (Formulation 9.3). Assume that each buffer in the library has an associated cost as does each arc (assume identical units). Hints:
   - Paths will be characterized by three parameters instead of two: cost b, capacitance c, and delay d.
   - It is helpful to order the priority queue first by the cost and second by capacitance (as a tie breaker) of the candidate configurations.
   - Adapt the data structure of Chapter 6 to efficiently determine if a candidate for $P(v)$ is dominated by a previously discovered path already in $P(v)$.
3. How might you modify an algorithm for either of the buffer insertion formulations to allow inverting buffers?
4. Two-pin bus routing: Suppose you are maze routing for a signal net that can be driven in either direction. What parameters would you use to characterize a path?

BIBLIOGRAPHY

[1] R. K. Ahuja, T. L. Magnanti, and J. B. Orlin, *Network Flows*, Englewood Cliffs, NJ, Prentice Hall, 1993.

[2] Optimal wire-sizing formula under the Elmore delay model, *Proc. ACM/IEEE Design Automation Conf.*, pp. 487–490.

[3] J. P. Fishburn, and C. A. Schevon, Shaping a distributed-RC line to minimize Elmore delay, *IEEE Trans. Circuits and Systems-I: Fundamental Theory and Applications*, 42(12); 1020–1022, Dec. 1995.

[4] C. Y. Lee, An algorithm for path connection and its applications, *IRE Transactions on Electonic Computers*, EC-10(3): 346–365, 1961.

[5] E. F. Moore, Shortest path through a maze, *Proceedings of the International Symposium on Switching Circuits*, Cambridge, MA, Harvard University Press, 1959. (In Annals of the Harvard Computing Laboratory, Volume 30, Part II.)

Index